D1195879

ROBUST VISION FOR VISION-BASED CONTROL OF MOTION

IEEE Press
445 Hoes Lane, P.O. Box 1331
Piscataway, NJ 08855-1331

IEEE Robotics and Automation Society, *Sponsor*
RA-S Liaison, Robert B. Kelley

Cover design: Caryl Silvers, *Silvers Design*

Technical Reviewers

Professor Thomas C. Henderson, *University of Utah*
Professor Jean-Claude Latombe, *Stanford University, CA*

Books of Related Interest from IEEE Press

ROBUST TRACKING CONTROL OF ROBOT MANIPULATORS
Darren M. Dawson and Zhihua Qu
1996 Hardcover 256 pp IEEE Order No. PP4218 ISBN 0-7803-1065-9

INDUSTRIAL APPLICATIONS OF FUZZY LOGIC AND INTELLIGENT SYSTEMS
John Yen
1995 Hardcover 376 pp IEEE Order No. PC4002 ISBN 0-7803-1048-9

ROBUST VISION FOR VISION-BASED CONTROL OF MOTION

Edited by

Markus Vincze
Technische Universitat Wien, Austria

Gregory D. Hager
The Johns Hopkins University

IEEE Robotics and Automation Society, *Sponsor*

SPIE/IEEE Series on Imaging Science & Engineering
Edward R. Dougherty, *Series Editor*

SPIE Optical Engineering Press
The International Society for Optical Engineering
Bellingham, Washington USA

The Institute of Electrical and Electronics Engineers, Inc., New York

This book and other books may be purchased at a discount
from the publisher when ordered in bulk quantities. Contact:

IEEE Press Marketing
Attn: Special Sales
445 Hoes Lane, P.O. Box 1331
Piscataway, NJ 08855-1331
Fax: +1 732 981 9334

For more information about IEEE Press products, visit the
IEEE Press Home Page: http://www.ieee.org/press

SPIE—The International Society for Optical Engineering
P.O. Box 10
Bellingham, Washington 98227-0010
Phone: (360) 676-3290
Fax: (360) 647-1445
Email: spie@spie.org
WWW: http://www.spie.org

Printed in the United States of America

10 9 8 7 6 5 4 3 2 1

An IEEE Press book published in cooperation with SPIE Optical Engineering Press

IEEE ISBN 0-7803-5378-1
IEEE Order No.: PC5403

SPIE ISBN 0-8194-3502-3
SPIE Order No.: PM73

Library of Congress Cataloging-in-Publication Data

Vincze, Markus, 1965–
Robust vision for vision-based control of motion / Markus Vincze,
Gregory D. Hager.
p. cm. — (SPIE/IEEE Series on Imaging Science & Engineering)
"IEEE Robotics and Automation Society, sponsor."
Includes bibliographical references (p.).
ISBN 0-7803-5378-1 (IEEE)
1. Robot vision. 2. Robots—Control systems. 3. Robots—Motion.
I. Hager, Gregory D., 1961– . II. Title. III. Series.
TJ211.3.R57 1999
629.8'92637—dc21 99–037155
CIP

SPIE/IEEE SERIES ON
IMAGING SCIENCE & ENGINEERING
Edward R. Dougherty, Series Editor

The SPIE/IEEE Series on Imaging Science & Engineering publishes original books on theoretical and applied electronic and optical imaging.

Titles in the Series

Robust Vision for Vision-Based Control of Motion
Markus Vincze and Gregory D. Hager, Editors

Digital Color Halftoning
Henry R. Kang

Fundamentals of Electronic Image Processing
Arthur R. Weeks, Jr.

Random Processes for Image and Signal Processing
Edward R. Dougherty

Nonlinear Filters for Image Processing
Edward R. Dougherty and Jaakko T. Astola, Editors

CONTENTS

PREFACE

Understanding how to make a "seeing robot" — this was the goal of the workshop *Robust Vision for Vision-Based Control of Motion* held at the 1998 IEEE International Conference on Robotics and Automation in Leuven, Belgium. As far-fetched as this goal might seem, the continuing development of computers, robotic hardware, and techniques for machine vision suggests that it inevitably will be reached.

It is clear that, given a sufficiently structured environment, today's robots can perform complex tasks with astonishing precision and speed. Hence it would seem that, from the standpoint of mechanics and control, robotics research has already advanced to the point where it could move out of the structured environment of the factory floor and into the "real" world. Why has this not happened? The limiting factor, simply stated, is the ability to perceive and react in such complex and unpredictable surroundings. In particular, vision — as the most versatile sense of perception — is required to aid the navigation, grasping, placing, steering, and the control of motion of a machine, be it a robot arm, a vehicle, or any other autonomous mechanism. Thus, it may be argued that *robust* vision is the key bottleneck to the development of autonomous robotic mechanisms.

For this reason, the use of visual sensing to motion control has drawn increasing interest over the last several years. The level of interest is documented by a series of Workshops and Tutorials at ICRA and other conferences as well as a marked growth in the number of sessions devoted to the topic at recent conferences. As a result, there have been many advances in the science and practice of vision-based control of motion. Yet, most laboratory vision-based control systems are constrained to use markers, special high contrast objects and background surfaces, or detailed models and good initialization — customizations often not possible in real applications.

This book, as well as the workshop from which it is derived, was motivated by the observation that fast, cheap, yet robust and reliable vision is still a major roadblock to industrial and service applications of vision-based control. Our objective is to present the most recent work dealing with this essential problem in visual feedback: *robustness*. As such, the collection contains a series of articles spanning a large range of issues including hardware design, system architecture, control, sensor data fusion, object modeling, and visual tracking. The 14 papers represent major research groups in Europe, America, and Japan. The following summarizes the contributions.

SUMMARY OF APPROACHES

The collection starts with two papers on integrating cues. D. Kragić and H. I. Christensen investigate the integration of color, cross correlation, edge tracking, disparity, and blob motion. Different voting schemes are evaluated using a stereo vision head and a robot at the Center for Autonomous Systems of the Stockholm Royal Institute of Technology. The results indicate that weighted consensus voting performs best, in particular in scenes with background clutter and partial occlusions. Another approach, presented by Y. Shirai et al., integrates optical flow and disparity to track persons using an active stereo head. Subdividing tracking windows into target blocks permits the system to determine if the two cues belong to one object. Therefore it becomes possible to track persons crossing each other. The occluded person is lost only for a short period but recovered once it becomes visible again.

K. Toyama et al. present a rather different approach to robust visual sensing. Using the principle of Incremental Focus of Attention (IFA), a hierarchy of trackers is established. Tracking begins from the coarsest level and tries to select and track the target. If tracking succeeds in one level, a more concise level is invoked. The advantage of this method is that the loss of tracking does not result in total system failure. Using the levels of attention, tracking can automatically reinitialize and therefore recover from loss of target.

Keeping track of a detailed model of the target object is another approach to robust visual processing. The approach by H. H. Nagel et al. exploits the model by filtering edges that are salient. Excluding ambiguous, short, or very close features increases the robustness of pose determination and therefore the control of the motion. K. Nickels and S. Hutchinson use a model to select templates for correlation-based feature tracking. The templates use a complete geometric model of the target object to derive the predicted appearance of a feature. A Kalman filter is used to optimally update the state estimates. Using a certainty measure for each feature, the reliability of pose determination is increased, as shown in the examples. Although the method presently does not operate at frame rate, the authors extrapolate that the steadily increasing computer speed will render the approach feasible in the near future.

An elegant method for fine-tuning a model to an object appearing in an image is presented by N. Giordana et al. Again, a wire-frame model is employed to determine the location of edges in the image. Using an energy minimization technique, a rough estimate of the model pose is then optimized. Examples show a very accurate fit of the object in the image to the model. Tracking can proceed by continuously fitting the model to a sequence of images.

An example of a mobile platform using stereo vision to navigate in an indoor environment is given in the work by D. Burschka et al. A specialized contour tracing method is used to accelerate the edge finding process. Edge data is then used to continuously update the portion of the model currently viewed. Adding exploration techniques permits the system to supervise the area in front of the platform, to detect unexpected objects, and to avoid collisions. A recognition module can be invoked to classify the object detected. In addition to navigating in a known or modeled environment, the stereo vision system can be also used to reconstruct the environment.

P. Rives and J. J. Borrelly report on a vision system to follow pipe systems under water. Focusing image processing on areas of interest reduces computation time. A DSP-based hardware system helps to obtain operation in real time — that is, at video frame rate. The technique has been already used to successfully control an underwater robot.

A number of vision techniques that have proven to be robust are presented in the work by K. Arbter et al. Hough Transform methods are used to detect lines and circles in order to steer the motion of a robot arm in a docking experiment for a space station. Optimizing the distance between color regions in hue-saturation space makes it possible to track the instruments in endoscopic operations. By combining line tracking with the model of an object and a module for hidden line removal, occlusions can be handled. Further improvement of robustness is achieved by integrating several miniature sensors all mounted at the robot hand.

Two works are concerned with the geometry of the visual sensor itself. A panoramic view of the scene can be obtained by using a convex mirror mounted over the lens, as introduced by A. Hicks et al. In this approach, vertical lines are transformed into radial lines, which then can be used to localize the platform carrying the camera. Using the special lens, it is shown that localization is inherently robust. The paper by N. Oshiro et al. analyzes the horopter of a stereo vision system. The results show that a zero disparity filter yields best results at reduced resolution and that a log-polar mapping of the image plane produces a more complete horopter map as compared to images using subsampling for reduction.

The next three chapters evaluate control using vision as feedback. J. P. Barreto et al. provide an in-depth study of the robustness of visual behaviors for a stereo vision head performing smooth pursuit control. The image-processing system and the control behavior are improved by analyzing step, ramp, parabolic, and sinusoidal response behavior of the system. W. J. Wilson et al. present a detailed study on fast feature extraction and robot control. Corner and hole features are extracted with directed search methods operating on grayvalue images. To isolate the feature, the prediction from pose determination is used. Pose estimation uses an Extended Kalman Filter and can switch between features, which allows stable control under partial occlusion. Finally, B. E. Bishop and M. W. Spong offer a highly dynamic air-hockey robot. Using an overhead camera, the motion of the puck is estimated and predicted. A fast direct drive robot intercepts the path of the puck toward the goal and tries to shoot toward the opponent's goal. The work shows the importance of closely coupling vision and control. A video derived from this work has been presented at the accompanying 1998 IEEE Conference on Robotics and Automation and has won the Best Video Award.

The final chapter presents the work of eight years of research in autonomous helicopter flying. A. Amidi and T. Kanade summarise the experience gained of building five helicopters that use sensors for robust control. They report that vision is successful when integrated with a GPS receiver and inertial sensors. The Autonomous Helicopter #2 is able to find target objects and to pick them up. The most noticeable work is the vision odometry, which is used for position control and stabilization.

CONCLUSION

The discussion at the workshop raised open questions as well as hinted at issues where consensus could be reached. An example is models, which are needed to relate the image information to three-dimensional object information. Models are also seen as a primary source to focus image processing. Better technology, in particular faster computers, are thought to solve many problems; however, there is a group of researchers that claims there are intrinsic problems, which speed will not solve. An issue touched by many authors and emphasized in the last paper by Bishop is the combination of vision and control. There has been the agreement that vision and control is more than adding vision to control.

Open issues are an experimental methodology to evaluate and compare the robustness of different approaches and an approach toward a general theory of robust systems.

In summary, the collection of work presented in this volume presents a comprehensive overview on the present state of the art in the field of Vision-Based Control of Motion. At this point, the editors would like to express their most sincere thanks to all authors. Without their effort and help neither the workshop nor this collection would have been possible. We hope the overview is of great help to application engineers and researchers in industry and academia. We are anticipating developments and improvements to continue the evolution of robust and reliable "seeing machines."

Markus Vincze
Technische Universitat Wein, Austria

Gregory D. Hager
The Johns Hopkins University

LIST OF CONTRIBUTORS

Omead Amidi
The Robotics Institute
Carnegie Mellon University
211 Smith Hall
Pittsburgh, PA 15213-3898

Helder Araujo
Institute of Systems and Robotics
Dept. of Electrical Engineering
University of Coimbra
Polo II – Pinhal de Marrocos
3030 Coimbra, PORTUGAL

K. Arbter
German Aerospace Center — DLR
Institute for Robotics and System Dynamics
D-82230 Wessling, GERMANY

R. Bajcsy
GRASP Laboratory
Department of Computer and Information Sciences
University of Pennsylvania
Philadelphia, PA 19104-6228

João P. Barreto
Institute of Systems and Robotics
Dept. of Electrical Engineering
University of Coimbra
Polo II – Pinhal de Marrocos
3030 Coimbra, PORTUGAL

Jorge Batista
Institute of Systems and Robotics
Dept. of Electrical Engineering
University of Coimbra
Polo II – Pinhal de Marrocos
3030 Coimbra, PORTUGAL

Bradley E. Bishop
Weapons and Systems Engineering
United States Naval Academy
Annapolis, MD 21402

Jean-Claude Bordas
DER-EdF
6, Quai Watier
78401 Chatou Cedex, FRANCE

Jean-Jacques Borrelly
2004 Route des Lucioles
06902 Sophia Antipolis Cedex, FRANCE

Patrick Bouthemy
Fabien Spindler
IRISA/INRIA Rennes
Campus Universitaire de Beaulieu
35042 Rennes Cedex, FRANCE

D. Burschka
Laboratory for Process Control and Real Time Systems
Technische Universität München
München, GERMANY

François Chaumette
Fabien Spindler
IRISA/INRIA Rennes
Campus Universitaire de Beaulieu
35042 Rennes Cedex, FRANCE

H. I. Christensen
Centre for Autonomous Systems
Numerical Analysis and Computing Science
Royal Institute of Technology
S-100 44 Stockholm, SWEDEN

K.S. Daniilidis
GRASP Laboratory
Department of Computer and
Information Sciences
University of Pennsylvania
Philadelphia, PA 19104-6228

Zachary Dodds
Department of Computer Science
Yale University
New Haven, CT 06520

C. Eberst
Laboratory for Process Controland Real
Time Systems
Technische Universität München
München, GERMANY

G. Färber
Laboratory for Process Controland Real
Time Systems
Technische Universität München
München, GERMANY

A. Gehrke
Institut für Algorithmen und Kognitive
Systeme
Universität Karlsruhe (TH)
Postfach 6980, D–76128 Karlsruhe
GERMANY

V. Gengenbach
Fraunhofer-Institut für Informations-und
Datenverarbeitung (IITB)
Fraunhoferstr. 1
D–76131 Karlsruhe, GERMANY

Nathalie Giordana
Fabien Spindler
IRISA/INRIA Rennes
Campus Universitaire de Beaulieu
35042 Rennes Cedex, FRANCE

Gregory D. Hager
Department of Computer Science
The Johns Hopkins University
3400 N. Charles St.
Baltimore, MD 21218

R.A. Hicks
GRASP Laboratory
Department of Computer and
Information Sciences
University of Pennsylvania
Philadelphia, PA 19104-6228

G. Hirzinger
German Aerospace Center — DLR
Institute for Robotics and
System Dynamics
D-82230 Wessling, GERMANY

Seth Hutchinson
Dept. of Electrical and Computer
Engineering
The Beckman Institute
University of Illinois
Urbana-Champaign, IL 61801

F. Janabi-Sharifi
Department of Mechanical Engineering
Ryerson Polytechnic University
Toronto, Ontario M5B 2K3
CANADA

Valéry Just
DER-EdF
6, Quai Watier
78401 Chatou Cedex, FRANCE

Takeo Kanade
The Robotics Institute
Carnegie Mellon University
211 Smith Hall
Pittsburgh, PA 15213-3898

D. Kragić
Centre for Autonomous Systems
Numerical Analysis and Computing
Science
Royal Institute of Technology
S-100 44 Stockholm, SWEDEN

J. Langwald
German Aerospace Center — DLR
Institute for Robotics and System
Dynamics
D-82230 Wessling, GERMANY

Ryan Miller
The Robotics Institute
Carnegie Mellon University
211 Smith Hall
Pittsburgh, PA 15213-3898

Fumio Miyazaki
Department of Systems and Human Science
Graduate School of Engineering Science
Osaka University
Osaka 560–8531, JAPAN

Th. Müller
Institut für Algorithmen und Kognitive Systeme
Universität Karlsruhe (TH)
Postfach 6980
D–76128 Karlsruhe, GERMANY

H.-H. Nagel
Institut für Algorithmenund Kognitive Systeme
Universität Karlsruhe (TH)
Postfach 6980 D–76128 Karlsruhe, GERMANY
Fraunhofer-Institut für Informations- und Datenverarbeitung (IITB)
Fraunhoferstr. 1
D–76131 Karlsruhe, GERMANY

Kevin Nickels
Dept. of Electrical and Computer Engineering
The Beckman Institute
University of Illinois
Urbana-Champaign, IL 61801

Atsushi Nishikawa
Department of Systems and Human Science
Graduate School of Engineering Science
Osaka University
Osaka 560–8531, JAPAN

Ryuzo Okada
Department of Computer-Controlled Mechanical Systems
Osaka University
2-1, Yamadaoka, Suita, Osaka 565-0871, JAPAN

Naoki Oshiro
Department of Mechanical Engineering
Faculty of Engineering
University of the Ryukyus
Okinawa 903–0213, JAPAN

Paulo Peixoto
Institute of Systems and Robotics
Dept. of Electrical Engineering
University of Coimbra
Polo II – Pinhal de Marrocos
3030 Coimbra, PORTUGAL

D.J. Pettey
GRASP Laboratory
Department of Computer and Information Sciences
University of Pennsylvania
Philadelphia, PA 19104-6228

Patrick Rives
INRIA
2004 Route des Lucioles
06902 Sophia Antipolis Cedex, FRANCE

C. Robl
Laboratory for Process Controland Real Time Systems
Technische Universität München
München, GERMANY

Yoshiaki Shirai
Department of Computer-Controlled Mechanical Systems
Osaka University
2-1, Yamadaoka, Suita
Osaka 565-0871, JAPAN

Mark W. Spong
Coordinated Science Laboratory
University of Illinois
Urbana, IL 61801

Kentaro Toyama
Vision Technology Group
Microsoft Research
Redmond, WA 98052

G.-Q. Wei
German Aerospace Center — DLR
Institute for Robotics and System Dynamics
D-82230 Wessling, GERMANY

C. C. Williams Hulls
Department of Electrical and Computer Engineering
University of Waterloo
Waterloo, Ontario N2L 3G1
CANADA

W. J. Wilson
Department of Electrical and Computer Engineering
University of Waterloo
Waterloo, Ontario N2L 3G1
CANADA

P. Wunsch
German Aerospace Center — DLR
Institute for Robotics and System Dynamics
D-82230 Wessling, GERMANY

Tsuyoshi Yamane
Department of Computer-Controlled Mechanical Systems
Osaka University
2-1, Yamadaoka, Suita
Osaka 565-0871, JAPAN

Chapter 1

CUE INTEGRATION FOR MANIPULATION

D. Kragić and H. I. Christensen
Royal Institute of Technology

Abstract

Robustness of vision is a notorious problem in vision, and it is one of the major bottlenecks for industrial exploitation of vision. One hypothesis is that fusion of multiple natural features facilitates robust detection and tracking of objects in scenes of realistic complexity. Use of dedicated models of objects of interest is another approach to the robustness problem. To provide some generality the use of natural features for estimation of the end-effector position is pursued in this work. The research investigates two different approaches to cues integration, one based on voting and another based on fuzzy logic. The two approaches have been tested in association with scenes of varying complexity. Experimental results clearly demonstrate that fusion of cues results in added robustness and increased estimation performance. The robustness is in particular evident for scenes with multiple moving objects and partial occlusion of the object of the interest.

1.1 INTRODUCTION

The potential of computational vision for control of manipulators is enormous. So far this potential has not been exploited. There are two major reasons for this: (1) only recently have off-the-shelf microprocessors offered adequate computational resources to allow real-time analysis of images, as needed to close the loop around a manipulator, and (2) robustness of visual cues has been a major obstacle, as it has been difficult to provide a sufficient performance to allow their use in real environments.

Over the last three decades a large number of efforts has tried to provide robust solutions to facilitate use of vision for real-world interactions. Most of these efforts have not been successful. Notable exceptions include Dickmanns [1, 2], and Nagel et al. [3, 4]. Both of these approaches have adopted specific models (in terms of the environment and/or the objects of interest, that is, cars). In terms of manipulation most approaches exploit markers on the object(s) of interest to simplify detection and tracking of cues. Examples of such work include Hager et al. [5, 6], Rives et al. [7, 8], Allen et al. [9]–[11], and Papanikolopoulos et al. [12, 13].

To enable vision to be used in real-world applications an added degree of robustness must be introduced to facilitate use of natural features for visual tracking. In a natural environment no single cue is likely to be robust for an extended period of time and there is thus a need for fusion of multiple cues to provide a system that is robust from a systems point of view. A lot of work has been reported on fusion of visual cues [14]–[17]. Most of the reported techniques are model based, where a specific model of the imaging process and feature extraction is used as the basis for fusion of cues. Good examples of model-based techniques can be found in [15]. One problem with many model based approaches is computational complexity. For use in real-time visual servoing it is necessary to use techniques that are suited for real-time implementation, which as a minimum implies upper-bounded computational complexity. To achieve this one often has to resort to 'simple' visual cues and fast execution. Redundancy on the other hand enables simplified tracking and temporary loss of individual features. In this chapter two different methods for real-time fusion of visual cues are investigated.

Initially, two methods for cue integration are introduced and the basic characteristics are explained. The set of visual cues (involving both monocular and binocular imaging) are then introduced in Section 1.3. Using the set of presented cues and the two integration methods an experimental system has been constructed. This system is introduced in Section 1.4. The constructed system has been used in a number of experiments to demonstrate the the utility of cue integration for visual manipulation. Section 1.5 presents the results for tracking of a manipulator in an environment with a significant amount of clutter and occlusions. Finally in Section 1.6 reflections on cue integration techniques are presented together with a number of issues to be pursued in future research.

1.2 CUE INTEGRATION

Many different methods for integration of information have been suggested in the literature. As, for example, described in [18] and [19], these method have primarily been used for pattern recognition. In computer vision the dominating method has been Bayesian estimation, where a probabilistic approach is formulated together with an optimization approach like simulated annealing or regularization. A good introduction to Bayesian-based integration can be found in [20] and [15].

The problem with model-based approaches is that the validity of the model often is limited and it might be impossible or difficult to verify the correctness of the model at run time. An alternative to use of strong models is model-free or weak models. Voting [21] is an example of a model free methods for fusion. Another method that has received a significant amount of attention in the robotics and control community is fuzzy fusion. The fuzzy approach has received a lot of attention since it was applied for the first time by Mamdani and Assilian [22]. Some authors argue that fuzzy techniques are suitable where a precise mathematical model of the process being controlled is not available [23, 24, 25].

These two methods are introduced below, and in Section 1.4 it is illustrated how the methods can be used for fusion of cues for manipulation.

1.2.1 Voting Methods

Voting methods have been used as a model-free approach to recognition. One of the major applications of voting has been redundant/robust computer systems for space and strategic applications. A good introduction to voting methods and their use for robust computing can be found in [21].

In voting-based integration there is a need for a common estimation/classification space. We will refer to the space as the voting domain, Θ. One can then think of each cue estimator v_i as a mapping:

$$v_i : \Theta \rightarrow [0; 1]. \tag{1.1}$$

The voting domain may for example be 3-space or the image plane. That is, we can estimate the presence of a particular feature at a given location in the image, as used in many image segmentation methods. Alternatively, the voting domain could be a control space like the joint space (R^6).

In terms of voting there are several possible methods which can be used. If each, of the n, cue estimators (c_i) produce a binary vote for a single class (i.e., present or not present) a set of thresholding schemes can be used ($c_i : \Theta \rightarrow \{0, 1\}$):

Unanimity: $\sum v_i(\theta) = n$
Byzantine: $\sum v_i(\theta) > \frac{2}{3}n$
Majority: $\sum v_i(\theta) > \frac{n}{2}$

If each cue estimator is allowed to vote for multiple classes, and the maximum vote is used to designate the final classification, the voting scheme is denoted consensus voting, that is, the class θ' is chosen according to

$$\delta(\theta') = \max\{\delta(\theta)|\theta \in \Theta\}, \tag{1.2}$$

where δ is the combination method, which for example could be simple addition of the votes, but it might also be a more complex weighting function that takes the relative reliability of the the cues into account; i.e.,

$$\delta(\theta) = \sum_{i=1}^{n} w_i * v_i(\theta)$$

A general class of voting schemes, known as *weighted consensus voting*, is defined by

Definition 1 [m-out-of-n voting] An m-out-of-n voting scheme, $V : \Theta \to [0, 1]$, where n is the number of cue estimators, is defined in the following way:

$$V(\theta) = \begin{cases} \Lambda(c_1(\theta), \ldots, c_n(\theta)) & \text{if } \sum_{i=1}^{n} v_i(\theta) \geq m; \\ 0 & \text{otherwise} \end{cases} \tag{1.3}$$

where

$$v_i(\theta) = \begin{cases} 1 & \text{if } c_i(\theta) > 0; \\ 0 & \text{otherwise} \end{cases} \quad \text{for } i = 1, \ldots, n, \tag{1.4}$$

is the voting function and $\Lambda : [0; 1]^n \to [0; 1]$ is a function for combining the confidence for each estimator.

A cue estimator can give a vote for a given class, θ, if the output of the estimator is > 0. If m or more cues vote for a given class θ the value is estimated using the fusion method Λ. As an example multiple cues may be used for estimation of the velocity of the end-effector. If more than m cues are compatible, a weighted estimate is produced by the structure function Λ. The motivation for not using simple averaging is that the different cues might have different levels of uncertainty associated, these can be taken into account by the fusion operator Λ.

1.2.2 Fuzzy Fusion

Fuzzy logic was initially introduced by Zadeh [26] as an extension to regular set theory. In traditional set theory membership is binary, i.e., a data point is a member of the set or it is not. In fuzzy logic, membership is defined by a membership function

$$\mu : \Theta \to [0; 1]$$

If the referential set is a finite set, membership values are discrete values defined in the range [0,1]. If the referential set is an infinite set, we can represent these values as a continuous membership function.

The most important operators in classical set theory are: complement, intersection and union. Their fuzzy counterparts are defined using membership functions.

For a two given sets **A** and **B** and their membership functions μ_A and μ_B, there are several proposed operators operating on the membership functions. Some of the operators most referred to in the literature are as follows:

Complement of A

$$\mu_{\bar{A}}(x) = \neg\mu_A(x) = 1 - \mu_A(x) \tag{1.5}$$

The symbol $\neg$ denotes the negation on the membership function.

Intersection of A and B

- *min*

$$\mu_{A\cap B}(x) = \mu_A \wedge \mu_B = \min[\mu_A(x), \mu_B(x)] \tag{1.6}$$

  The symbol $\wedge$ denotes the "fuzzy and" operator on the membership function.

- *algebraic product*

$$\mu_{A\cap B}(x) = \mu_A \wedge \mu_B = \mu_A(x)\mu_B(x) \tag{1.7}$$

- *bounded difference*

$$\mu_{A\cap B}(x) = \mu_A \wedge \mu_B = \max\{0, \mu_A(x) + \mu_B(x) - 1\} \tag{1.8}$$

Union of A and B

- *max*

$$\mu_{A\cup B}(x) = \mu_A \vee \mu_B = \max[\mu_A(x), \mu_B(x)] \tag{1.9}$$

  The symbol $\vee$ denotes the "fuzzy or" operator on the membership function.

- *algebraic sum*

$$\mu_{A\cup B}(x) = \mu_A \vee \mu_B = \mu_A(x) + \mu_B(x) - \mu_A(x)\mu_B(x) \tag{1.10}$$

- *bounded sum*

$$\mu_{A\cup B}(x) = \mu_A \vee \mu_B = \min\{1, \mu_A(x) + \mu_B(x)\} \tag{1.11}$$

A good overview on the additional operators can be found in [27]. There are several types of fuzzy reasoning and the most cited, in the literature, are the following:

Max Dot Method
The final output membership function for each output is the union of the fuzzy sets assigned to that output in a conclusion after scaling their degree of membership values to peak at the degree of membership for the corresponding premise (modulation by clipping) [28].

Min Max Method
The final output membership function is the union of the fuzzy sets assigned to that output in a conclusion after cutting their degree of membership values at the degree of membership for the corresponding premise (linear modulation). The crisp value of output is, most usually, the center of gravity of the resulting fuzzy set [29].

Tsukamoto's Method
The output membership function has to be monotonously non-decreasing [30]. Then, the overall output is the weighted average of each rule's crisp output induced by the rule strength and output membership functions.

Takagi and Sugeno's Method
Each rule's output is a linear combination of input variables. The crisp output is the weighted average of each rule's output [31].

1.3 CUES FOR VISUAL SERVOING

1.3.1 Color Segmentation

Color can be represented by RGB (red-green-blue) components with values between 0 and 255. However, this kind of representation is sensitive to variations in illumination. Therefore, the color detection of the robot's end-effector is based on the *hue* (H) and *saturation* (S) components of the color histogram values. The *saturation* is a measure of the lack of whiteness in the color, while the *hue* is defined as the angle from the red color axis.

$$H = acos\left[\frac{\frac{1}{2}[(R-G)+(R-B)]}{\sqrt{(R-G)^2+(R-B)(G-B)}}\right] \tag{1.12}$$

$$S = 1 - \frac{3}{(R+G+B)} min(R, G, B) \tag{1.13}$$

$$V = \frac{1}{3}(R+G+B) \tag{1.14}$$

The motivation for using the HSV space is found in experiments performed on monkeys and anthropological studies. Perez in [32] argues that the *hue* value gives better results in image segmentation because the material boundaries relate much better with the hue than intensity value. However, Enneser in [33] tested a wide number of different color spaces with different quantizations, and he argues that the RGB space performed "quite good."

To achieve real-time performance the color to be recognized has been selected a priori. This kind of color segmentation is known as supervised color segmentation. Color training can be done off-line; i.e., the known color is used to compute its color distribution in the H-S plane. In the segmentation stage all pixels whose hue and saturation values fall within the set defined during off-line training and whose brightness value is higher than a threshold are assumed to be object of interest.

See [32]–[35] for additional information about color-based segmentation.

1.3.2 Normalized Cross Correlation

The idea behind model-based matching is to track the object of interest in the entire image (or a part of the image) by searching for the region in the image that looks like the desired object defined by some mask or sub-image (template). The image template describes color, texture, and material that the object is made from. This kind of modeling includes assumptions about ambient lightning and background color that are not object's features and, therefore, will affect the performance of the cue.

There are several different techniques usually used for this purpose, based on minimization of sum of square differences. In all of them, pixels from the template and pixels from the image serve as basis vectors.

The main idea is that a reference template $T(m, n)$ is compared to all neighborhoods within some search region I(x,y). Normalized Cross Correlation (NCC) technique can be expressed as

$$T * I(x, y) = \frac{2\sum_{m,n} T(m, n) I(x+m, y+n)}{K_1 \sum_{m,n} T(m, n)^2 + K_2 \sum_{m,n} I(x+m, y+n)^2} \tag{1.15}$$

where

$$K_1 = \frac{\sum_{m,n} T(m,n)}{\sum_{m,n} I(x+m, y+n)^2}, \quad K_2 = \frac{1}{K_1} \tag{1.16}$$

The region providing the maximal similarity measure is selected as the location of the object in the image. To decrease the computation time the template matching algorithm may be initialized in the region where the object of interest was found in the previous frame.

In the present work, the template is initialized off-line. However, in the future work, the template will be dynamically changed based on the several frames. The reason is that during the motion of the manipulator there are changes in the end-effector configuration and in the ambient lightning. The size of the template should also be governed based on the values from the disparity map, i.e., depending on the distance between the end-effector and the camera, the size of the template should decrease if the distance is decreasing.

In addition, the following optimization techniques are used [36]:

- **Loop short-circuiting:**
  Because the NCC measure has to be performed over each pixel (x,y), during the execution of the loop we check the present NCC measure and the loop can be short-circuited as soon as the current sum exceeds the current reference value.
- **Heuristic best place search beginning and spiral image traversal pattern:**
  This optimization is based on the fact that the best place to begin the search is at the center of the region where the best match was found in the previous frame and to expand the search radially from this point (Figure 1.1).

This technique is sensitive to changes in object shape, size, orientation and changes in image intensities. Its main drawback is that it does not provide a general solution for view-point invariant object recognition [12], [37]–[39].

1.3.3 Edge Tracking

Ideally, edges are steep transitions between smoothly varying luminance areas. In real scenes transitions are not so steep hence their extraction requires careful processing. The edge detector is an interesting alternative for real-time applications since the amount of data can be significantly reduced by applying an edge detector. An edge detector can provide the fundamental data for some of the 3D reconstruction techniques.

Because at this point of research we are mainly interested in clusters of edges, we use the Sobel operator.

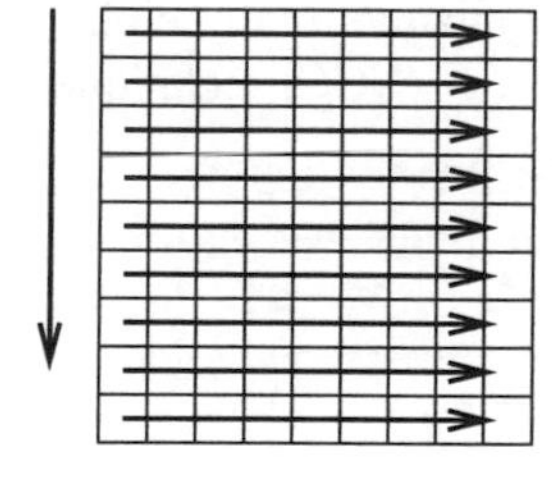

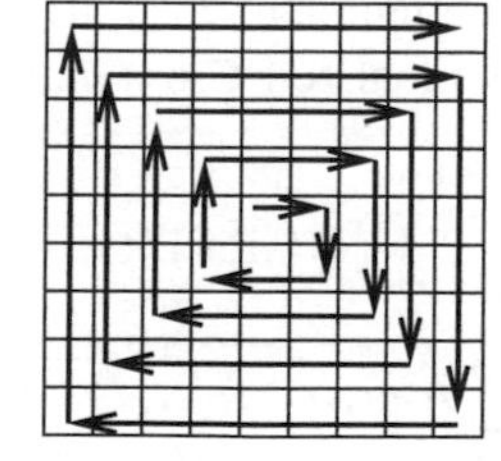

Figure 1.1 Traditional and spiral search patterns.

1.3.4 Disparity

For a particular feature in one image, there are usually several matching candidates in the other image. Although the problem of computing the binocular disparities has been studied in computer vision and photogrammetry community for more that 25 years, most of the methods existing today are far from useful in real-time applications. However, exceptions exist [40, 41, 42].

For computing the disparity map we used grey level values correlation based stereo technique. It is usually necessary to use additional information or constraints to assist in obtaining the correct match [43]. We have used the epipolar constraint, uniqueness and ordering constraint. Since we are using parallel stereo, we assume that the epipolar lines are near parallel so we search just along the y-axis in the image.

We implemented dynamic programming that is based on matching of windows of pixel intensities, instead of using windows of pixel intensities of gradient values. A maximum likehood cost function is used to find the the most probable combination of disparities along a scan line. This cost function takes into account the ordering and uniqueness constraints. The complexity of finding a unique minima is greatly reduced by dynamic programming. The size of the correlation mask is 9×9 pixels.

1.3.5 Blob Motion

For a system that is supposed to maintain the tracking of a moving target, the importance of motion cue is quite obvious. Lots of the existing work in tracking is based on deformable templates [44, 45] which assumes manual initialization. More complicated methods assume the computation of optical flow [46, 47, 48, 49].

In our implementation, the image differencing around a region of attention (image center) is performed as the absolute difference of the intensity component (I) of consecutive images:

$$M^{l,r}(\mathbf{X}) = \Theta(|I^{l,r}(\mathbf{x}, t) - I^{l,r}(\mathbf{x}, t-1)| - \Gamma) \tag{1.17}$$

where Γ is a fixed threshold and Θ is the Heavyside function.

By performing this processing we segment the scene into static and moving regions since only objects having a non-zero temporal difference change position between frames. Since motion cue responds to all moving image regions we have to compensate for the egomotion of the camera head itself before computing the motion cue. Egomotion estimation is based on encoder readings of the pan-tilt unit and inverse kinematics. Some of the implementations of motion in visual servoing can be found in [49, 50, 51, 52].

1.4 SYSTEM OUTLINE

1.4.1 System Setup

In this project, an external camera system is employed with a pair of color CCD cameras arranged for stereo vision (Fig. 1.2). The cameras view a robot manipulator and its workspace from a distance of about 2m. The camera pair is mounted on the pan-tilt unit with 2DOF and together they make up the "stereo head." The size of the original image was 320×240 pixels. In the experiments presented here, the object being tracked is a PUMA robotic arm. The implemented system is running on a regular 233 MHz Pentium with a Meteor framegrabber. The movement itself was under the external control.

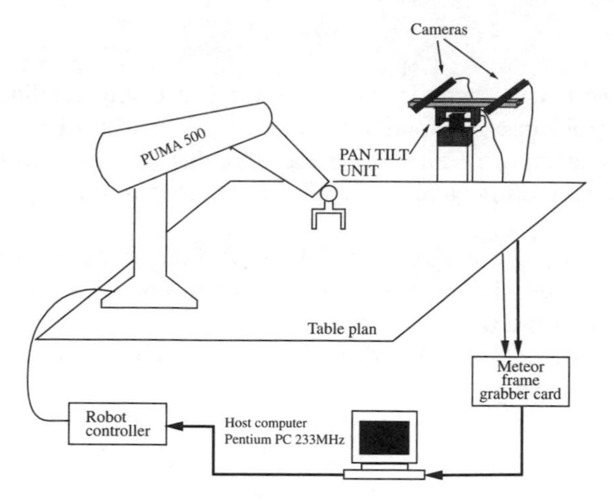

Figure 1.2 Robot configuration.

1.4.2 Implementation

Each cue estimator presented in Section 1.3 and Fig. 1.3 is a function that operates on a certain region (region of interest) and delivers the binary answer [0, 1] whether or not a certain pixel satisfies the conditions of the given function. In the case of voting schema, the voting space, Θ is the image plane (region of interest).

As presented in Fig. 1.4, in the case of the fuzzy fusion, we integrate the information from n-number of sample histograms. Here, n is the total number of cue estimators. Each cue estimator delivers a histogram where the values on the abscissa present the pixel number in the X horizontal image direction and the ordinata presents the sum of the pixel values from different cue estimators for a certain X in the Y vertical image direction.

1.4.2.1 Voting Schema. We have implemented *plurality voting*, which chooses the action that has received the maximum number of votes. This schema can be expressed as

$$\delta(\theta) = \sum_{i=1}^{n} c_i(\theta) \tag{1.18}$$

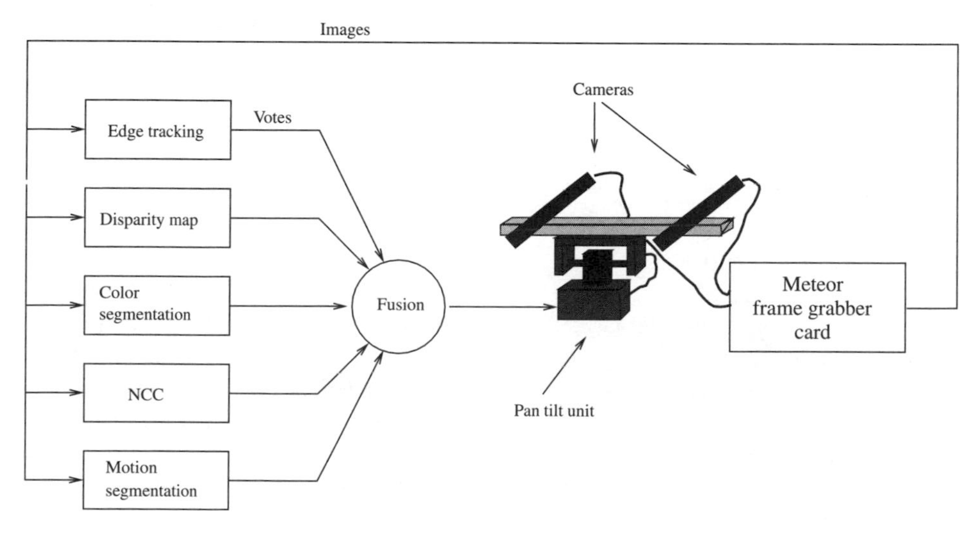

Figure 1.3 Fusion of the cue estimators. The voter in these experiments (denoted FUSION) implements a plurality approval voting scheme.

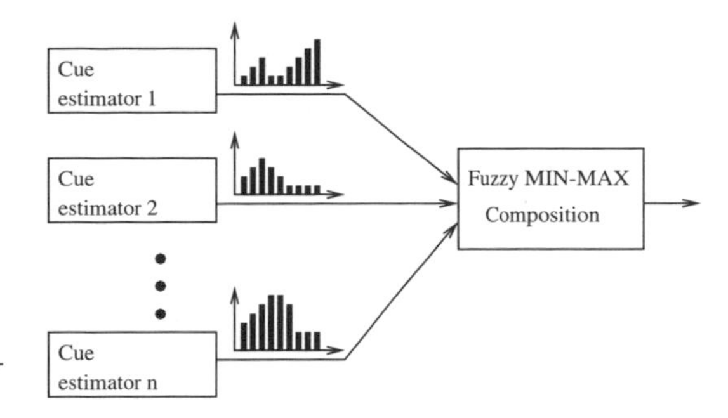

Figure 1.4 Schematic overview of the fuzzy control system.

where δ represents approval voting scheme for a team of cue estimators $c_1, c_2, \ldots, c_n$ and n is the number of cues. The most appropriate action is selected according to

$$\theta' = \max\{\delta(\theta) \mid \theta \in \Theta\} \tag{1.19}$$

1.4.2.2 Fuzzy Logic. In this approach we will consider that sample histograms are fuzzy relations defined on different product spaces [27] and we will combine them by the operation "composition." This operator was defined by Zadeh [53]:

$$m_{R_1 \circ R_2 \circ \ldots \circ R_n} = \max\{\min(m_{R_1}, m_{R_2}, \ldots m_{R_n})\} \tag{1.20}$$

where m_{R_i} are the outputs of the cue estimators.

1.5 EVALUATION

The aim of the experiments was to test each cue separately as well as the two implemented integration techniques in test scenes with different levels of clutter and to investigate the hypothesis that fusion of information from multiple sources can lead to improved overall reliability of the system. We have also investigated which of two different techniques of integration gives better results for our task. A series of tracking experiments has been carried out. The results from three of them are presented in Section 1.5.2. The three different scenarios can be seen in Fig. 1.5, Fig. 1.7, and Fig. 1.9.

1.5.1 Design

During the experiments the movement of the manipulator was as follows (the number of frames for each experiment was around 35):

Experiment 1. The manipulator was moving on the straight path for about 100 cm. The distance of the manipulator relative to the camera was increasing from about 75 cm to 110 cm. The rotation of the 6th joint was 20°. The background was not cluttered but all the objects in the scene had the same color properties as the end-effector. The aim was mainly to

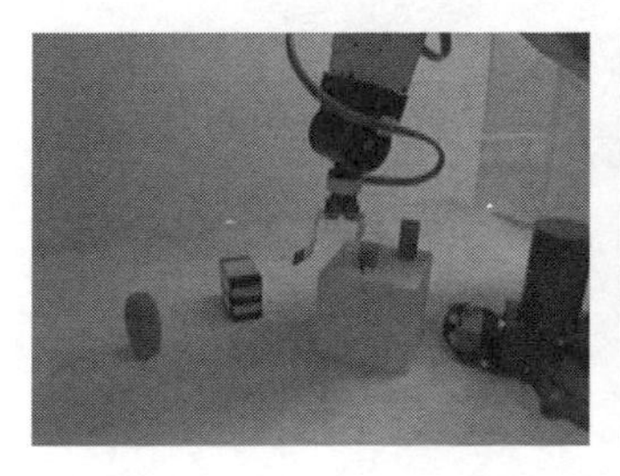 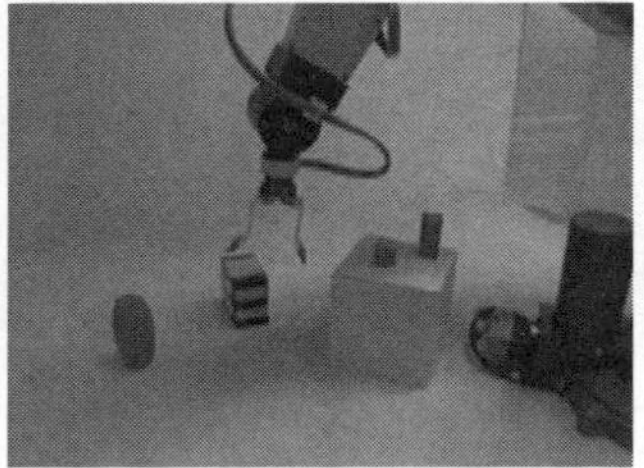 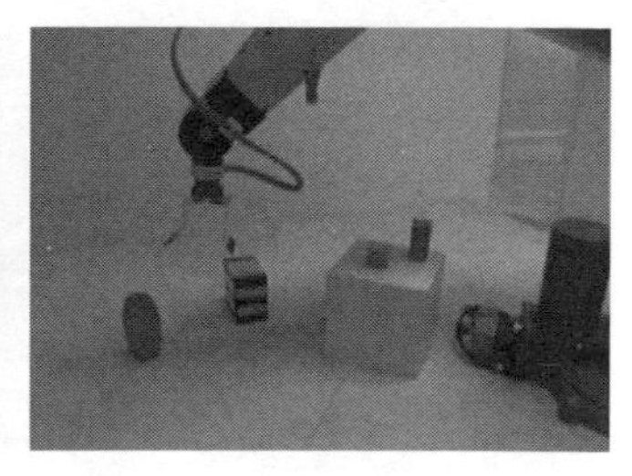

Figure 1.5 The configuration used in the Experiment 1.

test the color segmentation and estimation using normalized cross correlation to see whether the mentioned cues perform better in a rather simple setting.

Experiment 2. A complex background will introduce strong differences between image intensities in consecutive image frames. In addition to the first experiment, we wanted to investigate how clutter influences the edge detection as well as the motion estimation due to non constant illumination conditions. The manipulator was again moving on the straight path for about 100 cm. The distance of the manipulator relative to the camera was increasing from about 75 cm to 110 cm. The rotation of the 6th joint was 20°. The background was here highly complex.

Experiment 3. In this experiment more than one object was moving in the scene. During a few frames, the gripper was occluded. The idea was to investigate the robustness if one or two cue estimators totally fail to detect the end effector during a few frames.

1.5.2 Results

Three images from the first set of experiments are shown in Fig. 1.5. The homogeneous background and the similarly colored objects are clearly visible. The tracking results for each of the cue estimators and the integrated cues are shown in Fig. 1.6. The results from the experiments are summarized in terms of average error and standard deviation in Table 1.1. The results are discussed in more detail in Section 1.5.3. The results for experiments 2 and 3 are presented in a similar fashion on the following two pages.

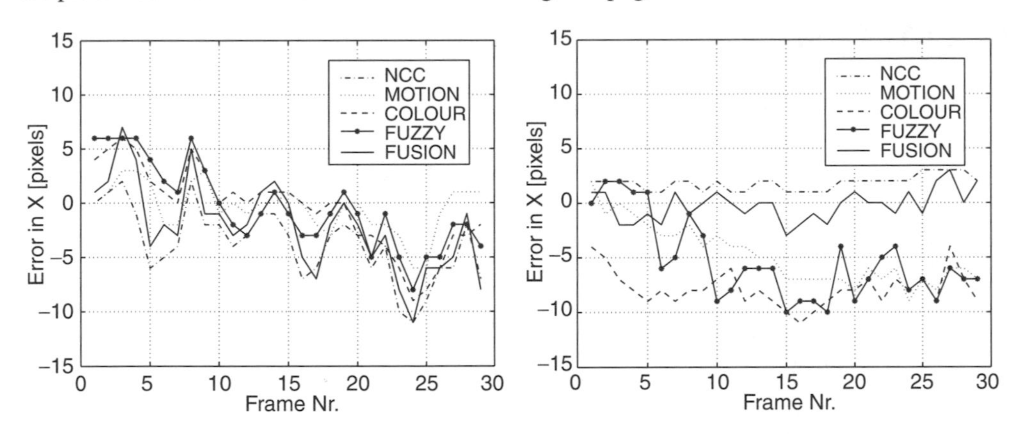

Figure 1.6 Distance error in the X-horizontal (left image) and Y-vertical (right image) direction for the modules during experiment. For clarity reasons, the results from the edge and disparity cues are not presented.

Figure 1.7 The configuration used in the Experiment 2.

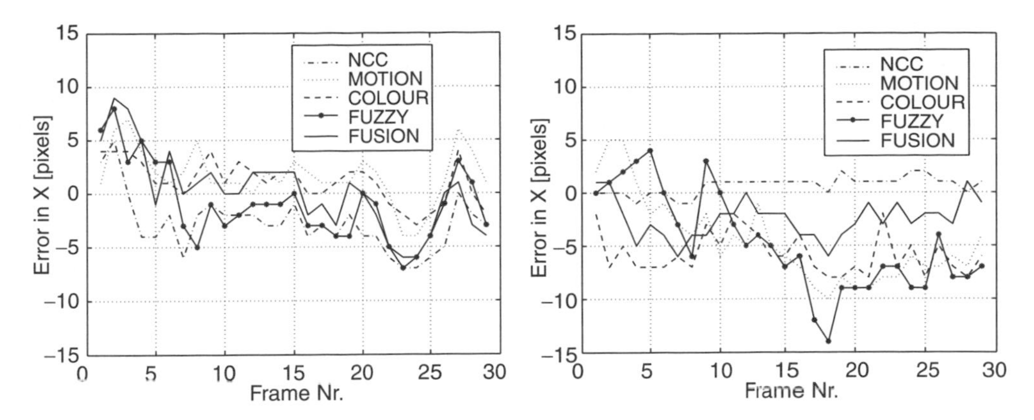

Figure 1.8 Distance error in the X-horizontal (left image) and Y-vertical (right image) direction for the modules during experiment. For clarity reasons, the results from the edge and disparity cues are not presented.

Figure 1.9 The configuration used in the Experiment 3.

1.5.3 Evaluation

In this section, we evaluate the performance of each cue estimator as well as the result obtained by their integration. To present the quality of tracking we used the measure of the *relative error*. To express the relative error we used the distance (in pixels) from the ground truth position values (that were obtained off-line) and the position values obtained by from the cue estimators and the integration modules. To estimate the ground truth, we chose the tool–center–point (TCP) to be our reference position.

The results presented in Table 1.1, Table 1.2 and Table 1.3 show the relative error separately for the X-horizontal and Y-vertical image components to determine which of these components gives stronger contribution to the overall result. We present the mean and the standard deviation of the distance error, along with the mean and the standard deviation of the X and Y image components.

1.5.3.1 Experiment 1. The first experiment demonstrates performance of the cues and their integration in a regular scene with few challenges. It is here evident that most most cues have a reasonable performance. The disparity cue is the only cue with significant deviations (due to the lack of texture on the target object). For the error graph for X values it is apparent that the cues and their integration all perform well. On the error graph for Y values it is apparent that the higher weight for the correlation implies that the voting scheme relies heavily on this particular cue. This implies that the correlation and votingfs-based fusion provides results on the same order. Overall tracking with a standard deviation of 2 pixels is very good. The bias trend in the errors for the X values is due to motion in depth. The fuzzy integration has an

TABLE 1.1 Relative error presented by the mean distance error and standard deviation. The main results are highlighted.

Module	Mean X	Std. Dev. X	Mean Y	Std. Dev. Y	Mean	Std. Dev.
color	−0.4483	3.8414	−7.8621	1.6415	**8.7411**	**1.5944**
Motion	−0.4483	2.2926	−4.8276	2.7264	**5.5291**	**2.3179**
Disparity	7.0345	4.4680	13.7931	2.3204	**15.9268**	**3.3047**
Edge track	−1.6552	2.3946	−4.1034	2.5403	**5.0062**	**2.5509**
NCC	−3.7586	3.3129	1.8276	0.6584	**4.7063**	**2.5613**
Voting schema	−2.1724	4.0976	−0.2069	1.4238	**4.0131**	**2.6590**
Fuzzy fusion	−0.3448	3.9304	−5.3448	3.6963	**7.2744**	**2.0035**

TABLE 1.2 Relative error presented by the mean distance error and standard deviation. The main results are highlighted.

Module	Mean X	Std. Dev. X	Mean Y	Std. Dev. Y	Mean	Std. Dev.
color	1.1379	2.0129	−5.7586	1.9394	**6.2685**	**1.6739**
Motion	1.6897	2.5369	−4.5172	4.0322	**6.2358**	**2.5517**
Disparity	7.4483	4.2560	−6.6552	3.4046	**10.7978**	**3.5042**
Edge track	0.7241	2.8017	−4.7241	2.8896	**5.7677**	**2.3244**
NCC	−2.7931	2.6777	0.6552	0.8140	**3.5408**	**1.8364**
Voting schema	0.2414	3.7288	−2.4828	1.8636	**4.3992**	**1.9436**
Fuzzy fusion	−0.8966	3.6679	−4.7586	4.7182	**7.0240**	**2.9836**

TABLE 1.3 Relative error presented by the mean distance error and standard deviation. The main results are highlighted.

Module	Mean X	Std. Dev. X	Mean Y	Std. Dev. Y	Mean	Std. Dev.
color	3.1034	3.6385	−5.5517	6.0568	**8.8805**	**3.1845**
Motion	1.8966	1.6550	−4.1724	3.7803	**5.5900**	**2.5342**
Disparity	11.3793	4.6323	−6.5172	3.2362	**13.7512**	**3.7664**
Edge track	1.3103	2.2056	−5.6552	3.4669	**6.5618**	**2.6815**
NCC	−3.2759	3.4836	0.1724	3.2190	**4.2967**	**3.8108**
Voting schema	1.2759	2.8772	−3.5517	2.1479	**4.9133**	**1.6249**
Fuzzy fusion	1.2414	4.3232	−5.0000	3.6839	**6.7763**	**3.4917**

equal weight for the different cues which implies that the estimator has a larger bias, but as most of the simple cues are in agreement the variation over the sequence is still very good.

1.5.3.2 Experiment 2. In this experiment significant background clutter is introduced into the scene, to determine handling of 'noise'. The distance errors in X-horizontal and Y-vertical direction are presented in Fig 1.8. Again the X-values, which have little variation provides a good result. From the Y error graph it is apparent that the simple cues like motion and edge tracking results in larger deviations. This in turn reflects on the integrated cue estimates. The fuzzy fusion is here very noisy and the end result is a large bias and a variation over the

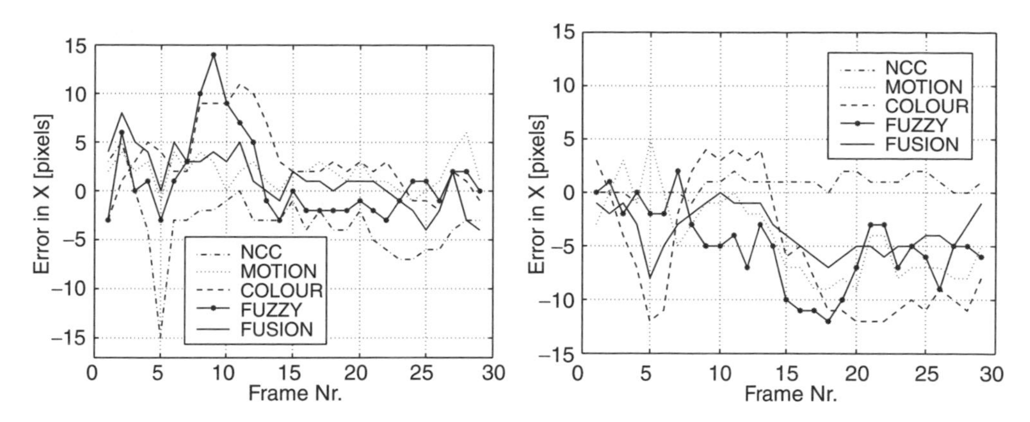

Figure 1.10 Distance error in the X-horizontal (left image) and Y-vertical (right image) direction for the modules during experiment. For clarity reasons, the results from the edge and disparity cues are not presented.

sequence which is worse that most of the cues alone. Fuzzy does no offer added robustness in this situation, due to variations between frames, caused by the background clutter. The weighted voting performs worse than the correlation due to the noise introduced by cues that are unable to handle background clutter. Weighted averaging is not an advantage in the presence of significant noise.

1.5.3.3 Experiment 3. The third experiment includes multiple moving objects and occlusion. The independent motion will confuse the simple motion detector, as it is used without a windowing function (obviously, it is possible to introduce validation gates to reduce this problem). The occlusion will at the same time result in failure for the correlation tracker, which demonstrates a case where cue integration is needed to ensure continuous tracking. The distance errors in X-horizontal and Y-vertical direction are presented in (Fig. 1.10). The error graphs clearly illustrate how correlation and motion cues fail. It is here evident that this failure is propagated almost unattenuated through the fuzzy tracker, which indicates that the simply fuzzy fusion used here is inappropriate for the type of cue integration pursued in this work. For the weighted consensus voting the consensus of most cues allow the system to cope with both types of error situations which is reflected in the limited variation over the image sequence and the best overall performance.

1.5.3.4 Summary. The above results clearly illustrate how integration of multiple cues facilitates robust tracking of objects in the presence of clutter and simple occlusion. It is here of particular interest to see how very simple cues, that are suitable for real-time implementation, can be used for tracking even though the would be unsuitable if applied individually.

1.6 SUMMARY

Vision is not in everyday use for manipulation of objects, due to cost and robustness considerations. The cost of computational systems is constantly dropping and systems with a reasonable performance are now within reach. To achieve robustness it is typically necessary to resort the use of markers on the objects to be manipulated. For a general use this is not a satisfactory solution. In this chapter the potential use of natural features combined with cue integration techniques has been pursued. The aim has been to use simple features that are amendable to real-time operation combined with model-free integration techniques. The choice of model free fusion methods was deliberate to avoid extensive training and use of models with limited validity.

The presented experimental results clearly indicate that our choice of a fuzzy integration technique was wrong. This does not indicate that fuzzy is the wrong choice for cue integration, but standard off the shelves techniques might not be adequate; an in-depth study of fuzzy rules, with specific models for cues might be needed to provide a robust solution.

Use of weighted consensus voting on the other hand performed extremely well, and the advantage is here a minimum need for explicit modeling of the different cues. The experiment demonstrated that weighted consensus voting is able to cope with failures in individual cues and handling of significant clutter that might corrupt output from many of the cues (but the corruption is uncorrelated between cues).

Overall a successful approach to cue integration has been presented. The experimental system was rather simple and there are thus a number of obvious problems for future research. First, little, if any, motion information was used for tracking of features throughout a sequence. It is here of interest to study use of robust estimation techniques for filtering of outliers, and definition of attention region that facilitate faster/simpler processing of individual images. The present systems always used information from all cue estimators. It is here of interest to define confidence measures that allow selective fusion of reliable cues, while discarding non-robust cues. Such a dynamic switching is likely to provide more robust results at a reduced computational cost.

1.7 ACKNOWLEDGMENTS

This work has been sponsored by the Swedish Foundation for Strategic Research through its contract with the Centre for Autonomous Systems.

REFERENCES

[1] E. D. Dickmanns, B. Mysliwetz, and T. Christians, "An integrated spatio-temporal approach to automatic visual guidance of autonomous vehicles," *IEEE Transactions on Systems, Man and Cybernetics*, vol. 37, pp. 1273–1284, November/December 1990.

[2] E. Dickmanns, "Vehicles capable of dynamic vision: a new breed of technical beings?," *Artificial Intelligence*, vol. 103, pp. 49–76, August 1998.

[3] D. Koller, K. Daniilidis, and H. H. Nagel, "Model-based object tracking in monocular image sequences of road traffic scenes," *International Journal of Computer Vision*, vol. 10, no. 3, pp. 257–281, 1993.

[4] H. Kollnig and H. Nagel, "3d pose estimation by directly matching polyhedral models to gray value gradients," *International Journal of Computer Vision*, vol. 23, no. 3, pp. 282–302, 1997.

[5] G. Hager, "Calibration-free visual control using projective invariance," *Proceedings ICCV*, pp. 1009–1015, 1995.

[6] G. Hager, W. Chang, and A. Morse, "Robot hand-eye coordination based on stereo vision," *IEEE Control Systems Magazine*, vol. 15, no. 1, pp. 30–39, 1995.

[7] B. Espiau, F. Chaumette, and P. Rives, "A new approach to visual servoing in robotics," *IEEE Transactions on Robotics and Automation*, vol. 8, no. 3, pp. 313–326, 1992.

[8] R. Pissard-Gibollet and P. Rives, "Applying visual servoing techniques to control a mobile hand-eye system," in *Proceedings IEEE Robotics and Automation, Nagoya, Japan*, pp. 725–732, October 1995.

[9] P. Allen, "Automated tracking and grasping of a moving object with a robotic hand-eye system," *IEEE Transactions on Robotics and Automation*, vol. 9, p. 152, 1993.

[10] P. Allen and B. H. Yoshimi, "Active, uncalibrated visual servoing," *IEEE ICRA*, vol. 4, pp. 156–161, May 1994.

[11] B. H. Yoshimi and P. Allen, "Closed-loop visual grasping and manipulation," *IEEE International Conference Robotics and Automation*, April 1996.

[12] N. P. Papanikolopoulous and P. K. Khosla, "Adaptive robotic visual tracking: theory and experiments," *IEEE Transactions on Automatic Controller*, vol. 38, no. 3, pp. 429–445, 1993.

[13] N. Papanilolopoulos, B. Nelson, and P. Khosla, "Six degree of freedom hand/eye visual tracking with uncertain parameters," *IEEE Transactions Robotics and Automation*, vol. 13, pp. 725–732, October 1995.
[14] J. Clark and A. Yuille, *Data fusion for sensory information precessing systems*. Kluwer Academic Publisher, 1990.
[15] J. Aloimonos and D. Shulman, *Integration of Visual Modules*. New York, Academic Press, 1989.
[16] P. Pirjanian, H. I. Christensen, and J. Fayman, "Application of voting to fusion of purposive modules: an experimental investigation," *Robotics and Autonomous Systems*, vol. 23, no. 4, pp. 253–266, 1998.
[17] J. Fayman, P. Pirjanian, and H. I. Christensen, "Fusion of redundant visual behaviours," *IEEE International Conference on Robotics and Automation 1997*, vol. 1, pp. 425–430, May 1997.
[18] I. Bloch, "Information combination operators for data fusion: a comparative review with classification," *IEEE Transactions on Systems, Man and Cybernetics, Part A:Systems and Humans*, vol. 26, no. 1, pp. 42–52, 1996.
[19] L. Lam and C. Suen, "Application of majority voting to pattern recognition: An analysis of its behaviour and performance," *IEEE Transactions on Systems, Man and Cybernetics, Part A: Systems and Humans*, vol. 27, no. 5, pp. 553–568, 1997.
[20] A. Blake and A. Zisserman, *Visual reconstruction*. Cambridge, Mass, MIT Press, 1987.
[21] B. Parhami, "Voting algorithms," *IEEE Transactions on Reliability*, vol. 43, no. 3, pp. 617–629, 1994.
[22] E. H. Mamdani and S. Assilian, "An experiment in linguistic synthesis with a fuzzy logic controller," *International Journal of Man-Machine Studies*, vol. 7, pp. 1–13, 1975.
[23] Y. F. Li and C. C. Lau, "Development of fuzzy algorithms for servo systems," in *IEEE International Conference Robotics and Automation*, Philadelphia, April 1988.
[24] H. Borotsching and A. Pinz, "A new concept for active fusion in image understanding applying fuzzy set theory," *Proceedings 5th IEEE International Conference on Fuzzy Systems*, 1996.
[25] A. Pinz, M. Prantl, H. Ganster, and H. Borotsching, "Active fusion–a new method applied to remote sensing image interpretation," *Pattern Recognition Letters*, 1996.
[26] L. Zadeh, "Fuzzy sets," *Information and Control*, vol. 8, pp. 338–353, 1965.
[27] J. Godjevac, *Neuro-Fuzzy Controllers; an Application in Robot Learning*. PPUR-EPFL, 1997.
[28] H. -J. Zimmermann, *Fuzzy Sets Theory-and Its Applications*. Kluwer Academic Publishers, 1990.
[29] C. C. Lee, "Fuzzy logic in control systems: fuzzy logic controller - part 1 and part 2," *IEEE Transactions on Systems, Man and Cybernetics*, vol. 20 (2), pp. 404–435, 1990.
[30] Tsukamoto, "An approach to fuzzy reasoning method," in *Advances in Fuzzy Set Theory and Applications* (M. Gupta, R. K. Ragade, and R. R. Yager, eds.), 1979.
[31] J. -S. R. Jang, "ANFIS, adaptive-network-based fuzzy inference systems," *IEEE Transactions on Systems, Man and Cybernetics*, 1992.
[32] F. Perez and C. Koch, "Towards color image segmentation in analog VLSI:Algorithm and hardware," *International Journal of Computer Vision*, vol. 12, no. 1, pp. 629–639, 1994.
[33] F. Enneser and G. Medioni, "Finding Waldoo, or focus of attention using local color information," *IEEE Transactions on Pattern Analysis Machine Intelligence*, vol. 17, no. 8, pp. 805–809, 1995.
[34] C. L. Novak and S. Shafer, "Color vision," *Encyclopedia of Artificial Intelligence*, pp. 192–202, 1992.
[35] C. Dejean, "Color based object recognition," Master's thesis, Computational Vision and Active Perception Laboratory, Royal Institute of Technology, Stockholm, Sweden, 1998.
[36] C. Smith, S. Brandt, and N. Papanikolopoulos, "Eye-in-hand robotic tasks in uncalibrated environments," *IEEE Transactions on Robotics and Automation*, vol. 13, no. 6, pp. 903–914, 1996.
[37] J. Crowley and J. Coutaz, "Vision for man machine interaction," *EHCI*, August 1995.
[38] J. Crowley and J. Martin, "Comparision of correlation techniques," *Conference on Intelligent Autonomous Systems, IAS'95*, March 1995.
[39] J. Crowley, F. Berard, and J. Coutaz, "Finger tracking as an input device for augmented reality," *IWAGFR*, June 1995.
[40] J. Little, "Robot partners: Collaborative perceptual robotic systems," *First Int. Workshop on Cooperative Distributed Vision*, pp. 143–164, 1997.
[41] A. Maki, *Stereo vision in attentive scene analysis*. Phd thesis, Dept. of Numerical Analysis and Comuting Science, KTH, Stockholm, Sweden, 1996.

[42] J. Malik, "On binocular viewed occlusion junctions," in *Proc. 4th ECCV* (B. Buxton and R. Cipolla, eds.), vol. 1064 of *Lecture Notes in Computer Science*, Springer Verlag, 1996.

[43] C. Eveland, K. Konolige, and R. Bolles, "Background modelling for segmentation of video–rate stereo sequences," *Proceedings Computer Vision Pattern Recognition*, June 1998.

[44] Y. Zhing, A. Jain, and M. Dubuisson-Jolly, "Object tracking using deformable templates," *Proc. 6th ICCV*, pp. 440–445, 1998.

[45] X. Shen and D. Hogg, "3D shape recovery using a deformable model," *Image and Vision Computing*, vol. 13, no. 5, pp. 377–383, 1995.

[46] M. Irani and S. Peleg, "Motion analysis for image enhacement: Resolution, occlusion and transparency," *Journal of Visual Communication and Image Representation*, vol. 4, no. 4, pp. 324–335, 1993.

[47] M. Black and A. Jepson, "Estimating optical flow in segmented images using variable–order parametric models with local deformations," *IEEE Transactions on Pattern Analysis Machine Intelligence*, vol. 18, no. 10, pp. 972–986, 1996.

[48] T. Uhlin, P. Nordlund, A. Maki, and J. -O. Eklundh, "Towards an active visual observer," Technical report, Department of Numarical Analysis and Computing Science, KTH, Stockholm, Sweden, 1995.

[49] P. Allen, A. Timcenko, B. Yoshimi, and P. Michelman, "Automated tracking and grasping of a moving object with a robotic hand-eye system," *IEEE Transactions on Robotics and Automation*, vol. 9, no. 2, pp. 152–165, 1993.

[50] Y. Shirai, R. Okada, and T. Yamane, "Robust visual tracking by integrating various cues," *Proceedings Conference Robotics and Automation'98 Workshop WS2*, 1998.

[51] K. Nickels and S. Hutchinson, "Characterizing the uncertainities in point feature motion for model-based object tracking," *Proceedings Workshop on New Trends in Image-Based Robot Servoing*, pp. 53–63, 1997.

[52] C. Smith and N. Papanikolopoulos, "Grasping of static and moving objects using a vision-based control approach," *Proceedings International Conference on Robotics Systems IROS*, vol. 1, pp. 329–334, 1995.

[53] L. Zadeh, "Outline of a new approach to the analysis of complex systems and decision processes," *IEEE Transactions on Systems, Man and Cybernetics*, vol. 3, no. 1, pp. 28–44, 1973.

Chapter 2

SPATIALLY ADAPTIVE FILTERING IN A MODEL-BASED MACHINE VISION APPROACH TO ROBUST WORKPIECE TRACKING

H. -H. Nagel
Universität Karlsruhe & Fraunhofer-Institut IITB Karlsruhe

Th. Müller
Universität Karlsruhe

V. Gengenbach
Fraunhofer-Institut IITB Karlsruhe

A. Gehrke
Universität Karlsruhe

Abstract

The introduction of model-based machine vision into the feedback loop of a robot manipulator usually implies that edge elements extracted from the current digitized video frame are matched to segments of a workpiece model which has been projected into the image plane according to the current estimate of the relative pose between the recording video camera and the workpiece to be tracked.

In the case in which two (nearly) parallel projected model segments are close to each other in the image plane, the association between edge elements and model segments can become ambiguous. Since mismatches are likely to distort the state variable update step of a Kalman-Filter-based tracking process, suboptimal state estimates may result which can potentially jeopardize the entire tracking process. In order to avoid such problems, spatially adjacent projected model segments in the image plane have been suppressed by an augmented version of a hiddenline algorithm and thereby excluded from the edge element association and matching step—see, e.g., Tonko et al. [1].

Here, we study an alternative approach towards increasing the robustness of a machine-vision-based tracker, exploiting the fact that a gradient filter can be adapted to the spatial characteristics of the local gray value variation: the filter mask is compressed in the gradient direction and enlarged perpendicular to the gradient direction, i.e., along the edge. We investigate the effect which such a spatially adaptive gradient filter for the extraction of edge elements from image frames exerts upon the association between edge elements and model segments, upon the subsequent fitting process in the update step, and thereby upon the robustness of the entire model-based tracking approach.

2.1 INTRODUCTION

Explicating the knowledge for a computer-based approach facilitates to study a solution at various stages of its implementation in a *systematic* manner. Provided knowledge has been *explicated*, failures which may be observed under certain conditions can be more easily attributed to particular 'chunks of knowledge' or to not having taken them into account. As a consequence, *incremental* improvement becomes feasible with less effort than in the case of a more heuristic approach based on *hidden assumptions*.

It appears, however, difficult to follow such insights when certain goals have to be achieved with insufficient (computing) resources: like in other real-life situations where one has to manage with insufficient resources, temptation is great to gamble 'that it just might work out.' Attempts to incorporate machine vision into the feedback loop of a robot manipulator provide ample examples.

In case of automatic *dis*assembly, one can not assume that pose, shape, and size of parts to be manipulated are provided with sufficient precision by a workpiece or product database. During extended use, parts may have been deformed, shifted to positions different from those assigned in the original construction, or even replaced by functionally equivalent ones of different appearance. For disassembly tasks, sensorial feedback thus needs to control robot manipulations. Machine vision offers enticing perspectives as a sensorial feedback channel.

The model-based approach of [1] toward tracking of tool and workpiece in video sequences of disassembly operations explicates knowledge of different kinds: for example, knowledge about objects in the form of 3D object models or knowledge about the imaging process by a camera model and its calibration. Knowledge about the short-time relative motion between a recording stereo camera configuration, a tool, and a workpiece is captured in a motion model which underlies a Kalman-Filter prediction and update step. Experience with this approach provides the background to illustrate the working hypothesis formulated in the introductory paragraph. In particular, we investigate the question whether and to which extent the use of an *adaptive filter* for a rather elementary operation—namely the extraction of edge elements from gray value images of a robotics scene—can contribute to a more robust *tracking* of tools and workpieces *based on machine vision*.

The next section will define the problem to be studied in more detail and outline the boundary conditions for our investigation. In subsequent sections we shall describe our approach, the experiments performed, and the results obtained thereby. The concluding discussion will also address related publications.

2.2 SPECIFICATION OF PROBLEM AND BOUNDARY CONDITIONS

The panels in the left column of Fig. 2.1 from [2] show approximately the same section from the first and last frame, respectively, of a video sequence recording the engine compartment of a used car while the mobile video-camera moved towards the battery. The panels in the right column of this figure show edge elements extracted by a special purpose computer ([3]) from the videoframes depicted in the corresponding left panel. A Kalman-Filter allows to track the workpiece—in this case the car battery—through the image sequence.

2.2.1 Outline of Our Machine-Vision-Based Tracking Process

The car battery is modeled as a quasi-polyhedron which includes the circular seals of the different battery sections. The recording camera is modeled by a central projection (pinhole model). The internal camera parameters have been determined in a separate calibration step and thus are known. The external camera parameters, i.e., the six degrees of freedom (DoF) for translation and rotation, constitute the components of a Kalman-Filter state vector which describes the current relative pose between video-camera and the engine compartment, in particular the battery.

In order to set the stage for subsequent discussions, the principal steps of the Kalman-Filter tracking loop are recapitulated here ([4, 2], see also [5, 6]):

- A special purpose computer (see [3]) digitizes and stores the current frame recorded by the moving video-camera.

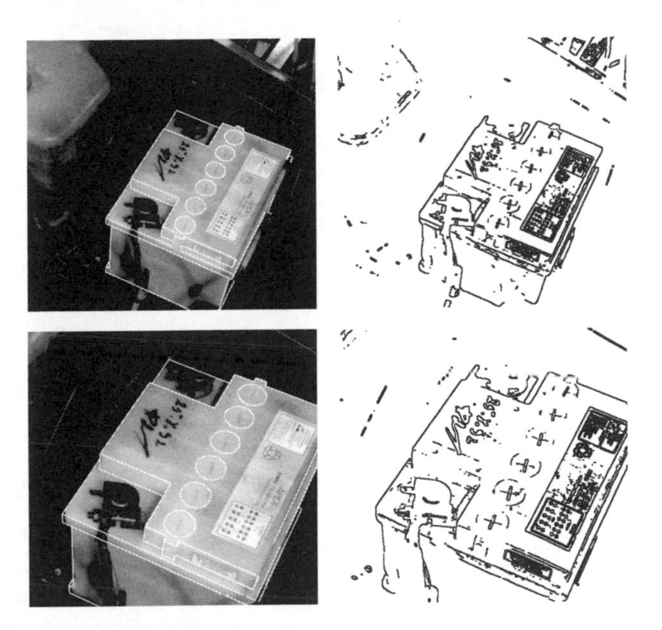

Figure 2.1 Visually servoed manipulation of car parts: Superimposed result of a pose estimation at the start (top left) and at the end (bottom left) of a purely translatory camera motion toward a partly occluded battery, which is still at its place of operation. Note that those parts of projected segments, which trigger association problems, have been removed in the projection step. The right column shows the corresponding edge elements, which were used as measurement data in the update step of the pose estimator. Despite the partial occlusion of the battery and the fact that it moves partially out of sight during the camera motion, the tracker relying on *edge elements* successfully estimates the object pose. A tracker which relies on *aggregated features* gets into trouble by the partial occlusion. (From [2].)

- Subsequently, this special purpose computer convolves the current digitized image with the digitized and quantized versions of the partial derivatives of a Gaussian. Data edge elements (DEEs) are then determined by searching for local maxima of the gray value gradient magnitude in the gradient direction. A DEE is represented by a three-tuple of image plane location and orientation $\boldsymbol{m} = (u, v, \phi)^T$.
- Based on the current estimate for the state vector describing the relative pose between the mobile video-camera and the battery to be manipulated, a model of the battery is projected (hidden lines removed) into the image plane of the camera—see the panels in the left column of Fig. 2.1. The visible part of a projected boundary between adjacent facets of the workpiece model are called 'Model Edge *Segment* (MES).' A MES $\boldsymbol{e}$ is represented by its center (u_m, v_m), orientation θ and length l: $\boldsymbol{e} = (u_m, v_m, \theta, l)^T$.
- DEEs $\boldsymbol{m}$ in a tolerance ribbon around a MES $\boldsymbol{e}$ are associated to this MES, provided a scalar distance function does not exceed a threshold—see [Heimes et al. 98] and Fig. 2.2. This distance function takes into account not only the perpendicular Euclidean distance between the DEE and the MES, but also the orientation difference Δ between the gradient direction of the DEE and the normal to the MES.
- The filter state is updated by minimizing the sum of such distances for all associated DEEs—see [4, 2].
- Based on a constant motion model, the filter state is predicted for the next image frame time point.

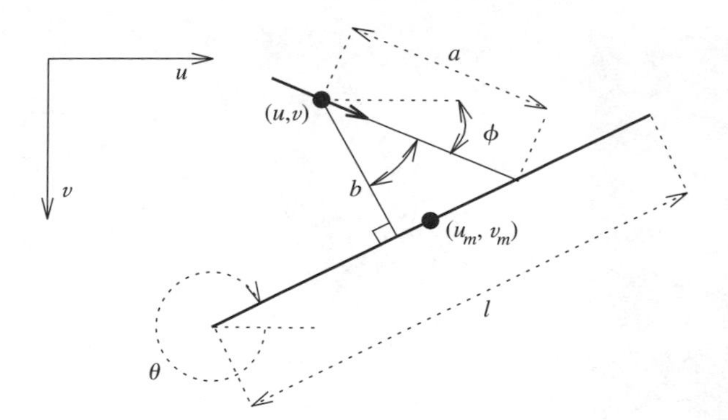

Figure 2.2 Association of a 'Data Edge *Element* (DEE)' to a visible 'Model Edge *Segment* (MES)': Angle Δ and distance b are used to determine a, which in turn determines the association of the DEE $(u, v, \phi)^T$ with the MES $(u_m, v_m, \theta, l)^T$. (From [2].)

2.2.2 Matching Data Edge Elements versus Matching Data Edge Segments

Usually, DEEs are aggregated by a data-driven process into data edge *segments* which in turn are matched to MESs—see, e. g., [8]. Such an approach is considerably faster than matching DEEs to MESs, since the number of data edge *segments* to be treated during tests for association with MESs and during the subsequent minimization of the filter state update is smaller than the number of DEEs. It is, however, noticeably less robust than the direct matching of DEEs to a MES as outlined in the preceding subsection.

The gain in robustness for a direct DEEs/MES-match can be attributed to the following fact: the location, length, and orientation of the MES—around which DEEs are collected for an association test—provide knowledge about which DEEs could constitute a data edge *segment*. In contradistinction, a *purely data-driven* aggregation process must rely on a general inter-DEE distance threshold in order to decide whether a DEE could form a candidate for the continuation of a current tentative data edge *segment*. The model-based tracking process thus *explicates* available knowledge about where edges are expected in the image plane. Exploitation of this knowledge then increases the robustness of DEE treatment. One price to be paid for this increase in robustness consists, of course, in the necessity to provide *more* computing power or—if the computing power remains the same—to wait longer than in the case of a purely data-driven, heuristic aggregation subprocess *outside* the Kalman-Filter update step.

2.2.3 Overcoming a Weakness of Model-Based Data Edge Element Collection

As can be seen in Fig. 2.1, MESs may be close neighbors in the image plane for certain relative poses between camera and workpiece, even if the corresponding boundaries between facets of the workpiece model are sufficiently separated in the scene space. The tolerance ribbons around two neighboring MESs may thus overlap with the consequence that the association of DEEs to a MES becomes ambiguous.

In order to avoid trouble which might be generated by such ambiguities, [4] modified the hidden line removal subprocess such that it suppresses MESs which are too close to each other in the image plane—see Fig. 2.3. The contributions of corresponding segments in the 3D boundary representation (BREP) of the modeled battery are suppressed likewise in the expression to be minimized during the update step. This suppression constitutes another example for the exploitation of explicated knowledge as a means to increase the robustness of machine vision in a robot control loop.

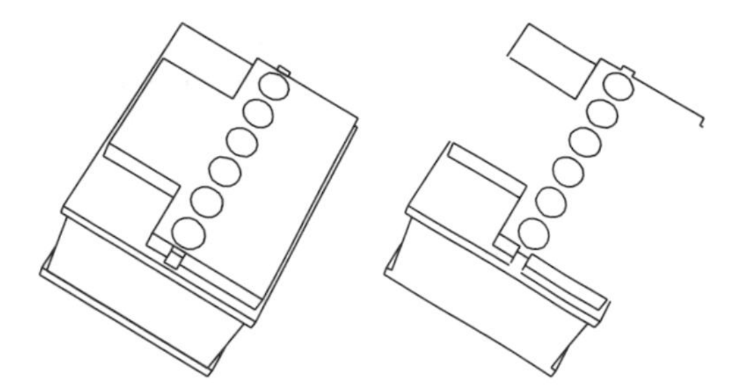

Figure 2.3 Tackling association problems: Most association problems are triggered by almost parallel visible edge segments which lie close to each other. Consequently, overlapping parts of almost parallel, neighboring visible model edge segments (MESs) as well as the corresponding parts of the respective BREP-segments are removed by a modified hiddenline algorithm. (From [4].)

2.2.4 An Alternative Solution Approach

Although experience with the approach outlined in the preceding subsection has been very encouraging ([4, 2, 1]), the following considerations motivate a search for an alternative to a simple suppression of potentially trouble-generating MESs.

In case the object to be tracked is partially occluded by other objects, removal of MESs may reduce the number of DEEs which can still be taken into account to a level where the remaining information from an image frame for the update step becomes dangerously low. Although we did not yet encounter such situations during our experiments so far, it nevertheless constitutes a possibility to be reckoned with.

Another consideration addresses the observation that DEEs which form two (almost) parallel lines in the image plane—see the edge lines for the top right battery contours in Fig. 2.1—still provide reliable information about the expected orientation of MESs, even if the association of individual DEEs to competing MESs might create mismatches and thereby negatively influence the tracking process. It would be advantageous if at least this information could be extracted from an image frame.

The challenge thus consists in formulating an alternative which on the one hand avoids or significantly reduces the possibility of mismatches between DEEs and MESs and, on the other hand, nevertheless allows to exploit a larger part of the information in the set of DEEs extracted from an image frame. In response to this analysis, it appears desirable to increase the reliability of the DEE-extraction process to a level where more reliable association and decision procedures may become acessible.

Based on experience in a different discourse domain—the tracking of vehicle images in video sequences of road traffic, see [9]—spatially adaptive filters for the computation of gray value gradients should be investigated.

2.3 SOLUTION APPROACH TO BE INVESTIGATED

So far, gradients $\nabla g(\mathbf{x})$ at pixel position $\mathbf{x} = (x, y)^T$ have been estimated by convolving the gray value function g($\mathbf{x}$) with the discretized version of the partial derivatives of a bivariate Gaussian:

$$G(\mathbf{x}) = \frac{1}{(2\pi)\sqrt{|\Sigma|}} e^{-\frac{1}{2}\mathbf{x}^T \Sigma^{-1} \mathbf{x}} \tag{2.1}$$

The covariance matrix Σ is initially set to

$$\Sigma_{\text{initial}} = \begin{pmatrix} \sigma_x^2 & 0 \\ 0 & \sigma_y^2 \end{pmatrix} \tag{2.2}$$

with equal values for the standard deviations σ_x in the x-direction and σ_y in the y-direction, chosen by the experimenter on the basis of a priori knowledge.

The spatial Gaussian introduced by (2.1) specifies a local environment around a location **x** which is sampled in order to compute a gradient at this location. In general, a position- and orientation-invariant environment does not appear to be optimal in image areas with inhomogeneous gray value variations. It should be preferable to adjust this sampling environment such that the low-pass action of the Gaussian is compressed in the direction of strong gray value transitions and extended in the direction parallel to gray value transition fronts. The implementation of such a postulate requires, however, that the prevalent direction of a local gray value variation is estimated at each location individually in a space-variant manner.

2.3.1 Estimation of Local Gray Value Characteristics

For this purpose, we introduce the *location-dependent* 'Gray Value-Local-Structure-Tensor (GLST)' as

$$GLST(\mathbf{x}) = \int_{-\infty}^{+\infty} d\boldsymbol{\xi} \frac{\nabla g(\boldsymbol{\xi} - \boldsymbol{x})\,(\nabla g(\boldsymbol{\xi} - \boldsymbol{x}))^T}{(2\pi)\sqrt{|2\Sigma|}} \cdot e^{-\frac{1}{2}\boldsymbol{\xi}^T(2\Sigma)^{-1}\boldsymbol{\xi}} \tag{2.3}$$

with twice the covariance (2Σ) in the Gaussian which determines the environment for weighting $\nabla g(\nabla g)^T$ than in the Gaussian used for estimating ∇g at each location **x**—see, e.g., [10].

By definition, the GLST is positive-semidefinite. $GLST_{\text{initial}}$ denotes a GrayValue-Local-Structure-Tensor computed by using Σ_{initial} as given by (2.2). Let $\mathbf{e}_{GLST_{\text{initial}},i}$ denote the ith eigenvector of $GLST_{\text{initial}}$ and $\lambda_{\text{initial},i}$ the corresponding eigenvalue, with $\lambda_{\text{initial},1} \geq \lambda_{\text{initial},2} \geq 0$. The matrix

$$U = \left(\mathbf{e}_{GLST_{\text{initial}},1} \;,\; \mathbf{e}_{GLST_{\text{initial}},2} \right) \tag{2.4}$$

denotes a 2D rotation matrix in the image plane which aligns the coordinate axes with the eigenvector directions of $GLST_{\text{initial}}(\mathbf{x})$. The latter can then be written in the form

$$GLST_{\text{initial}} = U(\mathbf{x}) \begin{pmatrix} \lambda_{\text{initial},1} & 0 \\ 0 & \lambda_{\text{initial},2} \end{pmatrix} (U(\mathbf{x}))^T \tag{2.5}$$

Following [9], we denote by $\sigma^2_{\text{minsize}}$ a parameter which forces a diagonal element in the covariance matrix for the Gaussian in (2.1) to remain compatible with the smallest admissible mask size. $\sigma^2_{\text{maxsize}}$ should be defined such that $\sigma^2_{\text{minsize}} + \sigma^2_{\text{maxsize}}$ determines the largest admissible mask size. Let I denote a 2×2 unit matrix. We use $I = U^T(\mathbf{x})U(\mathbf{x})$ and introduce a *locally adapted* covariance matrix $\Sigma(\mathbf{x})$ as

$$\Sigma(\mathbf{x}) \;=\; U(\mathbf{x})\; Diag(\mathbf{x})\; U^T(\mathbf{x}) \tag{2.6}$$

with—as one alternative studied—

$$Diag(\mathbf{x}) = \begin{pmatrix} \frac{\sigma^2_{\text{maxsize}}}{1+\sigma^2_{\text{maxsize}}\lambda_1(\mathbf{x})} & 0 \\ 0 & \frac{\sigma^2_{\text{maxsize}}}{1+\sigma^2_{\text{maxsize}}\lambda_2(\mathbf{x})} \end{pmatrix} + \sigma^2_{\text{minsize}}\; I \tag{2.7}$$

and $\lambda_i(\mathbf{x}) := \lambda_{\text{initial},i}(\mathbf{x}),\; i \in \{1, 2\}$. For our experiments to be discussed shortly, we have chosen $\sigma^2_{\text{minsize}} = 0.5$ and $\sigma^2_{\text{maxsize}} = 4.0$. Let us assume for the moment that $GLST_{\text{initial}}(\mathbf{x})$ has an eigenvalue $\lambda_{\text{initial},2}$ close to zero due to lack of gray value variation in the corresponding direction. In the limit of zero for $\lambda_{\text{initial},2}$, the second eigenvalue of $\Sigma(\mathbf{x})$ will become $\sigma^2_{\text{minsize}} + \sigma^2_{\text{maxsize}}$, thereby restricting a digitized version of the resulting Gaussian to the chosen maximal mask size. In case of a very *strong* straight line gray value transition front in the vicinity of image location **x**, the first eigenvalue $\lambda_{\text{initial},1}(\mathbf{x})$ of $GLST_{\text{initial}}(\mathbf{x})$ will be very large and thus will force the first contribution to the corresponding eigenvalue of $\Sigma(\mathbf{x})$ in (2.7) to be small, delimiting the sum of both terms from below by a suitably chosen $\sigma^2_{\text{minsize}}$.

In [10] (for the estimation of depth cues from stereo image pairs) and [9] (for the estimation of optical flow) the knowledge about the local spatiotemporal gray value characteristics at image (sequence) location $\mathbf{x}$ provided by the 'Gray Value-Local-Structure-Tensor' was exploited. The gradient is *recomputed*, but this time using the locally adapted $\Sigma(\mathbf{x})$ given by (2.6) in the Gaussian of (2.1) instead of the constant Σ given by (2.2).

Equation (2.6) only yields acceptable results for the eigenvalues of $\Sigma(\mathbf{x})$ and therefore practical adaptive masks, if the eigenvalues of $GLST_{\text{initial}}$ range between $\sigma^2_{\text{minsize}}$ and $\sigma^2_{\text{maxsize}}$. In order to limit the effects of large *absolute* variations of the initial eigenvalues, a normalization of these eigenvalues by $\frac{1}{2}\text{trace}(GLST_{\text{initial}}(\mathbf{x}))$—i.e., setting $\lambda_i(\mathbf{x}) := \frac{\lambda_{\text{initial},i}(\mathbf{x})}{\frac{1}{2}\text{trace}(GLST_{\text{initial}}(\mathbf{x}))}$, $i \in \{1, 2\}$—forces the eigenvalues of $\Sigma(\mathbf{x})$ to vary between reasonable limits (see also [9]).

2.3.2 Expected Improvements by Using Locally Adapted Gradient Estimates

Although locally adaptive gradient estimation is computationally much more expensive then gradients computed on the basis of a location-invariant Gaussian weight function, it appears worthwhile to explore its effect in the context of an operational system. This allows us to assess the contribution of such an approach on the overall performance and to optimize its effect by a suitable selection of parameters.

We expect that such an experiment should yield the following results:

- Closely adjacent, almost parallel gray value transition fronts should be separated better since the low-pass action of the adapted Gaussians should be weakened in the gradient direction, i.e., neighboring stronger gray value transitions should be less 'smeared' into a broader, but less pronounced one. As a consequence, separation between parallel edge elements should be improved and, thereby, the possibilities for a clean association between DEEs and MESs.
- We may as well recompute the GLST based on the locally adapted $\Sigma(\mathbf{x})$. This again should emphasize the local gray value characteristics. In the case of a dominant straight line gray value transition front, the eigenvalue associated with the eigenvector in the dominant gradient direction will be much larger than the eigenvalue in the perpendicular direction tangential to the gray value transition front: in the latter direction, the gray value should essentially remain constant, i.e., the related eigenvalue will be determined mostly by noise.

  We thus expect that both eigenvalues will become of comparable size around the endpoints or intersection of line segments where the gradient direction may change significantly over short distances. Around such locations, gradient estimates will be less well defined and thus should preferentially be excluded from matching to MESs.
- On the contrary, one might even think about matching locations with local extrema of roughly equal GLST eigenvalues to intersections of two or more MESs and, thereby, re-enforce the localisation of the model match.
- In image areas where two or more nearly parallel MESs almost merge (i.e., approach each other closer than a tolerance distance), one should define a *virtual MES* as the average of sufficiently close MESs and associate DEEs to this virtual MES. If the current filter state and the object model adequately describe the actually observed image, only one (kind of averaged) DEE should be extracted from the digitized image. This local prediction could be tested by inspection of the GLST eigenvalues. If corroborated, it would make sense to associate such 'joint' DEEs to 'merged' MESs. In this manner, it is expected to provide a better approach than suppressing the contributions of both DEEs and MESs when the latter are getting to close to each other.

2.4 RESULTS

In order to facilitate a systematic investigation of these various effects and their interaction, we recorded image sequences while tracking an object with the program version described by [4, 2, 1]. We then rerun tracking experiments with a version modified according to the suggestions discussed above.

In the first experiment, parts of model edge segments (MESs), which are visible and almost parallel with an intersegment distance below 5 pixel in the image plane, have been removed prior to the association process (Fig. 2.3). This value has emerged to be appropriate for stable battery tracking results under varying lighting conditions when working with fixed filter masks of size 5 × 7 pixel corresponding to $\sigma_x = \sigma_y = 1.0$. For lower minimal intersegment distance values, considerable tracking problems are observed (see the misaligned workpiece position overlayed in the upper left section of Fig. 2.5). Integrating the described adaptive filtering technique into the tracking process allows a minimal intersegment distance

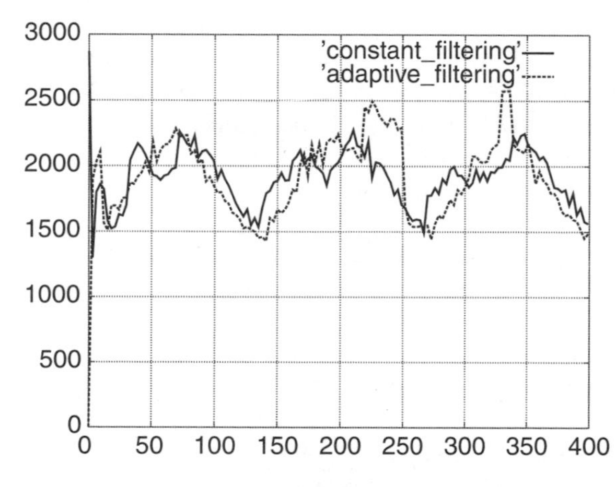

Figure 2.4 Comparison of the number of 'Data Edge *Elements* (DEEs)' associated with 'Model Edge *Segments* (MESs)' in the association process over a time interval of 400 half-frames when working with the constant and the adaptive filtering technique. One can observe two sections (half-frames 220 to 250 and 330 to 340), where the number of associated DEEs is significantly higher when using adaptive instead of constant filtering.

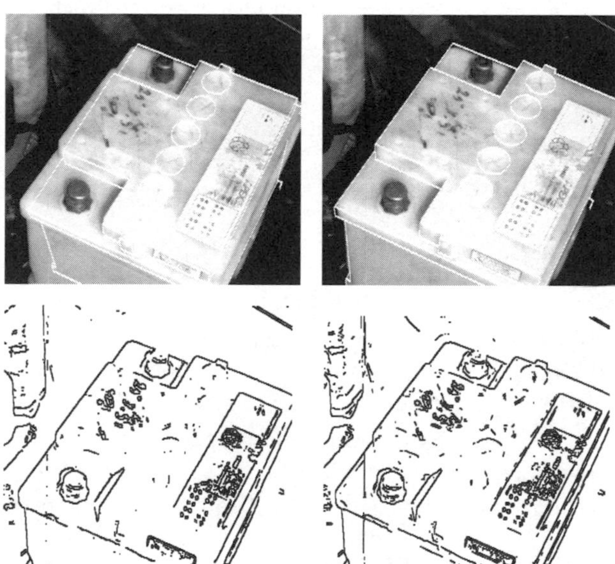

Figure 2.5 Result of pose estimation at image frame 201 (left) and calculated 'Data Edge *Elements* (DEEs)' (right) in the constant (upper) and the adaptive (lower) filtering experiment, respectively. In both experiments, a minimal intersegment distance of 2.75 pixels is used.

of 2.75 pixel and nevertheless yields substantially better tracking results than those observed with constant filter masks and a minimal intersegment distance of 5 pixel.

Figure 2.4 shows the number of DEEs used by the association process when working with constant and adaptive filter masks, respectively. In most of the image frames, the number of associated DEEs is comparable in both experiments, although there are sections in which significantly more DEEs are associated in the adaptive filtering experiment. In addition, the half-frame to half-frame variation of the number of associated DEEs in the adaptive filter experiment is smaller than in the constant filter experiment. Figure 2.5 shows pose estimations, extracted DEEs and associated DEEs at half-frame 201 for these two experiments.

2.5 DISCUSSION AND OUTLOOK

First experimental results show that the proposed adaptive filtering approach increases the robustness of our tracking process. More detailed model knowledge can be used due to the improved spatial separation between parallel edge segments. The proposed method to choose a locally adaptive covariance matrix depending on the underlying gray value structure allows further variations. Considering the real-time characteristic of the tracking process, we could reduce computation time by choosing the appropriate filtering mask from a fixed set of precalculated filtering masks. Further experiments have to be performed in order to investigate the influence of this restriction in the generation process of the filter masks. If we are able to show that a limited set of filter kernels would suffice, a real-time software solution seems to be possible with off-the-shelf computers. Depending on the number of filter kernels, we can even challenge a dedicated hardware solution for the filtering process, where the on-chip storage for filter kernels is always expensive.

REFERENCES

[1] M. Tonko, J. Schürmann, K. Schäfer, H. -H. Nagel: *Visually Servoed Gripping of a Used Car Battery*. In C. Laugier (ed.): Proceedings IEEE/RSJ International Conference on Intelligent Robots and Systems, Grenoble, France, 7–11 September 1997, INRIA – Domaine de Voluceau B.P. 105, Le Chesnay, France, 1997, pp. 49–54.

[2] M. Tonko, K. Schäfer, F. Heimes, and H. -H. Nagel: *Towards Visually Servoed Manipulation of Car Engine Parts*. In Proc. IEEE Int. Conf. on Robotics and Automation, Albuquerque, NM, 20–25 April 1997, R. W. Harrigan (ed.), Los Alamitos, CA, IEEE Computer Society Press, pp. 3166–3171.

[3] V. Gengenbach: *Einsatz von Rückkopplungen in der Bildauswertung bei einem Hand-Auge-System zur automatischen Demontage*, Dissertation, Fakultät für Informatik der Universität Karlsruhe (TH), Juli 1994. Published in the series 'Dissertationen zur Künstlichen Intelligenz,' Vol. DISKI 72, infix Verlag: Sankt Augustin, Germany, 1994 (in German).

[4] M. Tonko: *Zur sichtsystemgestützten Demontage am Beispiel von Altfahrzeugen*. Dissertation, Fakultät für Informatik der Universität Karlsruhe (TH), Juni 1997. Published in the series 'Dissertationen zur Künstlichen Intelligenz,' Vol. DISKI 166, infix Verlag: Sankt Augustin, Germany, 1997 (in German).

[5] A. Gelb (Ed.): *Applied Optimal Estimation*. Cambridge, MA, MIT Press, 1974.

[6] Y. Bar-Shalom, T. E. Fortmann: *Tracking and Data Association*. Orlando, FL, Academic Press, 1988.

[7] F. Heimes, H. -H. Nagel, T. Frank: *Model-Based Tracking of Complex Innercity Road Intersections*. Mathematical and Computer Modelling, Vol. 27, 9–11 (1998) 189–203.

[8] D. Koller, K. Daniilidis, and H. -H. Nagel: *Model-Based Object Tracking in Monocular Image Sequences of Road Traffic Scenes*. International Journal of Computer Vision, Vol. 10, No. 3 (1993) 257–281.

[9] H. -H. Nagel and A. Gehrke: *Spatiotemporally Adaptive Filtering for Estimation and Segmentation of Optical Flow Fields*. Proc. Fifth European Conference on Computer Vision (ECCV '98), June

2–6, 1998, Freiburg/Germany; H. Burkhardt and B. Neumann (Eds.), Lecture Notes in Computer Science LNCS 1407 (Vol. II), Springer-Verlag Berlin, Heidelberg, New York, 1998, pp. 86–102.

[10] T. Lindeberg and J. Gårding: *Shape-Adapted Smoothing in Estimation of 3-D Depth Cues from Affine Distortions of Local 2-D Brightness Structure*. Proc. Third European Conference on Computer Vision ECCV '94, May 2–6, 1994, Stockholm, S; J. -O. Eklundh (Ed.), Lecture Notes in Computer Science 800 (Vol. I). Springer-Verlag, Berlin Heidelberg, New York, 1994, pp. 389–400.

Chapter 3

INCREMENTAL FOCUS OF ATTENTION: A LAYERED APPROACH TO ROBUST VISION AND CONTROL

Kentaro Toyama
Microsoft Research

Gregory D. Hager and Zachary Dodds
Yale University

Abstract

Robust sensing is a crucial ingredient for robust robotics. This chapter considers an architecture called Incremental Focus of Attention (IFA) for real-time robot vision that robustly recovers from failures to track a target object. IFA is a state-based architecture for combining different tracking algorithms into a single system that is cognizant of its own degree of success.

The states of an IFA system are used to issue *action prohibitions*, in which certain robot actions are suppressed when sensory information required to perform the actions is unavailable. Action prohibition is an intuitive way to integrate robust, failure-cognizant vision systems with almost any type of robot planning scheme. We present three examples of robot tasks with IFA and action prohibition, where the robots implement a range of planning paradigms, from reactive agents to more centrally controlled planners.

3.1 INTRODUCTION

A careful study of robustness is required to make robotics applicable to worlds less structured than the strictly controlled environments of today's industrial robots. In the real world, robots must monitor themselves and act in acceptable ways regardless of the environment. A poorly behaved robot not only fails at the task it is expected to perform, it also threatens to damage surrounding equipment and may even endanger human workers.

Accurate perception is the minimal requirement for correct, or at least innocuous, robot action. A trivial option for a robot which can detect an irregularity in its perception of the environment is to shut down, perhaps signaling a human operator as it does so. More preferable, however, is an execution system which is able to detect and classify relevant sensing and/or environmental anomalies and to continue with its task to the degree that its (now degraded) perceptual inputs allow.

In this chapter, we apply Incremental Focus of Attention (IFA), a state-based framework for robust real-time vision, to robust robotics. IFA serves as a mechanism for performing *action prohibition*, where robot agents use cues from sensor modules to suppress those actions which require unavailable sensory data. Action prohibition simply vetoes certain actions based on deficiencies in sensing, and therefore, it has the advantage that it can be flexibly integrated with any robot control model, whether the model is reactive or driven by complex plans.

3.2 ROBUST TRACKING

The goal of *vision-based object tracking* is the recovery of a relevant visual event. We view vision-based tracking as a repeated search and estimation process, where the search occurs in the configuration space of the target, denoted $\mathcal{X}$, *not* in the image. The input to a tracking algorithm is an *input configuration set*, or a set of candidate target states, $\mathbf{X}^{\text{in}} \subseteq \mathcal{X}$, together with an image, I. The output at each time step consists of an *output configuration set*, $\mathbf{X}^{\text{out}} \subset \mathcal{X}$, such that $\mathbf{X}^{\text{out}} \subset \mathbf{X}^{\text{in}}$, where $\mathbf{X}^{\text{out}}$ includes $\hat{\mathbf{x}}$, the estimated configuration. Note that for actual implementation, neither input sets nor output sets are necessarily explicit. An algorithm may simply return $\hat{\mathbf{x}}$, but such output together with the known margin of error can be interpreted as a set of configurations.

The explicit margin of error allows us to precisely define several relevant terms. We say that tracking is *accurate* if and only if $\mathbf{x}^* \in \mathbf{X}^{\text{out}}$. *Mistracking*, or tracking *failure* occurs when the actual configuration, $\mathbf{x}^*$ is in $\mathbf{X}^{\text{out}}$. *Precision* is related to a measure of the size of the error margin, denoted $|\mathbf{X}^{\text{out}}|$. The simplest formulation of precision is that it relates inversely with the volume of configurations occupied by the error margin. Under this formulation, algorithms which return large margins of error are less precise than algorithms with smaller margins of error. In addition, the dimensionality of the configuration information computed by an algorithm affects its precision. For example, a tracking algorithm which can determine the position *and* orientation of a face is more precise than an algorithm which determines only the position: the former has a smaller output set than the latter, which includes sets ranging over all possible orientations.

3.2.1 IFA System Construction

IFA is a state-based framework for robust vision-based tracking in which the states encode semantic information both about the target and the system's degree of knowledge of target configuration. Transitions between states cause the system to dynamically swap different algorithmic components, called *layers*, as appropriate to the situation.

3.2.2 Layer Elements

The raw materials for IFA layers are tracking algorithms and visual search heuristics. In general, the more algorithms are available for finding and tracking a target, the better the final system, so it is advantageous to explore and exploit every visual uniqueness of the target.

Given a set of visual searching and tracking algorithms, they must first be classified as either *trackers* or *selectors*. For the sake of correctness a clear distinction between trackers and selectors must be enforced. Trackers should be algorithms which almost always generate an output set that contains the target state given an input set which contains the target. If a tracking algorithm often tracks non-target objects, it may be more suitable as a selector. Conversely, if an attentional heuristic is based on a cue that is known to be unique within its input sets, it may be better suited as a tracker. An attention algorithm focusing on flesh-colored "blobs" could be a successful tracking algorithm if guaranteed that its input will only contain one flesh-colored object. The sensitivity of the surrounding task to mistracking and accuracy will also come into consideration.

In the following, let $\overline{\mathcal{X}}$ denote the set of all subsets of $\mathcal{X}$ and let $\mathcal{I}$ denote the set of all images. We model trackers and selectors as functions from configuration sets and images to configuration sets. For a given function f, $\text{Dom}(f) \subseteq \overline{\mathcal{X}}$ and $\text{Rng}(f) \subseteq \overline{\mathcal{X}}$ denote the domain of input configuration sets and range of output configuration sets of f, respectively. In all cases, we assume that $\text{Dom}(f)$ covers $\mathcal{X}$.

3.2.2.1 ***Trackers.*** Formally, a tracker is defined as follows:

Definition 1 An *idealized tracker* is a function, $f: \overline{\mathcal{X}} \times \mathcal{I} \mapsto \overline{\mathcal{X}}$, such that for all $\mathbf{X} \in \text{Dom}(f)$ and $I \in \mathcal{I}$,

1. $f(\mathbf{X}, I) \subset \mathbf{X}$.
2. If $\mathbf{x}^* \in \mathbf{X}$, then either $\mathbf{x}^* \in f(\mathbf{X}, I)$ or $f(\mathbf{X}, I) = \emptyset$.
3. If $\mathbf{x}^* \not\in \mathbf{X}$, then $f(\mathbf{X}, I) = \emptyset$.

Together, Properties 2 and 3, which we refer to as the *filter criterion*, are a formalization of partial correctness. They state that trackers may return occasional false negatives, where a target which is present goes undetected, but never produces false positives, where a nontarget is *hallucinated* although none exists. Thus, trackers must monitor their performance and report tracking failure. Usually, geometric constraints or thresholds can be set so that a tracker will report failure when appropriate. A tracker based on a wire-frame object model, for example, might report failure when a certain percentage of its model edges lack correspondences in the image. Although precise self-assessment is optimal, for the correctness of the algorithm, it is better for trackers to err on the side of conservativeness in their own estimation of success.

3.2.2.2 ***Selectors.*** Selectors are attention-focusing algorithms which are heuristic in nature and hence prone to returning sets not containing the target configuration:

Definition 2 An *idealized selector* is a randomized function, $g : \overline{\mathcal{X}} \times \mathcal{I} \mapsto \overline{\mathcal{X}}$, such that for all $\mathbf{X} \in \text{Dom}(f)$ and $I \in \mathcal{I}$,

1. $g(\mathbf{X}, I) \subset \mathbf{X}$.
2. There is some $\epsilon_g > 0$ such that if $\mathbf{x} \in \mathbf{X}$, there is finite probability ϵ_g that $\mathbf{x} \in g(\mathbf{X}, I)$.

The purpose of selectors is to output manageable configuration sets for higher layers with a possible bias toward certain geometric or image-based constraints. For instance, a selector for face tracking will prefer *but not insist on* returning output sets which include configurations consistent with detected motion or flesh color. Thus, selectors return different output sets over time such that different portions of the configuration space are passed up the hierarchy with each call.

Finally, associated with each selector, g_i, are an *iteration index*, σ_i, and an iteration threshold, σ_i^{MAX}. The iteration index counts the number of times a selector is executed and elicits selector *failure* when it surpasses the iteration threshold (Subsection 3.2.5 describes when indices are incremented or reset). Monitoring of the iteration index prevents repeated, unsuccessful attempts to find the target in a region of the configuration space not containing the target. Actual σ_i^{MAX} values will depend on the reliability of the attention heuristic.

3.2.2.3 ***Set Dilation.*** One last adjustment is made to both trackers and selectors. Since targets are moving even as computation in layers takes place, the configuration sets output by layers must be adjusted in order to accommodate target movement. The union of potential adjusted output sets must be guaranteed to include the target configuration if the input set included it. In practice, this seemingly strict requirement is satisfied by *set dilation*, in which the initial output sets of a layer are simply expanded to include neighboring configurations as well. For the remainder of this article, $\mathbf{X}^{\text{out}'}$ will be taken to mean the output set after set dilation.

3.2.3 Layer Composition

In order to construct the system, all layers are sorted by output precision. Because greater precision often requires more processing, this means that faster algorithms will tend to occur at the bottom layers. Algorithms which are superseded by others in both speed and precision should be discarded.

Next, additional selectors are inserted wherever one layer's output is not precise enough to satisfy the input constraints of the layer above it. These selectors may be specially designed for the system and the environment, or they may be brute-force search algorithms which systematically partition a configuration space into subsets of the appropriate size. If the bottommost layer does not take the full configuration set as its input, a selector is added at the bottom. Any selectors that are more precise than the most precise tracker are discarded or converted into trackers.

Layers are then labeled from 0 to $n - 1$ such that the topmost layer is $n - 1$, and the bottommost is 0.

3.2.4 State Transition Graph

At the heart of an IFA system lies a deterministic finite state automaton (Fig. 3.1(right)), in which transitions occur based on the success of layers. States are split into `search` states which perform a search for a lost target and `track` states which output at least part of the configuration of the target. The state automaton is constructed as follows:

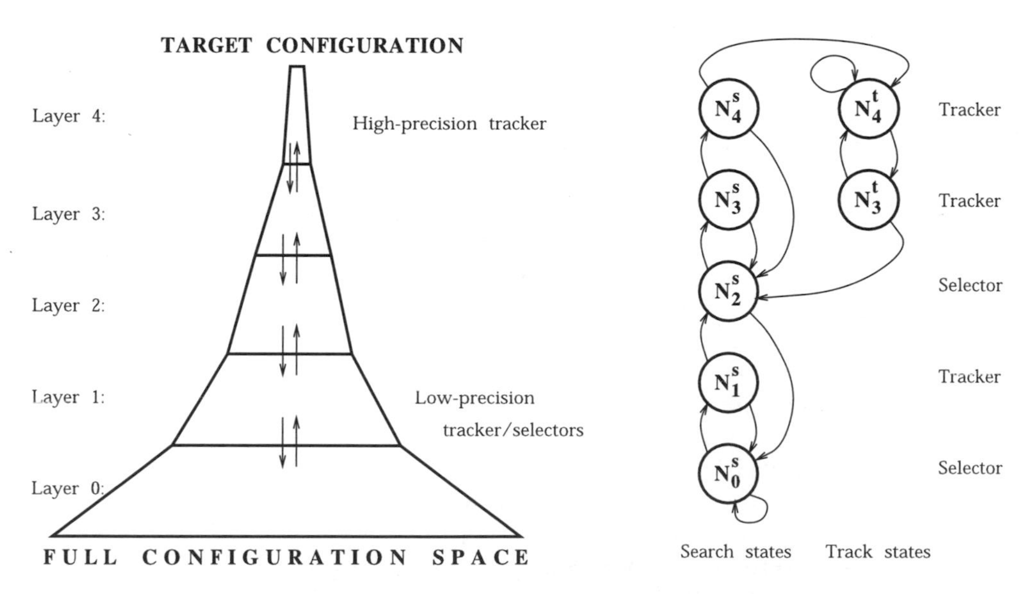

Figure 3.1 Incremental Focus of Attention. On the left, a five-layer IFA system. On the right, the corresponding finite state transition graph for an example with three trackers and two selectors. Subscripts indicate corresponding layer number; superscripts indicate whether the state is in `search` or `track` mode. Outbound edges at the top left of a state are taken when a layer is successful; outbound edges at the bottom right are taken when a layer fails.

- For each layer k $(0 \leq k < n)$, we create a node N^{s}_k.
- Let m be the index of the first selector that occurs above a tracker. We create nodes N^{t}_j, for each layer j $(m+1 \leq j < n)$. In Fig. 3.1, $n = 5$, $m = 2$, layers 0 and 2 are selectors, layers 1, 3, and 4 are trackers, `search` nodes are on the left, and `track` nodes are on the right. The superscripts indicate the obvious correspondence of states with the two modes.
- For each k, $0 < k < n$, we create a *success link* (a directed edge) from node N^{s}_{k-1} to N^{s}_k and a *failure link* from node N^{s}_k to node N^{s}_l, where l is the first layer below layer k that is a selector. One more failure link is created from node N^{s}_0 to itself.
- From N^{s}_{n-1}, the `search` node for the topmost tracker, we add a success link to N^{t}_{n-1}.
- For each j, $m < j < n$, we add a failure link from node N^{t}_j to N^{t}_{j-1}; and similarly, we add a success link from node N^{t}_{j-1} to N^{t}_j. One more success link is created from node N^{t}_{n-1} to itself.
- Finally, we add a failure link from node N^{t}_{m+1} to N^{s}_m.

Since each state represents a unique combination of tracking layer and operating mode (i.e., `search` or `track`), the internal knowledge of the system in terms of tracking mode and precision is represented in its entirety by the current state of the state automaton.

3.2.5 Algorithm

In describing the IFA algorithm (Fig. 3.2), we use the following variables: N represents the current node in the state transition graph, $i = L(N)$ represents the layer associated with N, l_i represents the algorithm at layer i, and $\mathbf{X}^{\mathrm{in}}_i$ and $\mathbf{X}^{\mathrm{out}}_i$ denote the input and output sets of layer i. The functions $S(N)$ and $F(N)$ return the destination nodes of transition edges out of node N for success and failure, respectively.

Once initialized, the algorithm spends its time in the main loop, repeatedly executing the currently active layer and evaluating the results. Layers cause a straightforward transition

Initialization: Set all iteration indices, σ_i, to 0, set N to N^{s}_0 $(i = 0)$, and set $\mathbf{X}^{\mathrm{in}}_0 \leftarrow \mathcal{X}$, the entire configuration space.

Loop: Compute $\mathbf{X}^{\mathrm{out}}_i = l_i(\mathbf{X}^{\mathrm{in}}_i, I(t))$, perform set dilation (resulting in $\mathbf{X}^{\mathrm{out}'}_i$), and evaluate the following conditional:

1. If layer i is a tracker and $\mathbf{X}^{\mathrm{out}}_i = \emptyset$, update $N \leftarrow F(N)$, update $i \leftarrow L(N)$, and increment σ_i if Layer i is a selector. Go to Loop.
2. If layer i is a selector and $\sigma_i > \sigma^{\mathrm{MAX}}_i$, then reset σ_i to 0, reset the selector, set $N \leftarrow F(N)$, and set $i \leftarrow L(N)$. Go to Loop.
3. Otherwise, set $N \leftarrow S(N)$, $i \leftarrow L(N)$, and set $\mathbf{X}^{\mathrm{in}}_i \leftarrow \mathbf{X}^{\mathrm{out}'}_{i-1}$. If $N = N^{\mathrm{t}}_{n-1}$, reset all selectors, set all iteration indices to 0, and set $\mathbf{X}^{\mathrm{in}}_i \leftarrow \mathbf{X}^{\mathrm{out}'}_i$. Go to Loop.

Figure 3.2 The IFA algorithm.

based on their success, moving control up a layer when successful and down to the next tracker or selector, if unsuccessful.

Some additional bookkeeping happens for these transitions, both to keep iteration indices updated and to send the proper configuration sets to the next executing layer. A selector's iteration index is incremented each time its corresponding node is visited and reset to 0 either when the selector fails or when top layer tracking is achieved. Output configuration sets at a particular layer are only used as input sets of the next processing layer when the layer in question is successful. If a layer is unsuccessful, then the selector to which control falls uses the same input set as it did the previous time it was called. In Fig. 3.2, Steps 1 and 2 are concerned with tracker and selector failure, respectively, and Step 3 handles success for both trackers and selectors.

3.3 ACTION PROHIBITION IN ROBOT CONTROL

The state of an IFA system encapsulates its cognizance of tracking success. Therefore, IFA state can be used directly to activate or inhibit robot actions as appropriate. For wide applicability, we adopt a robot action prohibition paradigm where the vision system merely disables certain actions in a robot's repertoire based on availability of perceptual information.

Consider the scenario of a robot welder welding two sheets of metal at a corner. Tracking might involve stereo cameras tracking the edge formed by the corner. Most of the time, tracking is successful and welding proceeds at full speed. As sparks fly from welding, robust algorithms may be called to handle difficult tracking conditions—full speed welding is prohibited due to the slowdown in the tracking process, and welding now crawls to a slower pace. In some instances, smoke may completely obscure the welded surface, making tracking impossible. All welding actions are now prohibited by the deterioration in tracking, and the robot welder removes the tool from the surface. When the smoke clears, tracking recovers, the prohibition is lifted, and the robot continues welding.

In this instance, robustness is achieved because cognizant failure of the sensing prohibits the robot from actions which are likely to result in task failure; action prohibition limits the robot to innocuous actions. When the prohibitions are lifted, the robot continues with its task.

3.3.1 Perception Formalism

Let a perceptual specification, $S = (\Pi, \varepsilon)$, be a vector space of perceptual inputs, Π, together with a compact, convex, estimation error set, $\varepsilon \subset \Pi$. For example, for the welding task, the input may require the coordinates of the point requiring welding in some 3D Euclidean world frame within a 1mm sphere about the actual point. Thus, S can be written $(\Re^3, \{\mathbf{x} : \mathbf{x} \in \Re^3, \|\mathbf{x}\|_2 <= 0.5 \text{ mm}\})$, where $\|\cdot\|_2$ is the L_2 norm.

Let $\mathcal{X}$ denote a sensor sample space. If Π is a subspace of $\mathcal{X}$ (i.e., we can write $\mathcal{X} = \Pi \times \mathcal{K}$ for some set $\mathcal{K}$) then we define $\text{Im} : \mathcal{X} \rightarrow \Pi$ to be the corresponding projection operation. A perceptual event, $\mathbf{X}(t) = (\hat{\mathbf{x}}(t), \epsilon(t))$, consists of a perceptual estimate, $\hat{\mathbf{x}}(t) \in \mathcal{X}$, together with a margin of error set, $\epsilon(t) \subset \mathcal{X}$ (we will drop the explicit dependence on time, t, when unnecessary in context). Note that the margin of error, ϵ, is a *set* of possible alternate estimates and could correspond, for example, to the configuration sets of IFA. Now, let $\hat{\mathbf{x}}' = \text{Im}(\hat{\mathbf{x}})$, and let $\epsilon' = \{\mathbf{e} : \mathbf{e} = \text{Im}(\mathbf{x}), \forall \mathbf{x} \in \epsilon\}$. Then, an event, $\mathbf{X}$, *satisfies* S if

- Π is a subspace of $\mathcal{X}$ and
- if $\forall \mathbf{e} \in \epsilon', \mathbf{e}' = \mathbf{e} - \hat{\mathbf{x}}'$ is in ε.

Continuing with the welding example, a perceptual event may include a 5D estimate, $\hat{\mathbf{x}} = [x_0\ y_0\ z_0\ \theta_0\ \phi_0]^T$, indicating a 3D coordinate of the welding point and the 2D direction in which welding should proceed, together with a margin of error $\epsilon = \{[x\ y\ z\ \theta\ \phi]^T : \|[x\ y\ z]^T - [x_0\ y_0\ z_0]^T\| \leq 0.2\text{ mm}, |\theta - \theta_0| \leq 0.1, |\phi - \phi_0| \leq 0.1\}$. The perceptual specification of the welding task requires only the 3D position of the welding point, so $\hat{\mathbf{x}}' = [x_0\ y_0\ z_0]^T$ and $\epsilon' = \{[x\ y\ z]^T : \|[x\ y\ z]^T - [x_0\ y_0\ z_0]^T\| \leq 0.2$. Since elements of ϵ' translated by $\hat{\mathbf{x}}'$ are all in ε, $\hat{\mathbf{x}}$ satisfies S.

We also note that there is a partial order on perceptual specifications where the relation '$\prec$' (which we will read "is more precise than") applies as follows:

$$S_1 \prec S_2 \quad \text{if } \forall \mathbf{X}\ (\mathbf{X} \text{ satisfies } S_1) \Rightarrow (\mathbf{X} \text{ satisfies } S_2)$$

In general, high-layer IFA trackers, when compared with lower layers, will produce perceptual events that satisfy perceptual specifications that are more precise.

3.3.2 Action Prohibition

Let $\{A_i\}$, for $0 < i \leq n$ be a set of robot actions. Let the perceptual specification for an action, A_i, be denoted by S_i. For convenience, we will represent all of the perceptual input available at a time t as a single perceptual event, $\mathbf{X}(t)$.

A_i is prohibited from execution at time t, if $\mathbf{X}(t)$ does not satisfy S_i. Prohibition implies just that: An action is prohibited from execution based on insufficient perceptual data to proceed with the action. Prohibitions are temporary in principle; they are lifted as soon as the requisite perceptual data becomes available.

The action prohibition formalism places no explicit constraints on the underlying robot action architecture. Robots are free to plan, react, backtrack, learn, and do otherwise, just as long as they do not execute prohibited actions. Indeed, action prohibition can apply to the gamut of agent types, from perception-driven, reactive robots, to centrally controlled robots that act upon complex plans.

Reactive control through action prohibition occurs if, for example, actions have a strict ordering of preference. Assume that A_i is always preferred over A_j for $i < j$. Then, for any given perceptual event, $\mathbf{X}$, the action to execute is the action indexed by

$$\arg \min_{0<i\leq n} (\mathbf{X} \text{ satisfies } S_i)$$

In other words, the most preferred action not vetoed by the perceptual system is executed at any given time. Such a system is purely reactive, since the available percpetual information drives action.

Centralized planning and robot control can be integrated with action prohibition by adding run-time checks and additional preconditions into a planning framework. Run-time checks simply require accurate sensing of relevant portions of the world prior to operation execution. The additional preconditions state that particular perceptual inputs must be available for certain operations to be legal.

Initially, planning would proceed as usual, with assumptions that all perceptual conditions are met. Then, at run time, certain actions may fail because of faulty sensing and enforced action prohibitions. At that point the agent may choose to wait out the sensing deficiencies, perform actions designed to acquire the desired sensory data, or it may replan and backtrack by first eliminating those operations which require the unavailable sensory input.

3.4 EXAMPLES

Below, we illustrate two examples of robust robot action using vision subsystems based on IFA.

3.4.1 Face Tracking

In this task, the "robot" is a simple pan-tilt head, and the goal is to maintain a target face in the center of the image. Although there are several face tracking systems described in the literature [1, 2, 3], few are able to consistently maintain track of a single, specified individual given occlusions, distraction by other people in view, illumination changes, and so forth.

The underlying IFA system consists of seven different algorithms which can be broadly classified into three types. Listing these in ascending layer order, we have (1) search algorithms which find those regions of the image exhibiting motion and skin color, (2) fast tracking algorithms for tracking face position based on color, and (3) slow tracking algorithms for tracking face position precisely based on templates (based on [4]. During tracking, (2) takes over when (3) is too slow to track (layers in (3) are still necessary to insure that the proper target is being tracked).

The robot action vocabulary consists of (A) scanning, in which the pan-tilt head pans from side to side, (B) standstill, (C) full-speed pan-tilt to a specified angular coordinate, and (D) slower and smoother target pursuit.

The action prohibitions are straightforward. During tracking in layers in (3), there are no prohibitions. During (2), smooth target pursuit is prohibited. During (1), both (C) and (D) are prohibited because there is no tracked target to follow. We allow the system to be largely reactive, with one exception: when the IFA system enters (1), action (B) is performed until either a prespecified amount of time elapses and action (A) begins or until tracking recovers.

Figure 3.3 shows the face tracking system in various stages of operation. The implementation is on a 266 MHz Pentium II PC running Windows NT using a Matrox Meteor framegrabber and a Sony single-CCD color camera with pan-tilt head.

The behavior is as expected: tracking proceeds by smooth pursuit while the subject remains in view and motion is restricted [Fig. 3.3(a)]; high-speed motions cause tracking to fall to color-based tracking, which prohibits smooth pursuit and results in fast, jerky pans to follow a subject significantly off-center (b); when visual tracking is lost, the pan-tilt head first sits idle as a software search occurs, and if the latter is unsuccessful, begins panning left and right in search of the target (c). When the correct target is found, smooth pursuit resumes. We note that non-target faces are *not* tracked by the system.

3.4.2 Doorknob Homing

For indoor mobile robots, one important capability is navigation through doorways, whether the doors are open or closed. In order to successfully accomplish this task when doors are closed, doorknobs in the environment must be identified and tracked prior to grasping of the doorknob. We have implemented the vision module for this subtask, for a mobile robot platform used in the Siemens robotics lab [5].

The IFA system consists of several search layers and one additional layer at the top which tracks doorknobs as an edge-based polygon. The vision system was implemented on a Sun Sparc 20 equipped with a K2T-V300 digitizer and Sony single-CCD color camera. Figure 3.4 shows doorknob search at different layers. For more information on the details of the layers, see [5].

(a) (b) (c)

Figure 3.3 Face tracking: (a) high precision tracking, smooth pursuit; (b) low precision tracking, rapid pursuit; (c) tracking lost, pan and search.

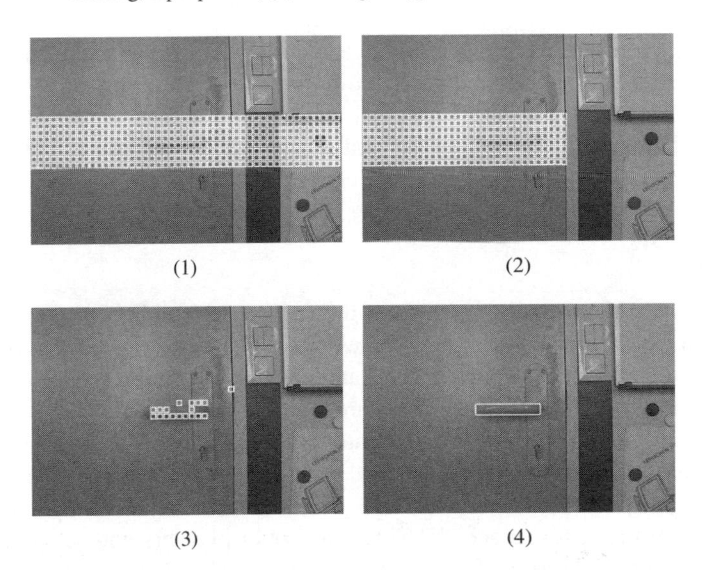

(1) (2) (3) (4)

Figure 3.4 Layers in doorknob search. Regions marked by white indicate remaining regions of the search space projected onto the image. (1) position constraint, (2) color search, (3) horizontal edge search, (4) polygon tracking.

The relevant robot actions are simply (A) standstill, and (B) servoing toward the doorknob. The obvious action prohibition is to suppress (B) when the IFA system is not in the sole `track` state.

Again, the behavior of the robot during the doorknob homing subtask is as expected. The robot servoes toward the door when visual tracking is reliable and stops when doorknob positional information is unavailable. In the current implementation, unavailability of doorknob position is assumed to occur because of temporary occlusions or jerky robot motions which cause temporary mistracking. The robot waits out these periods of sensory deprivation and resumes its task when the vision module recovers track of the doorknob. We note that more complex reasoning processes may take an extended prohibition on doorknob servoing as an indication of a more permanent situation, in which case the robot may proceed to look for other doors, move to refind the same door, or take on another task.

3.5 RELATED WORK

The combination of IFA with action prohibition is related to several prominent notions in robotics.

Perhaps the most similar in conception is the subsumption architecture, in which hardware modules organized in layers inhibit execution of modules below them [6]. In fact, IFA was inspired by the robustness of subsumption robots [7], and it shares similar structure. There is a major difference, however: subsumption emphasizes complete perception-action modules whereas the action prohibition concept is free of bias toward any single robot planning paradigm. Consequently, action prohibition can be laid on top of an existing system, whether reactive or centrally controlled.

Adaptive control considers dynamic adjustment of different control algorithms based on observed state [8]. These adjustments are not strictly prohibitions on actions, however, as they are modifications of operating parameters. In addition, we view adaptive control as a reactive, low-level form of robot control. Action prohibition can apply to higher level planners equally well.

One interesting example of adaptive control is *sequential composition*, in which different control algorithms are applied based on the phase of an observed object [9]. This work is complementary to ours. Where we focus on the robustness of vision, sequential composition assumes that input from vision is assured and focuses instead on robust control architectures. Both are required for truly robust robotics.

3.6 DISCUSSION

The power of IFA systems come from the framework's ability to unite search and tracking algorithms such that recovery from tracking failure is not an ad hoc process but a natural outcome of the layered hierarchy. The framework makes vision *post-failure robust*, or efficient at recovering from noncatastrophic failures [10].

Of course, for the robotics and automation community, robust vision is simply one of many requirements for reliable robotics. One lesson that transfers from efforts in real-time vision is that post-failure robustness is as crucial for real-world operation as the ability to avoid failure in the first place. In particular, robot planners and actuators must be designed to recover from the various failure modes that can take place during task execution. Robust vision is worth little by itself unless there is corresponding robot action to handle undesirable circumstances.

3.7 ACKNOWLEDGMENTS

Work by Greg Hager and Zachary Dodds was supported by NSF IRI-9420982. Work by Kentaro Toyama was supported in part by a grant from Siemens Central Research.

REFERENCES

[1] S. Birchfield and C. Tomasi. "Elliptical head tracking using intensity gradients and color histograms." *Proceedings of the IEEE Computer Society Conference on Computer Vision and Pattern Recognition*, pp. 232–237, Santa Barbara, June 1998.

[2] M. Collobert, R. Feraud, G. Le Tourneur, O. Bernier, J. E. Viallet, Y Mahieux, and D. Collobert. "LISTEN: a system for locating and tracking individual speakers." *Proceedings of the International Conference on Automatic Face and Gesture Recognition*, pp. 283–288, Killington, VT, 1996.

[3] N. Oliver, A. Pentland, and F. Berard. "LAFTER: Lips and face real-time tracker." *Proceedings of the IEEE Computer Society Conference on Computer Vision and Pattern Recognition*, pp. 123–130, Puerto Rico, June 1997.

[4] G. Hager and P.N. Belhumeur. "Efficient Region Tracking with Parametric Models of Geometry and Illumination." *IEEE Transactions on Pattern Analysis and Machine Intelligence*, vol. 20, no. 10, pp. 1025–1039, October 1998.

[5] W. Feiten, G. D. Hager, J. Bauer, B. Magnussen, and K. Toyama. "Modeling and control for mobile manipulation in everyday environments." *Proceedings of the 8th International Symposium on Robotics Research*, Kanagawa, Japan, October 1997.

[6] R. A. Brooks. "A robust layered control system for a mobile robot." *IEEE Transactions on Robotics and Automation*, vol. 1, no. 1, pp. 24–30, March 1986.

[7] K. Toyama and G. Hager. "Incremental Focus of Attention for robust visual tracking." *Proceedings of the IEEE Computer Society Conference on Computer Vision and Pattern Recognition*, pp. 189–195, San Francisco, June 1996.

[8] K. Watanabe. "Adaptive Estimation and Control." Englewood Cliffs, NJ: Prentice Hall, 1992.

[9] R. Burridge, A. Rizzi, and D. Koditschek. "Sequential composition of dynamically dexterous robot behaviors." *International Journal of Robotics Research*, vol. 18, no. 6, pp. 534–555, June 1999.

[10] K. Toyama and G. Hager. "If at first you don't succeed..." *Proceedings of the Fourteenth National Conference on Artificial Intelligence* (AAAI-97), pp. 3–9, Providence, RI, August 1997.

Chapter 4

INTEGRATED OBJECT MODELS FOR ROBUST VISUAL TRACKING

Kevin Nickels and Seth Hutchinson,
University of Illinois at Urbana-Champaign

Abstract

The robustness of visual tracking, or following the movement of objects in images, can be improved with an explicit model for the objects being tracked. In this paper, we investigate the use of an object model in this way. An object geometric model can tell us what feature movements to expect, and what those movements reveal about object motion. We characterize the tracking problem as one of parameter estimation from incomplete feature tracking data, and apply the Extended Kalman Filtering algorithm to the situation. Having an object model integrated into the tracking system overconstrains feature trackers, so that erroneous tracking results are selectively ignored and feasible tracking results are used to optimally update the object configuration estimate.

4.1 INTRODUCTION

Visual object tracking is the following of the movements of objects in a scene by analysis of computer images of that scene. The tracking of features on an object is a necessary precursor to many tasks in vision-based control, such as positioning the gripper in a predefined orientation with respect to a known object in an unknown configuration in the workspace. Object tracking is helpful when the shape and geometry of an object are known, but the current configuration is not known. This is particularly useful when kinematically complex objects, such as those with internal degrees of freedom, are involved.

In object tracking, information about the movement of individual features in images is used to update the *state* of the tracked object. The state is defined to be the position, orientation, and any internal degrees of freedom of the object. In feature tracking, each feature is assigned a state (here, a state would include size, orientation, and position of the *feature*). One advantage of object tracking is that the overconstrained nature of the problem may be used to reject faulty feature tracking results. Hashimoto et al. [1] have shown that increasing the number of redundant features increases the robustness of visual servoing.

The problem of determining object state from image feature locations has been extensively studied. The earliest approach to this problem was direct analysis of the transformation from object state (typically, position and orientation of a rigid object) to image feature locations, and study of the relationships involved. Nagel gives a review of the early work in [2]. This type of analytical work typically assumed that the 2D image feature measurements were given [3], [4]. More recent work in this area has focused on the analysis of uncertainty and the propagation of error through this transformation [5], [6], [7], [8]. In the 1980s, researchers began using an algorithm called Kalman Filtering to estimate object state by modeling the system structure. Kalman Filtering is an optimal state estimation process applied to a dynamic system [9], and is described in Section 4.5. Gennery [10] uses an algorithm very similar to the Kalman Filter to estimate the position and rotation of a rigid object. In [11], Broida and Chellappa employ a Kalman Filter to estimate the 2D rotation and 1D translation of an object from a sequence of 1D images. Dickmanns and Graffe [12] presented a seminal work describing the relationship of the modeled world (object state) to the real world and measured world (feature measurements), and describing the use of the framework in guiding autonomous vehicles. Since then, this dynamic machine vision approach has been used in many situations, to estimate the depth of points in a scene [13], [14], to estimate the state of objects in the world [15], [16], and for visual servoing [17], [18], [19], [20].

In addition to its use as an algorithm for incorporating feature measurements into a state estimate, many researchers have benefited from the use of the Kalman Filter as a purely spatiotemporal filter [21], [22], [23], [24]. In this way, spurious measurements and noisy data can be handled in a numerically robust manner. The filtered measurements can then be used in an analysis of the scene to generate more robust estimates for object states.

In this paper, we present a method for the use of object models for complex articulated objects, instantiated in an EKF framework, for the visual tracking of these objects. Furthermore, we describe how the overconstrained nature of the problem can be exploited to optimally update the object state, given the results from the tracking of the object features. We utilize explicit models for the articulated object and the imaging system to achieve this. Finally, we present some tracking results for the case where the object of interest is a three degree of freedom robotic arm.

4.2 MODELS USED

In this paper, five kinds of models will be used: the overall system model, the object geometric model, the object appearance model, the imaging model, and the object dynamic model. In this section, we will describe these models and discuss the possible assumptions that can be made for each.

4.2.1 System Model

We utilize a general nonlinear model in our tracking system. The other models we will describe relate to assumptions about object geometry, appearance, and dynamics, as well as the the imaging system, and are realized by instantiating functions in this system model. In this model, the *state vector* at time $k+1$, $\mathbf{x}_{k+1}$ is given by a vector valued function $\mathbf{f}_k$ of the state vector at time k, $\mathbf{x}_k$ with additive noise. Observations of the system are given by another vector valued function $\mathbf{h}_k$ of the state vector at time k, $\mathbf{x}_k$ with additive noise. The equations for the nonlinear model used for the system are of the form

$$\mathbf{x}_{k+1} = \mathbf{f}(\mathbf{x}_k) + \mathbf{w}_k \tag{4.1}$$

$$\mathbf{z}_k = \mathbf{h}_k(\mathbf{x}_k) + \mathbf{v}_k \tag{4.2}$$

See Table 4.1 for a notational summary. We make the usual (in the Kalman Filtering literature) assumptions with respect to the correlation of the noise and initial conditions:

$$E(\mathbf{w}_k\mathbf{w}_l^T) = \mathbf{Q}_k\delta_{kl}$$

$$E(\mathbf{v}_k\mathbf{v}_l^T) = \mathbf{R}_k\delta_{kl}$$

$$E(\mathbf{w}_k\mathbf{v}_l^T) = E(\mathbf{w}_k\mathbf{x}_0^T) = E(\mathbf{v}_k\mathbf{x}_0^T) = 0$$

4.2.2 Object Geometric Model

The object geometric model describes the size and shape of each link of the articulated object under consideration, and how the links move with respect to one another. This model is a superset of the standard kinematic model used in robotics, that only defines the interrelation of the coordinate systems of each link, but not the shapes of the links [25]. Thus, the object

TABLE 4.1 Notation Used in this Section

Symbol	Definition
$\mathbf{x}_k$	state (at the kth time instant)
$\mathbf{f}_k$	system function
$\mathbf{w}_k$	white process noise
$\mathbf{z}_k$	measurement
$\mathbf{h}_k$	measurement function
$\mathbf{v}_k$	white measurement noise
$\mathbf{R}_k$	covariance matrix for $\mathbf{v}_k$
$\mathbf{Q}_k$	covariance matrix for $\mathbf{w}_k$
$\hat{\mathbf{x}}_k = \hat{\mathbf{x}}_{k\vert k}$	state estimate at time k given data $\mathbf{z}_0 \ldots \mathbf{z}_k$
$\hat{\mathbf{x}}_{k\vert k-1}$	state estimate at time k given data $\mathbf{z}_0 \ldots \mathbf{z}_{k-1}$

geometric model will determine the position of all features in the world coordinate frame, given the robot's configuration parameters. The use of complex explicit geometric models in object tracking has increased in recent years [26], [27], [28]. We assume that the object model is known a priori, and does not change.

4.2.3 Object Appearance Model

The object appearance model describes the color, texture, and materials used on each link described by the object geometric model. The combination of the object appearance model and the object geometric model defines the position and appearance of every point on the object. Specifically, this allows us to recover the 3D appearance of the area immediately surrounding a feature, given the feature location and the object configuration [29]. The method used to do this is described in Section 4.4.

In addition to direct modeling of object attributes, this model includes assumptions such as ambient lighting and background color that are not strictly object feature, but will affect the appearance of features.

4.2.4 Imaging Model

The imaging model mathematically describes the camera used to image the scene. This is a superset of the perspective or orthogonal projection models, including any lens modeling (such as vignetting, lens distortion, or focusing effects) for the camera. We include the position of the camera in the scene in this model. A fixed camera position could also be assumed, by including the object position relative to the camera in the object geometric model. This model is assumed to be known and constant.

4.2.5 Object Dynamic Model

The dynamic model describes the assumptions about motion through the filter's *state space*. In our case, $\mathbf{x}_k$ contains the joint angles $\mathbf{q}$ and joint velocities $\dot{\mathbf{q}}$, so the object dynamic model describes movement through the robot's configuration space. For an in-depth description of configuration space concepts, see [25] or [30]. Intuitively, the dynamic model $\mathbf{f}_k(\mathbf{x}_k)$ is the best estimate (from a modeling standpoint) of the next state of the system given the current state $\mathbf{x}_k$.

For instance, if we know that the robot is being controlled such that it maintains a constant velocity in configuration space, we adopt a "constant velocity" dynamic model

$$\mathbf{f}_k(\mathbf{x}_k) = \mathbf{q}_k + \Delta\dot{\mathbf{q}}_k$$

where Δ is the sampling interval. If we wish to avoid explicit modeling of motion, we can assume a "constant position" dynamic model,

$$\mathbf{f}_k(\mathbf{x}_k) = \mathbf{x}_k$$

Note that since $\mathbf{x}_{k+1} = \mathbf{f}_k(\mathbf{x}_k) + \mathbf{w}_k$, this is not the same as assuming that the object never moves, but rather that the motion is completely unpredictable. Specifically, deviations from the motion model are assumed to be zero-mean, Gaussian, and white. If "no dynamic model" were assumed, the implicit assumption would be what we have described as a constant position dynamic model. In this case, the state vector is simply the object configuration parameters,

$$\mathbf{x}_k = \mathbf{q}_k$$

If constant velocity is assumed, the velocity through configuration space needs to be estimated as well, so

$$\mathbf{x}_k = \begin{bmatrix} \mathbf{q}_k \\ \dot{\mathbf{q}}_k \end{bmatrix}$$

Motion models $\mathbf{f}$ could be also formulated to model constant workspace end-effector velocities or a number of other feasible dynamic models.

4.2.6 Errors in Models

Each of these models affects the robustness of the visual tracking process, and errors in each model will degrade the performance in different ways. Errors in the appearance and imaging models will affect the appearance of an feature template, and will degrade the feature tracking portion of the algorithm in as much as the appearance of the features in the input images no longer matches the expected appearance. As long as the "correct" location of a feature is most similar to the template, feature tracking will occur without errors. Since the our correlation measure does not account for changes in illumination [31], errors in the modeling of illumination would change the object appearance, and would affect feature tracking in much the same way.

Errors in the object geometric model would affect the system in a different way. First, the forward kinematics of the object describe the mapping between the object configuration space and the workspace. Section 4.3 describes this process. Thus, errors in this mapping will initialize the feature trackers at incorrect positions in the image, positions different from the actual locations of the features. Second, the forward kinematics as well as the partial derivatives of this map are used to compute a linearization of this nonlinear map for the purpose of assimilating feature tracking results into an updated object state. Section 4.5 describes this process. The implication of an incorrect geometric model on the process is that a minimum least squares fit to the object feature locations is computed, and the estimates of the object parameters would be incorrect in so far as the placement of the features is incorrect with respect to the object due to the geometric errors. Luckily, since the same geometric map is used for the forward projection of feature locations and the assimilation of feature tracking results, small errors in the geometric map result in proper feature tracking at the expense of absolute object configuration estimates. If feature location prediction and object dynamic models were used to a greater extent, such that only the most likely feature locations in the image were searched, errors in the geometric map might be more problematic.

Errors in the dynamic model will draw the state estimate off during the time projection portion of the system. This means that the feature locations will be incorrect to a certain extent. As long as the actual feature locations still fall within the search regions, the only result is that the correction portion of the prediction-correction update equation (see Section 4.5 for an explanation of the Extended Kalman Filter) will be larger. If too many features fall outside the search region due to an incorrect dynamic model, the feature trackers will not be able to bring the full state estimate, and tracking will fail.

4.3 OBSERVATION FUNCTION

Of great importance in our object tracking system is the map from the object state $\mathbf{x}_k$ to the location in the image plane of each feature $\mathbf{m}_k^f$. We call this map the observation function $\mathbf{h}_k(\mathbf{x}_k)$, a vector valued function with two elements for each feature of interest (the u and v location in the image plane of the image of that feature). In our case, the observation function is the composition of the forward kinematics of the robot and the projection equations used. Since we use a PUMA robotic arm with the proximal three joints free to move, and we assume perspective projection with focal lengths f_x and f_y, this reduces to

$$\mathbf{m}_k^f = CT_0^{c_f} p_f = C\Pi_{i=1}^{c_f} A_i p_f$$

$$\mathbf{h}_k^f = \begin{bmatrix} -\frac{\mathbf{m}_k^f[1]}{f_x \mathbf{w}_k^f[3]} \\ -\frac{\mathbf{m}_k^f[2]}{f_y \mathbf{w}_k^f[3]} \end{bmatrix}$$

where $\mathbf{m}_k^f$ is the position of feature f in the camera coordinate frame. The 4×4 transformation matrix C determines the mapping from the world coordinate frame to the camera coordinate frame, and $T_0^{c_f}$ is the standard homogeneous transformation from the coordinate frame in which feature f is specified, c_f, to the world coordinate frame. Transformation $T_0^{c_f}$ is dependent on $\mathbf{x}_k$ and can be decomposed into homogeneous transforms A_i corresponding to each link of the robot. See [25] for more discussion on the Denevit-Hartenberg convention for assigning coordinate transformations.

Since we have closed form expressions for each component of $\mathbf{h}_k(\mathbf{x}_k)$, we can compute the partial derivatives of $\mathbf{h}$ with respect to each component of $\mathbf{x}$ algebraically. These functions will play an important role in Section 4.5 when we describe the assimilation of feature tracking results into the state estimate update.

4.4 FEATURE TRACKING

Given the object appearance model and the imaging model, each link can be rendered using standard computer graphics techniques [32]. If a geometric model and object configuration is assumed, the links can be rendered in the appropriate position, and a complete *synthetic scene* can be rendered. This synthetic scene represents the best estimate, given the models and object configuration estimate, of what the input image will look like. An example synthetic scene is shown in Fig. 4.1. This scene is then used for several computations.

The assumptions used to render the scene can also be used to compute the visibility of each point on the robot from the given viewpoint [33]. Specifically, points on the robot specified as *features* are checked for visibility at each step. A feature is defined to be a three-dimensional point defined in a given link coordinate system. Features f determined to be non-visible at a given object configuration $\hat{\mathbf{x}}_k$ have the algebraic observation

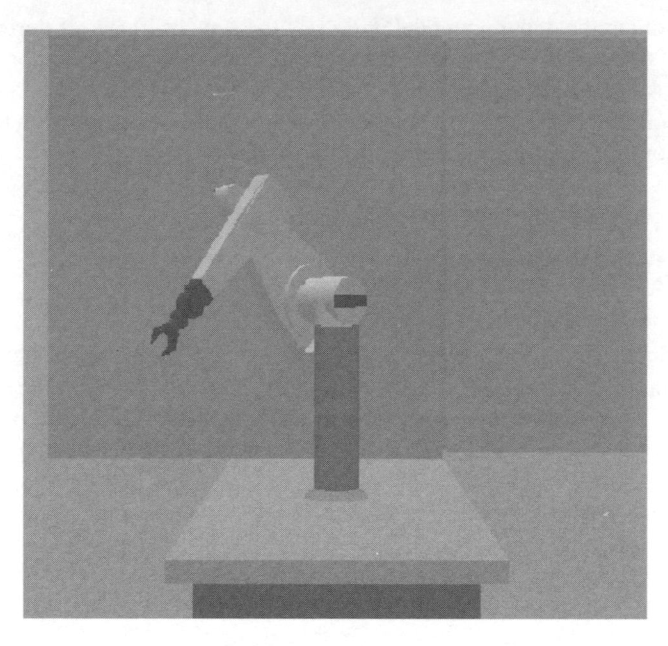

Figure 4.1 A synthetic scene.

function $\mathbf{h}_k^f(\mathbf{x}_k) \triangleq [0\ 0]^T$; otherwise the observation function described in Section 4.3 is used.

Visible features have an expected feature location $\mathbf{h}_k^f(\mathbf{x}_k)$ defined by the modeling assumptions. Since the synthetic scene represents the best estimate of the appearance of the object at the given configuration, it also represents the best estimate for the appearance of the features at that configuration [29]. Therefore, we extract the region surrounding the expected feature location for use as a feature appearance template. Currently, we use a fixed sized rectangular template for each feature.

A correlation measure, in our case the Sum of Squared Differences (SSD) correlation measure, is used to compare the template with the region in the input image surrounding the expected feature location. As this similarity measure is used on each pixel in a region, the scores make up what we term the SSD Surface (SSDS). The shape of the SSDS can tell us how well the template matches the image content surrounding the expected feature location [34], [35]. The amount of drop off in the SSDS, or equivalently the degree to which the template discriminates the most likely feature location from the surrounding area, is an important contributor to increasing the robustness of the tracking system as a whole. If the system can detect when an individual feature tracker is failing, the results from that tracker can be selectively disregarded. This is a different concept from outlier rejection, which although feasible in the context of the EKF [36], is not implemented in the tracking system at the current time. In this case, we are looking strictly at the shape of the SSDS to determine the spatial discrimination power of the template. Figure 4.2 illustrates what the shape of the SSDS can reveal about feature tracking results. In the top row, vertical movement of the feature in the image cannot be detected. Since SSD scores in the vertical direction do not drop off, the SSDS reveals this fact. In contrast, the second row shows a feature discriminating well in two directions. The SSDS for this case shows SSD score drop offs in every direction away from the peak.

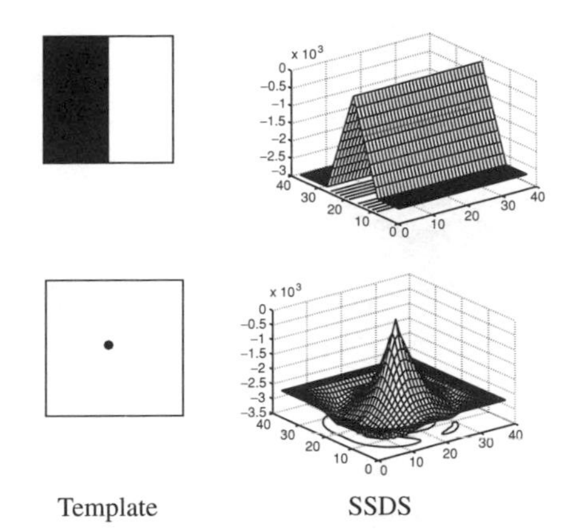

Figure 4.2 SSDS shape.

We follow [14] and interpret the SSDS probabilistically. The feature tracking reveals that the feature is most likely at the point of maximum similarity (the peak in Fig. 4.2), and that the probability of the feature location being other places is proportional to the SSD score at that location. Thus the SSDS, scaled appropriately, can be used as an estimate of the probable location of a feature in the image plane. More importantly, the local shape of the SSDS can be used to derive a measure of the spatial certainty of that location.

As we will see in Section 4.5, the location of each feature in the image is modeled as a two dimensional Gaussian random vector (GRV). The statistics of a GRV are completely determined by the mean $\boldsymbol{\mu}$ and covariance $\mathbf{R}$ of the GRV. We next present a method for estimating the mean and covariance for the GRV we use to represent the feature location in the image. We denote the measurement as the vector $\mathbf{z}_k$ and the covariance as the matrix $\mathbf{R}_k$.

The peak of the SSDS surface is determined, and this is taken to be the mean of the GRV.

$$\boldsymbol{\mu} = \text{argmin } SSD(u, v)$$

where argmin returns the u and v that minimize $SSD(u, v)$. Then the variance in the u and v directions, as well as the covariance of u and v are numerically computed from the scaled SSDS. These measures are used to compute the covariance matrix for the GRV.

$$\begin{aligned}\mathbf{R} &= \begin{bmatrix} Var^2(u) & Cov(u, v) \\ Cov(u, v) & Var^2(v) \end{bmatrix} \\ &= \begin{bmatrix} \sigma_u^2 & \rho_{uv}\sigma_u\sigma_v \\ \rho_{uv}\sigma_u\sigma_v & \sigma_v^2 \end{bmatrix}\end{aligned}$$

The geometric interpretation of this approximation with respect to the SSDS is shown in Fig. 4.3.

In this section, we will denote individual feature measurements by $\mathbf{z}_k^f$ and $\mathbf{R}_k^f$. Once $\mathbf{z}_k^f$ and $\mathbf{R}_k^f$ have been computed for each feature, the information from the individual feature trackers needs to be combined to make an observation vector and covariance matrix for the system as a whole. The combined measurements are denoted by $\mathbf{z}_k$ and $\mathbf{R}_k$.

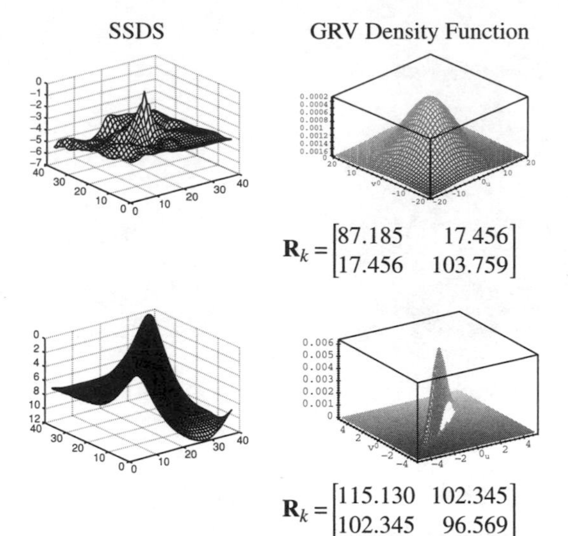

Figure 4.3 SSDS and GRV density function for two features.

To make the measurement vector, we concatenate the means of the individual vectors,

$$\mathbf{z}_k = \begin{bmatrix} \mathbf{z}_k^1 \\ \vdots \\ \mathbf{z}_k^F \end{bmatrix}$$

where there are F features. If we make the assumption that the movement of each feature is independent of the movement of the other features, the covariance of $\mathbf{z}_k$ can be computed by constructing a block-diagonal matrix from the individual covariance matrices,

$$\mathbf{R}_k = \begin{bmatrix} \mathbf{R}_k^1(1,1) & \mathbf{R}_k^1(1,2) & \cdots & 0 & 0 \\ \mathbf{R}_k^1(2,1) & \mathbf{R}_k^1(2,2) & & 0 & 0 \\ \vdots & & \ddots & & \vdots \\ 0 & 0 & & \mathbf{R}_k^F(1,1) & \mathbf{R}_k^F(1,2) \\ 0 & 0 & \cdots & \mathbf{R}_k^F(2,1) & \mathbf{R}_k^F(2,2) \end{bmatrix}$$

where the individual $\mathbf{R}_k^f$ are as described above. Modeling the interactions between the uncertainties of the individual feature trackers is a nontrivial task, and is a topic for future research. The interactions of the locations of the features are modeled, in that the movement of each feature plays a role in updating the state estimates $\hat{\mathbf{x}}_k$, as described below.

4.5 EXTENDED KALMAN FILTER

This system utilizes the Extended Kalman Filter (EKF), an extension of the standard Kalman Filter [37], [38] to the case of nonlinear observation and dynamic functions. Many tracking researchers use KF or EKF to track individual features or pixels, thus estimating the motion of the feature in the image plane or the workspace [13], [22], [21]. After tracking, feature location and movement information are gathered into an object estimate. In contrast, we use the object configuration as our state vector, and model the forward kinematics of the robot and the projection of the imaging system as part of the EKF. This allows us to incorporate feature tracking results within the context of the object model.

The state space for our case is the configuration space of the robot $\mathbf{x}_k = \mathbf{q}$. We have presented the two main models associated with our modeling of the system, the observation model $\mathbf{h}_k(\mathbf{x}_k)$ and the object dynamic model $\mathbf{f}_k(\mathbf{x}_k)$. The complete nonlinear model used for the system is as given in (4.1) and (4.2). Table 4.1 summarizes the notation used in the EKF.

The following set of equations is generally collectively called the *extended Kalman filter* [39]:

$$\mathbf{P}_{0,0} = Var(\mathbf{x}_0)$$

$$\hat{\mathbf{x}}_0 = E(\mathbf{x}_0)$$

$$\mathbf{P}_{k,k-1} = \left[\frac{\partial \mathbf{f}_{k-1}}{\partial \mathbf{x}_{k-1}}(\hat{\mathbf{x}}_{k-1})\right] \mathbf{P}_{k-1,k-1} \left[\frac{\partial \mathbf{f}_{k-1}}{\partial \mathbf{x}_{k-1}}(\hat{\mathbf{x}}_{k-1})\right]^T + \mathbf{Q}_{k-1}$$

$$\hat{\mathbf{x}}_{k|k-1} = \mathbf{f}_{k-1}(\hat{\mathbf{x}}_{k-1})$$

$$\mathbf{K}_k = \mathbf{P}_{k,k-1}\left[\frac{\partial \mathbf{h}_k}{\partial \mathbf{x}_k}(\hat{\mathbf{x}}_{k|k-1})\right]^T \times \left[\left[\frac{\partial \mathbf{h}_k}{\partial \mathbf{x}_k}(\hat{\mathbf{x}}_{k|k-1})\right] \mathbf{P}_{k,k-1} \left[\frac{\partial \mathbf{h}_k}{\partial \mathbf{x}_k}(\hat{\mathbf{x}}_{k|k-1})\right]^T + \mathbf{R}_k\right]^{-1}$$

$$\mathbf{P}_{k,k} = \left[I - \mathbf{K}_k \left[\frac{\partial \mathbf{h}_k}{\partial \mathbf{x}_k}(\hat{\mathbf{x}}_{k|k-1})\right]\right] \mathbf{P}_{k,k-1}$$

$$\hat{\mathbf{x}}_{k|k} = \hat{\mathbf{x}}_{k|k-1} + \mathbf{K}_k(\mathbf{z}_k - \mathbf{h}_k(\hat{\mathbf{x}}_{k|k-1}))$$

4.6 EXPERIMENTAL RESULTS

Laboratory experiments were performed to validate the tracking performance of the system. In the experiments presented here, the object being tracked is a PUMA robotic arm. Thc three proximal joints of the PUMA are free to move in these tests. The movement of the PUMA is under external control, but the speed is restricted sufficiently to prevent features from escaping the fixed search areas in a single sampling period.

As the arm moves, some subset of the full feature set is tracked as described in Section 4.4. The feature tracking results are used as input to the EKF algorithm as described in Section 4.5, resulting in state estimates. For the purposes of experimental validation, the actual joint angles are available and are shown below. Note that these values are not available to the tracking system.

In this experiment, the arm moves from a parking position down towards the table, as if picking up a block from the table. The arm then moves up away from the table, swings right (from the point of view of the camera), and moves down toward the table again. All three joints move significantly with constant velocity motion in joint space. Note that this fact is not exploited at the current time. The actual joint angles commanded, and the estimation of the joint angles by our filter, are shown in Fig. 4.4. The error in the tracking is shown in Fig. 4.5.

One feature of the system is the characterization and use of the uncertainties in the kinematic and imaging chain, as modeled by our object and imaging models. See [40] for an in-depth discussion on this. Figure 4.6 illustrates the uncertainty present in the system during the experiment described above. The variance of the estimate for each joint angle is shown.

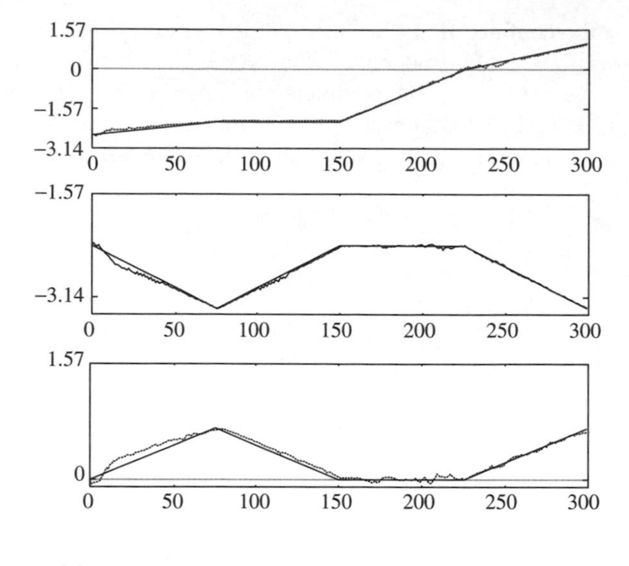

Figure 4.4 Exercising three degrees of freedom. Top: x_0 and $\hat{x}_0$, middle: x_1 and $\hat{x}_2$, bottom: x_2 and $\hat{x}_2$.

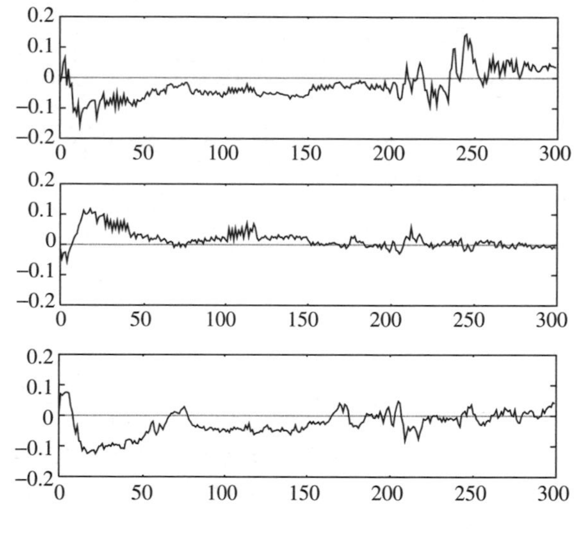

Figure 4.5 Tracking error. Top: $x_0 - \hat{x}_0$, middle: $x_1 - \hat{x}_1$, bottom: $x_2 - \hat{x}_2$.

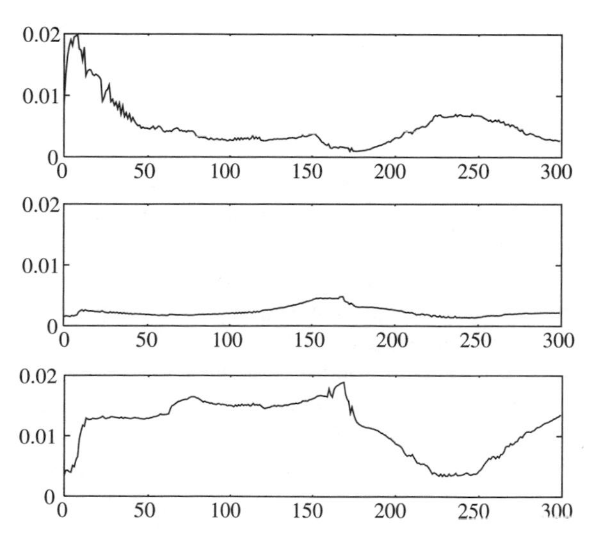

Figure 4.6 Uncertainty in the tracking system. Top: $P[0, 0]$, middle: $P[1, 1]$, bottom: $P[2, 2]$.

The time scale is the same as in Fig. 4.4. See [41] for a detailed analysis of the tracking behavior in this experiment. The web site `http://www-cvr.ai.uiuc.edu` contains other demonstrations of the tracking behavior, and animations showing tracking results along with the relevant input images.

4.7 CONCLUSIONS

We have described an object tracking system that uses object modeling and online confidence measures to improve the robustness of tracking. The system uses Extended Kalman Filtering to optimally update the current state estimate for an integrated object model using uncertain feature tracking results.

By having an object model integrated into the EKF framework and analyzing feature tracking results in the context of this model, individual feature tracker failures will not cause mistracking. If Kalman Filters are used on individual feature trackers, knowledge about the kinematics of the object and the resultant information about expected motions in the image plane are not directly available.

By evaluating the spatial certainty of the feature tracking results online, templates that are not doing a good job of discriminating the most likely location of a feature can pass this information along to the object tracking algorithm. Thus, the information that the feature tracking is obtaining can be used to its maximum effectiveness. A system that only extracts the location of the minimum SSD score will not have this information available, and will implicitly assign the same confidence to every feature tracking result.

Areas for future research in the context of this system include the extension of this framework to track the full 6 degrees of freedom possible for our PUMA arm, and investigating tracking in situations where the geometric and appearance models for an object are incompletely specified.

REFERENCES

[1] K. Hashimoto, A. Aoki, and T. Noritsugu, "Visual servoing with redundant features," *Proceedings of the 35th Conference on Decision and Control*, Kobe, Japan, December 1996.

[2] H. H. Nagel, "Overview on image sequence analysis," in *Image Sequence Processing and Dynamic Scene Analysis*, T. S. Huang (Ed.), pp. 2–38, Berlin, Germany, Springer-Verlag, 1983.

[3] J. Q. Fang and T. S. Huang, "Some experiments on estimating the 3D motion parameters of a rigid body from two consecutive image frames," *IEEE Transactions on Pattern Analysis and Machine Intelligence*, vol. 6, pp. 545–554, IEEE, September 1984.

[4] A. Z. Meiri, "On monocular perspective of 3D moving objects," *IEEE Transactions on Pattern Analysis and Machine Intelligence*, vol. 2, pp. 582–583, IEEE, Piscataway, NJ, November 1980.

[5] D. Lowe, "Fitting parameterized three-dimensional models to images," *IEEE Transactions on Pattern Analysis and Machine Intelligence*, vol. 13, pp. 441–450, IEEE, Piscataway, NJ, May 1991.

[6] D. Lowe, "Robust model-based motion tracking through the integration of search and estimation," *International Journal of Computer Vision*, vol. 8, pp. 113–122, Kluwer Academic Publishers, August 1992.

[7] D. P. Huttenlocher, J. J. Noh, and W. J. Rucklidge, "Tracking non-rigid objects in complex scenes," *Proceedings of Fourth International Conference on Computer Vision*, pp. 93–101, Berlin, Germany, May 1993.

[8] A. Gee and R. Cipolla, "Fast visual tracking by temporal consensus," *Image and Vision Computing*, vol. 14, pp. 105–114, New York, Elsevier, March 1996.

[9] C. K. Chui and G. Chen, Kalman Filtering with Real-Time Applications. Springer Series in Information Sciences, Berlin Heidelberg, Springer-Verlag, 1987.

[10] D. B. Gennery, "Tracking known three-dimensional objects," *Proceedings of the National Conference on Artifical Intelligence*, pp. 13–17, Pittsburgh, PA, August 1982.

[11] T. J. Broida and R. Chellappa, "Estimation of object motion parameters from noise images," *IEEE Transactions on Pattern Analysis and Machine Intelligence*, vol. 8, pp. 90–99, IEEE, Piscataway, NJ, January 1986.

[12] E. E. Dickmanns and V. Graefe, "Dynamic monocular machine vision," *Machine Vision and Applications*, vol. 1, pp. 223–240, New York, Springer-Verlag, 1988.

[13] L. Matthies, T. Kanade, and R. Szeliski, "Kalman filter-based algorithms for estimating depth from image sequences," *International Journal of Computer Vision*, vol. 3, pp. 209–238, Kluwer Academic Publishers, September 1989.

[14] A. Singh and P. Allen, "Image flow computation: An estimation-theoretic framework and a unified perspective," *Computer Vision Graphics and Image Processing: Image Understanding*, vol. 56, pp. 152–177, New York, Academic Press, September 1992.

[15] N. A. Andersen, O. Ravn, and A. T. Sorensen, "Real-time vision based control of servomechanical systems," *Proceedings of 2nd International Symposium on Experimental Robotics*, June 1991.

[16] J. J. Wu, R. E. Rink, T. M. Caelli, and V. G. Gourishankar, "Recovery of the 3D location and motion of a rigid object through camera image (an extended Kalman filter approach)," *International Journal of Computer Vision*, vol. 2, pp. 373–394, Kluwer Academic Publishers, April 1989.

[17] A. Singh, "Incremental estimation of image flow using a Kalman filter," *Journal of Visual Communication and Image Representation*, vol. 3, pp. 39–57, New York, Academic Press, 1992.

[18] W. J. Wilson, "Visual servo control of robots using Kalman filter estimates of robot pose relative to work-pieces," in *Visual Servoing: Real-Time Control of Robot Manipulators Based on Visual Sensory Feedback*, K. Hashimoto (Ed.), Ch. 3, pp. 71–104, Singapore, World Scientific, 1993.

[19] J. W. Lee, M. S. Kim, and I. S. Kweon, "A Kalman filter based visual tracking algorithm for an object moving in 3D," *Proceedings of IEEE/RSJ International Conference on Intelligent Robots and Systems*, pp. 342–347, Pittsburgh, August 1995.

[20] W. J. Wilson, "Relative end-effector control using cartesian position based visual servoing," *IEEE Transactions on Robotics and Automation*, vol. 12, pp. 684–696, Piscataway, NJ, IEEE, October 1996.

[21] D. Metaxas and D. Terzopoulos, "Shape and nonrigid motion estimation through physics-based synthesis," *IEEE Transactions on Pattern Analysis and Machine Intelligence*, vol. 15, pp. 580–591, Piscataway, NJ, IEEE, June 1993.

[22] P. Matteucci, C. S. Regazzoni, and G. L. Foresti, "Real-time approach to 3-D object tracking in complex scenes.," *Electronics Letters*, vol. 30, pp. 475–477, IEE, March 1994.

[23] K. Stark and S. Fuchs, "A method for tracking the pose of known 3D objects based on an active contour model," *Proceedings of International Conference on Pattern Recognition*, pp. 905–909, Vienna, Austria, 1996.

[24] P. Wunsch and G. Hirzinger, "Real-time visual tracking of 3D objects with dynamic handling of occlusion," *Proceedings of International Conference on Robotics and Automation*, pp. 2868–2873, Albuquerque, NM, April 1997.

[25] M. Spong and M. Vidyasagar, Robot Dynamics and Control. New York, John Wiley & Sons, 1989.

[26] A. Hauck, S. Lanser, and C. Zierl, "Hierarchical recognition of articulated objects from single perspective views," *Proceedings of Conference on Computer Vision and Pattern Recognition*, pp. 870–883, San Juan, PR, June 1997.

[27] J. E. Lloyd, J. S. Beis, D. K. Pai, and D. G. Lowe, "Model-based telerobotics with vision," *Proceedings of International Conference on Robotics and Automation*, Albuquerque, NM, April 1997.

[28] H. A. Rowley and J. M. Rehg, "Analyzing articulated motion using expectation-maximization," *Proceedings of Conference on Computer Vision and Pattern Recognition*, pp. 935–941, San Juan, PR, June 1997.

[29] R. Lopez, A. Colmenarez, and T. S. Huang, "Vision-based head and facial feature tracking," *Advanced Displays and Interactive Displays Federated Laboratory Consortium*, Annual Symposium, January 1997.

[30] J. C. Latombe, Robot Motion Planning. Norwell, MA, Kluwer Academic Publishers, 1991.

[31] G. Hager and P. Belhumeur, "Real-time tracking of image regions with changes in geometry and illumination," *Proceedings of Conference on Computer Vision and Pattern Recognition*, pp. 403–410, San Fransisco, CA, June 1996.

[32] M. Woo, J. Neider, and T. Davis, The OpenGL Programming Guide. Reading, MA, Addison-Wesley, 1996.

[33] A. Watt, 3-D Computer Graphics. Reading, MA, Addison-Wesley, 2nd ed., 1995.

[34] P. Anandan, "A computational framework and an algorithm for the measurement of visual motion," *International Journal of Computer Vision*, vol. 2, no. 3, pp. 283–310, Kluwer Academic Publishers, 1989.

[35] N. P. Papanikolopoulos, "Selection of features and evaluation of visual measurements during robotic visual servoing tasks," *Journal of Intelligent and Robotic Systems*, vol. 13, pp. 279–304, Kluwer Academic Publishers, July 1995.

[36] A. Zolghadri, "An algorithm for real-time failure detection in Kalman filters," *IEEE Transactions on Automatic Control*, vol. 41, Piscataway, NJ, IEEE, October 1996.

[37] P. S. Maybeck, Stochastic Models, Estimation, and Control, vol. 141-1 of Mathematics in Science and Engineering, New York Academic Press, 1979.

[38] F. L. Lewis, Optimal estimation with an introduction to stochastic control theory. New York, John Wiley & Sons, 1986.

[39] R. G. Brown, Introduction to Random Signal Analysis and Kalman Filtering. New York, John Wiley & Sons, 1983.

[40] K. Nickels and S. Hutchinson, "Characterizing the uncertainties in point feature motion for model-based object tracking," *Proceedings Workshop on New Trends in Image-Based Robot Servoing*, pp. 53–63, Grenoble, France, 1997.

[41] K. Nickels and S. Hutchinson, "Weighting observations: The use of kinematic models in object tracking," *Proceedings of International Conference on Robotics and Automation*, Leuven, Belgium, May 1998.

Chapter 5

ROBUST VISUAL TRACKING BY INTEGRATING VARIOUS CUES

Yoshiaki Shirai, Ryuzo Okada, and Tsuyoshi Yamane
Osaka University

Abstract

This chapter describes methods of tracking of moving objects in a cluttered background by integrating optical flow, depth data, and/or uniform brightness regions. First, a basic method is introduced in which a region with uniform optical flow is extracted as the target region. Then two methods are described which integrate optical flow and another visual cue. In the first method, a target region is extracted by Baysian inference in term of optical flow, depth and the predicted target location. In the second method, considering uniform brightness regions where neither optical flow nor depth is obtained, a target region is extracted from optical flow and/or uniform brightness regions. Real-time human tracking is realized for real image sequences by using a real-time processor with multiple DSPs.

5.1 INTRODUCTION

Object tracking has been a popular research theme. If an object has a different brightness from the background, the object can be easily detected. This assumption, however, does not hold for scenes with a complex background.

Another simple method is to extract objects from the difference of an image from the background image. This method does not work if a camera moves during tracking. Moving objects can better extracted from apparent motion in images (optical flow). However, optical flow alone is not enough to discriminate overlapping two persons with similar motions. An effective cue to solve this problem is the depth information which is obtained by a stereo pair of cameras. By integrating optical flow and depth, a target object is discriminated from another overlapping object with similar motion or similar depth.

Another problem is the lack of information in uniform brightness regions where neither optical flow nor depth is obtained. In order to solve the problem, the uniform region itself is used as a cue to find a target object. By integrating optical flow and uniform brightness region, a moving object is more reliably tracked.

Experiments with the above methods are shown for real-time human tracking in cluttered background.

5.2 OPTICAL FLOW EXTRACTION

Most methods of optical flow extraction try to eliminate errors caused by noises, disregarding the computational cost. Horn and Schunck's method [1], for example, requires iterations for regularization.

For a limited number of specified points in an image the optical flow can be extracted by correlation. One problem with the correlation method is the computation cost, which make it difficult to extract dense flows, say 10,000 flows.

Optical flow is calculated by the generalized gradient method [2] based on spatio-temporal filtering. Although the obtained flow is noisy, this method requires less computation.

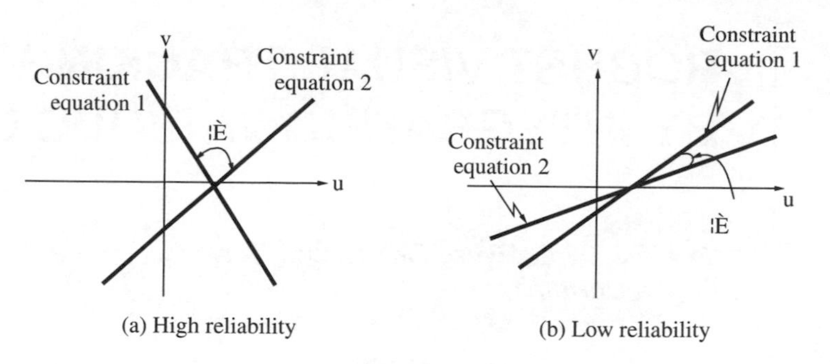

Figure 5.1 Reliability of optical flow.

Let $f(x, y, t)$ denote the brightness at a point (x, y) in an image at time t. The constraint equation of the gradient method is

$$f_x(x, y, t)u + f_y(x, y, t)v + f_t(x, y, t) = 0 \tag{5.1}$$

where f_x denotes partial derivative with x, and (u, v) is the flow vector.

In order to obtain u and v, two different spatial filters g, h are applied to the input image $f(x, y, t)$ and the following two constraint equations are derived:

$$\begin{pmatrix} (g * f)_x & (g * f)_y \\ (h * f)_x & (h * f)_y \end{pmatrix} \begin{pmatrix} u \\ v \end{pmatrix} = \begin{pmatrix} (g * f)_t \\ (h * f)_t \end{pmatrix} \tag{5.2}$$

where $*$ represents convolution. The flow vector (u, v) is obtained by solving these simultaneous equations.

The reliability of optical flow is defined as the angle between two lines corresponding to (5.2) as shown in Fig. 5.1. The reliability R_v is given as

$$R_v = |sin\theta| = \frac{|\mathbf{G} \times \mathbf{H}|}{|\mathbf{G}||\mathbf{H}|} \tag{5.3}$$

where $\mathbf{G} = ((g * f)_x, (g * f)_y)$ and $\mathbf{H} = ((h * f)_x, (h * f)_y)$.

Another reliability R'_v is for the contrast of the brightness:

$$R'_v = (g * f)_x^2 + (g * f)_y^2 + (h * f)_x^2 + (h * f)_y^2 \tag{5.4}$$

The flow vector is not calculated at a point where either reliability R_v or R'_v is low.

If more than two spatial filters are available, the constraint equations are solved by the least square method. In this case, more reliable flow is obtained and the reliability can be represented by the residual errors.

5.3 TRACKING WITH OPTICAL FLOW

We propose an efficient object tracking method that distinguishes a target object flow vectors from the others [3]. An object is tracked by updating a rectangular window which circumscribes an object region in the image sequences. The neighborhood of the window is searched for the next object region. This method is not subject to the influence of the flow vectors that are far from the target object.

Before tracking, we have to determine a moving object to track in a scene. Our current system tracks the first object that moves in the image. In order to find a moving object, the camera direction is fixed and the window size is initialized to the image size. When enough flow vectors are obtained, we consider them as belonging to a moving object. Once a moving object is obtained, it is tracked by the following procedure.

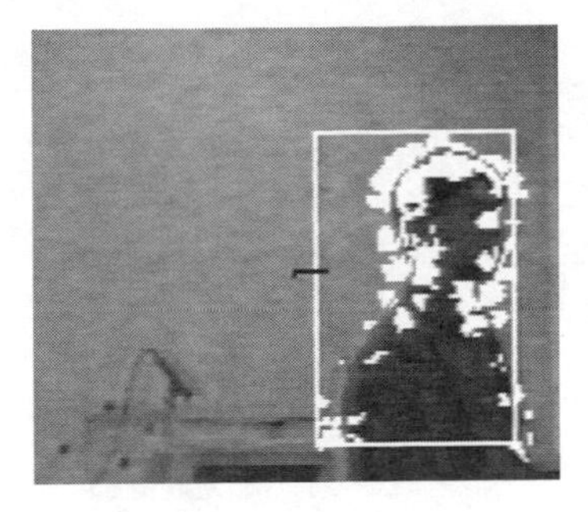
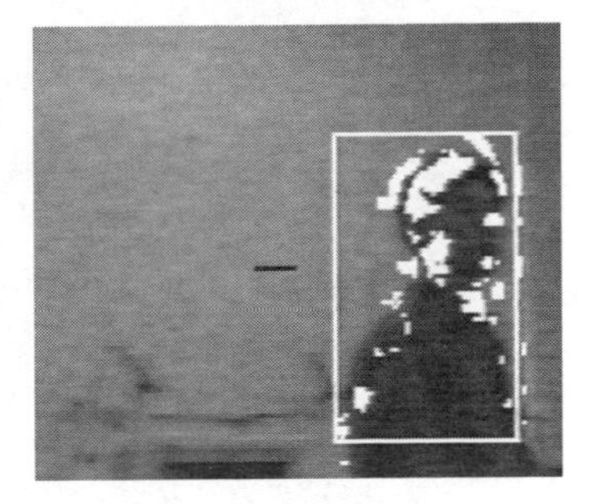

Figure 5.2 Tracking a person with camera motion.

1. Calculate optical flows in the image.
2. If the number of pixels with a flow vector in the window is more than a threshold, go to 3. Otherwise, go to 1, to detect a moving object again.
3. Calculate the mean flow vector in the window.
4. Search the window and the neighborhood of the window for the pixels whose flow vector are similar to the mean flow. A set of these pixels is the object region in the next frame.
5. Update the window to circumscribe the object region. If the number of pixels of the region is less than a threshold, the window is not updated.
6. Based on the position of the object region and the flow vector, the camera direction is controlled to capture the object in the image center.
7. Return to 1.

Figure 5.2 shows an example of tracking of a person walking from left to right in the image. The direction of the camera is controlled from left to right. The black lines at the image center represent the vertical and the horizontal velocities of the target object.

5.4 TRACKING TWO PERSONS

The tracking method described above can be extended to multiple objects tracking [4]. Real-time tracking of two objects is realized as follows. Initially a moving object is found and tracked in the same way as above. Meanwhile the rest of the image is searched for another object. When another object is found, they are tracked independently.

If two objects overlap, it must be determined which is in the foreground so that windows are correctly updated. Figure 5.3 shows a situation where window A and window B overlap. Each overlapping window is divided into two regions: "overlapping region" and "non-overlapping region." The decision is made as to which window is in the foreground

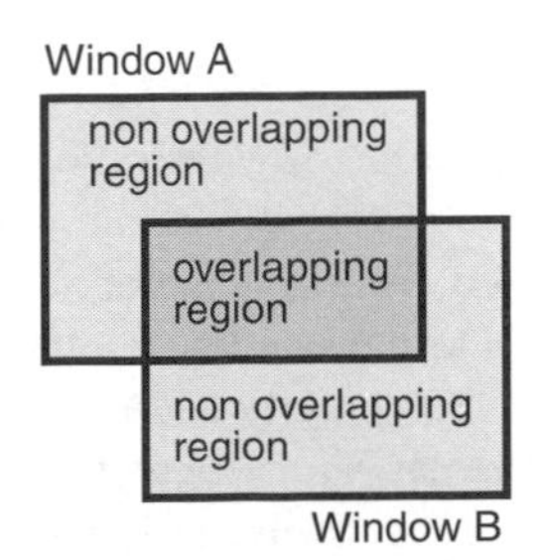

Figure 5.3 Overlapping windows.

Figure 5.4 Tracking two persons.

based on the reliable optical flow in these regions. Figure 5.4 shows an example of an experiment for two persons crossing in the image.

5.5 TRACKING WITH OPTICAL FLOW AND DEPTH

Tracking with velocity data alone is difficult if the velocity of other objects are similar to the target. On the other hand, tracking with depth data alone [5], [6] is also difficult if the target and other objects have similar depth data.

This problem is solve by using two sensor data: optical flow and disparity of a stereo pair of images [7]. At each frame, the optical flow and the disparity are calculated. The target velocity and the disparity are estimated using the optical flow and the disparity inside the predicted target region. The region which has the similar velocity and disparity to the predicted ones is extracted as the target region. Occlusion of the target is detected from the abrupt disparity change in the target region and the motion is estimated based on the past record.

5.5.1 Disparity Extraction

The disparity is derived from the one-dimensional optical flow between a pair of stereo images. We use the vergence motor of active camera head (turning the right camera inside) to see close objects. The reliability of disparity depends on the contrast. We define the reliability of the disparity as

$$R_d = ({f^R}_x)^2 + ({f^L}_x)^2 \tag{5.5}$$

where f^L and f^R denote the left and the right image, respectively, and d denotes disparity. If R_d is small for a point, then the disparity of the point is not calculated.

5.5.2 Tracking

Initially a target object is detected as a moving region or it is specified by a user. The region is approximated by a rectangle called *target window*. As shown in Figure 5.5, the candidate region of the target is predicted from the target window of the previous frame (initially, the candidate region is the initial target window itself).

We approximate the distribution of the velocity and disparity in the target region with Gaussian distribution. The mean target velocity (μ_u, μ_v) and the disparity μ_d are determined from the optical flow and the disparity, respectively, in the temporal target window. The

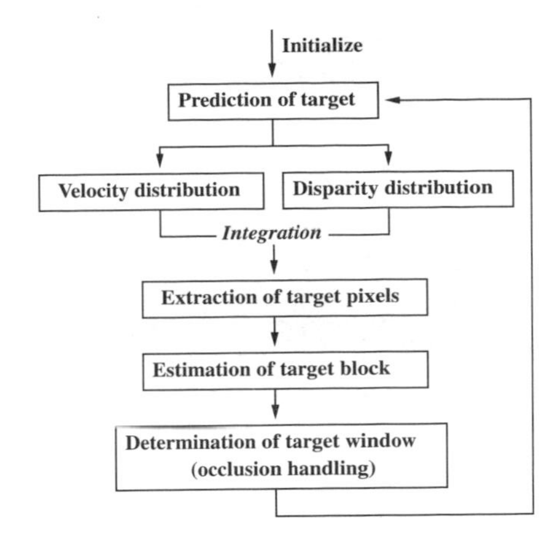

Figure 5.5 Outline of tracking.

standard deviation σ is estimated to be linear to the average:

$$\sigma_i = \alpha|\mu_i| + \beta$$

where the suffix i represents the optical flow or the disparity, and α and β are constants.

The probability of the pixel belonging to the target is calculated based on the distributions of the target velocity, that of the target disparity and the pixel attributes in the predicted candidate region of the target. The pixels with high probability are extracted and are called *target pixels*. The rectangle circumscribing the target pixels is called a *temporal target window* (Fig. 5.6).

Because the extracted optical flow and disparity are noisy, the final decision is made with a sequence of images. When a target region is found, the target window is divided into small fix-sized rectangles (called *block*). The block moves at the same velocity as that of the target window. If enough target pixels are extracted in a block in many frames, the block is determined to be a part of the target and called *target block*.

If the mean disparity in the target block becomes abruptly large, the region is determined to be occluded. This block is called an *occluded block*. The next *target window* is a rectangle circumscribing the target pixels, the target blocks, and the occluded blocks. In the next frame, the target region is predicted as the target window shifted by the mean target velocity (μ_u, μ_v).

5.5.2.1 Prediction of the Target Location. The candidate target region is determined by enlarging the predicted target region (see Fig. 5.7). We define the prior probability of the pixel belonging to the target region as follows. The center region has the probability of 1.0, and the probability is reduced linearly toward the edge of the candidate target region.

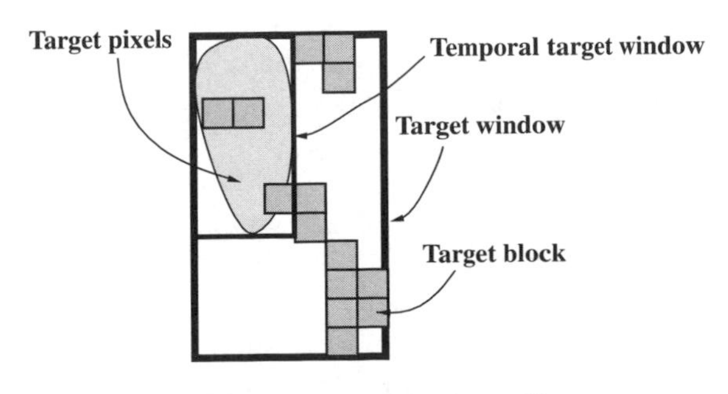

Figure 5.6 Terminologies for tracking.

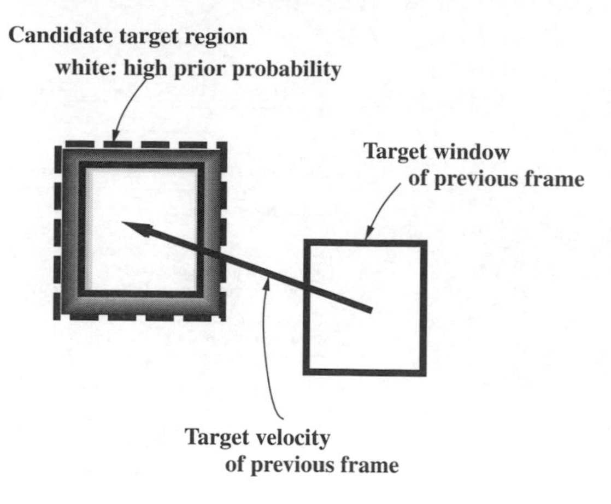

Figure 5.7 Candidate target region and prior probability (the darker the shading, the higher prior probability).

5.5.2.2 Extraction of Target Pixels. The target pixels are extracted in and around the temporal target window shifted by the mean target velocity of the previous frame. Let observation vector $\mathbf{F}(x, y)$ be at (x, y) be defined as

$$\mathbf{F}(x, y) = [u(x, y), v(x, y), d(x, y)]\,. \tag{5.6}$$

The probability of the pixel (x, y) being on the target object $P(obj|\mathbf{F})$ is calculated as follows:

$$P(obj|\mathbf{F}) = \frac{P(\mathbf{F}|obj)P(obj)}{P(\mathbf{F})} \tag{5.7}$$

where $P(\mathbf{F})$ is constant. Let us assume that u, v and d in a target region are independent of one another. Then, $P(\mathbf{F}|obj)$ is represented as

$$P(\mathbf{F}|obj) = P(u|obj)P(v|obj)P(d|obj) \tag{5.8}$$

where $P(u|obj)$, $P(v|obj)$, $P(d|obj)$ are likelihoods of the observation vector (the distributions are previously defined). $P(obj)$ is the prior probability of the pixel (x, y) (described previously).

The pixels which have high probability $P(obj|\mathbf{F})$ are extracted as the target pixels.

5.5.3 Estimation of the Target Block

Because the optical flow and depth are noisy, the target block is estimated not from an instantaneous input images but from images of a certain period. The target window is divided into small fix-sized *blocks*. Each block is assigned one of the following states:

- *Target*: belonging to the target and not occluded
- *Occluded*: belonging to the target and occluded
- *Not_Target*: not belonging to the target

The state transition is shown in Fig. 5.8. All the initial states of blocks are *Not_Target*. If there are enough target pixels in a block, the block gets one vote. Enough number of votes changes the state *Not_Target* to *Target*. If the target pixels are not extracted in a block of *Target* and the average disparity in the block is much larger than the mean target disparity d_obj, then the state *Target* changes to *Occluded* (because it is in the occluded target region). If the average disparity is not much larger than the mean target disparity d_{obj}, the block gets one

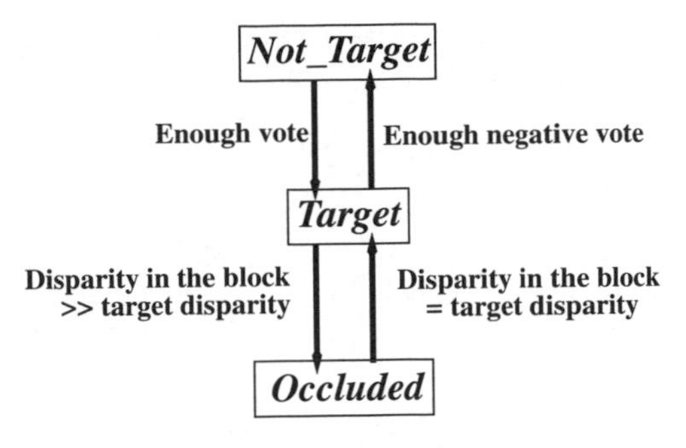

Figure 5.8 State transition of block.

negative vote. A sufficient number of negative votes changes the state *Target* to *Not_Target*. If the average disparity in a block of state *Occluded* becomes similar to the mean target disparity d_{obj}, the state of the block changes again to *Target*.

5.5.3.1 Determination of the Target Window. Determination of the target window depends on occlusion state of the target. We decide that the target is completely occluded when either of the following two conditions is satisfied.

1. Few target pixels are extracted, and there are the occluded blocks.
2. The mean target disparity is much larger than that of the previous one.

Unless the target is completely occluded, the target window is determined as the rectangle circumscribing the target blocks, occluded blocks and the target pixels. If the target is completely occluded, the target velocity is kept constant. If the disparity in target window abruptly decreases, we decide that the target appears again.

5.5.4 Experimental Result

Figure 5.9 shows some of our experimental result using 100 images (in about 6.7 seconds). The resolution of each image is 160×120 pixels. Persons walk from right to left. In image No. 1, the left person is the target person and the right one is nearer to the camera than the target person. While the right person overtakes the target person (image No. 25 to No. 40), tracking using only the optical flow might fail because the optical flow is similar.

After the target object is overtaken, he passes by the standing person. Because the distance of the standing person is similar to the target person, the tracking using only depth data might also fail.

The target person was successfully tracked in our method. Note that the feet of the target person are not extracted because they violated the assumption that the target velocity is uniform in the target region.

5.5.5 Real-Time Tracking

The hardware configuration of our system is illustrated in Fig. 5.10. The image processor processes the sequence of stereo images, and puts the result of tracking out to the monitor. The image processor also sends the motor commands to the active camera head controller.

The image processor has many DSPs (Digital Signal Processor). Each DSP is mounted on a DSP board which has memories and interface for data transfer. The connections between DSP boards can be changed according to the image processing algorithm. The functions of the DSP boards and the connections between the DSP boards for our method are also depicted in Fig. 5.10.

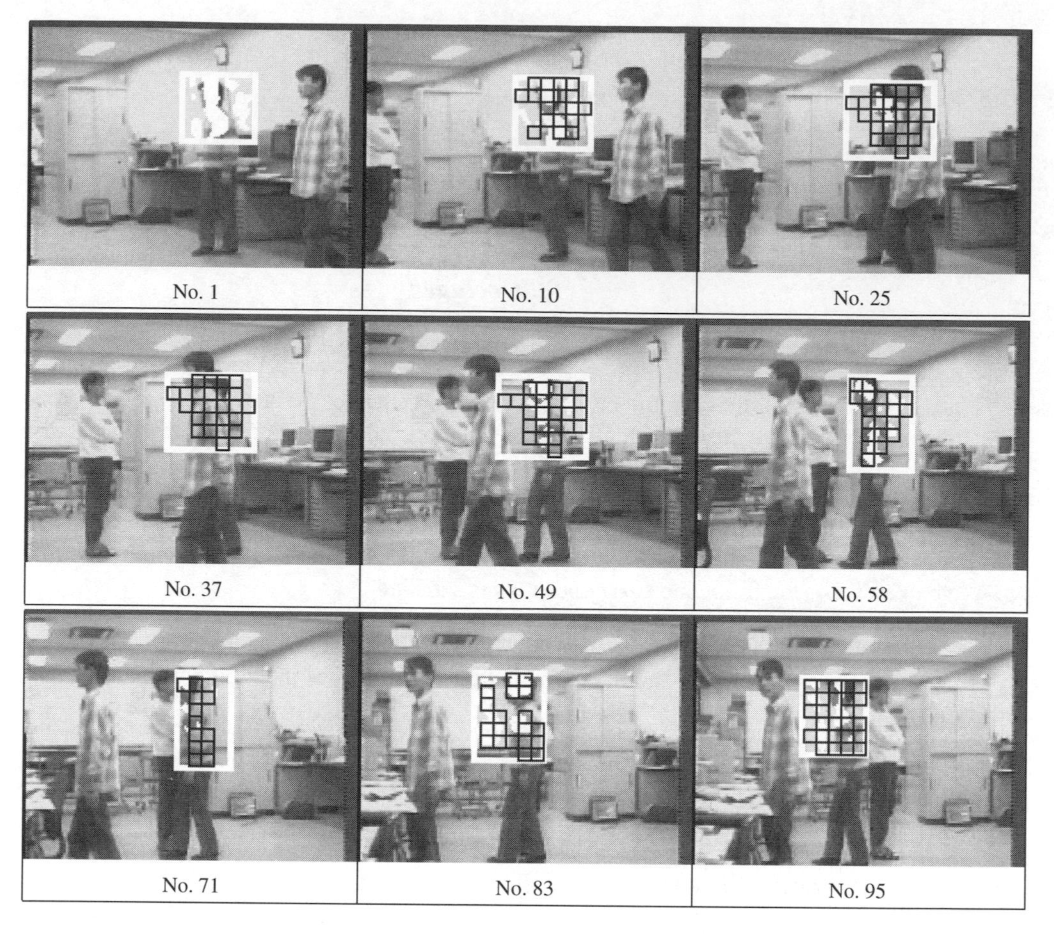

Figure 5.9 Experimental result (white region: target pixels; black rectangle: target blocks and occluded blocks; and white rectangle: target window).

5.6 TRACKING WITH OPTICAL FLOW AND UNIFORM BRIGHTNESS REGIONS

Optical flow and the depth are not obtained reliably in uniform brightness regions. In order to solve this problem, the uniform region itself is used as the cue of the target [8].

Initially, a moving person is detected with optical flow. The area is circumscribed by a rectangle (called *motion window*). Uniform brightness regions are extracted near the motion window and circumscribed by a rectangle (called *brightness window*). In tracking, the motion and the brightness windows are updated at each frame and mean optical flow vector is calculated in the motion window. Then the motion and the brightness windows are predicted from the mean optical flow vector. The target motion region is extracted around the predicted position and the new motion window is set to circumscribe it. The target brightness regions are extracted around the predicted position and the new brightness window is set to circumscribe all of them.

We integrate optical flow and uniform brightness regions in order to track the target person reliably even if one of them does not offer correct information.

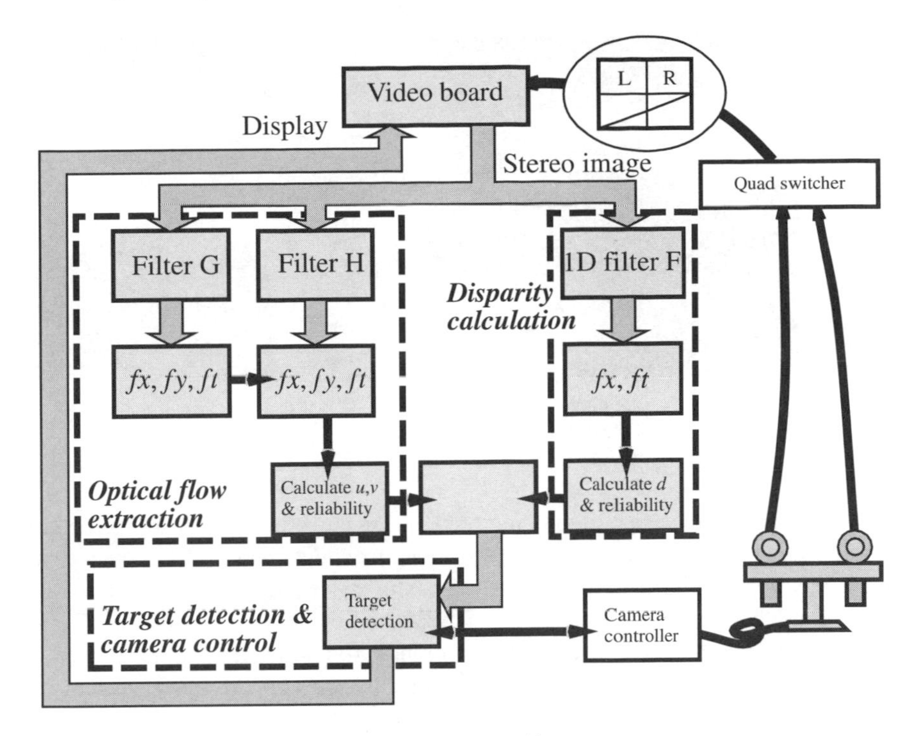

Figure 5.10 Real-time tracking system.

5.6.1 Target Motion Region

Extraction of the target motion region is the same as in the previous sections. The method, however, cannot discriminate the target from other objects with similar flow. In Fig. 5.11, the motion window (the white rectangle) circumscribes both persons.

5.6.2 Target Brightness Region

A uniform brightness region is defined as the region in which optical flow cannot be obtained. Fig. 5.12 shows an example of uniform brightness regions.

After removing small regions, each region is approximated by the rectangle. All rectangles are circumscribed by the *brightness window*.

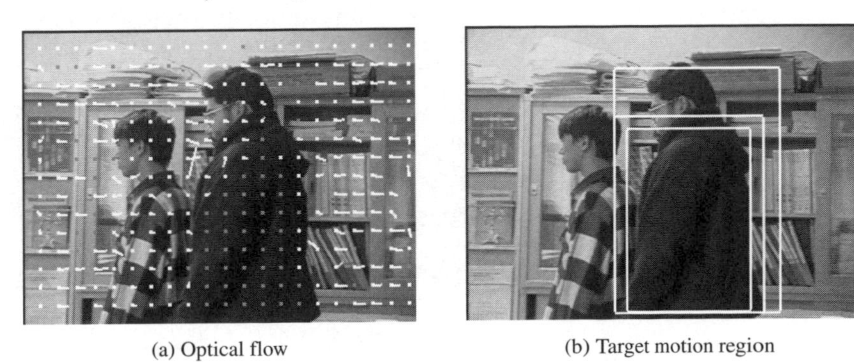

(a) Optical flow (b) Target motion region

Figure 5.11 Two persons in a target motion region.

(a) Input image

(b) Uniform brightness regions

Figure 5.12 Extraction of uniform brightness regions.

For each brightness region being tracked, the corresponding region is determined in the current image. First, a search area is determined from the previous brightness window and the mean flow vector. Uniform brightness regions in the search area are extracted as candidates. Correspondence is made between the candidate and the predicted brightness regions.

This method, however, cannot discriminate the target from other objects with similar brightness. In Fig. 5.13(a) the black clothing region and the black curtain region merge into one uniform brightness region. In Fig. 5.13(b), the brightness window (outer rectangle) circumscribes both regions.

5.6.3 Integration of Motion and Brightness

The width of the brightness window and that of the motion window are similar. If the width of one window is much larger than that of the other, the larger one is less reliable and the width is reduced to that of the smaller one. If the motion and/or brightness regions are lost during tracking, a search area is determined and the lost area is searched for again.

5.6.3.1 Modification of Brightness Window by Motion. Some of uniform brightness regions may be background regions. If the center of gravity of the region is outside the motion window, we remove it as a background. Background regions are tracked in Fig. 5.14(a). The regions are removed in Fig. 5.14(b).

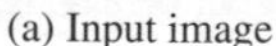

(a) Input image

(b) Target brightness region

Figure 5.13 Background and a person in a target brightness region.

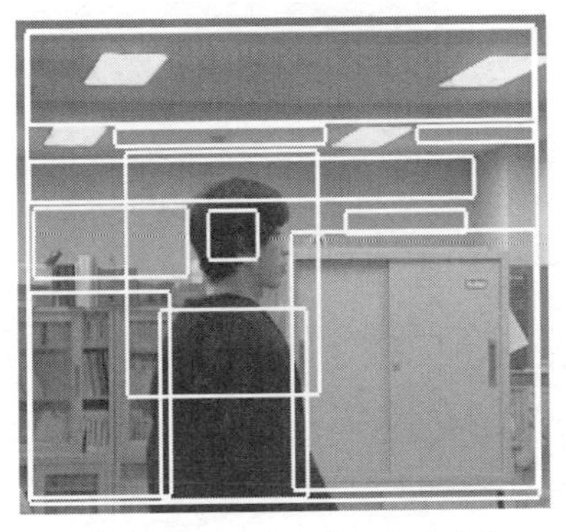

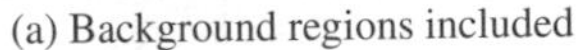

(a) Background regions included

(b) Background regions removed

Figure 5.14 Removing background regions by motion.

Some of the uniform brightness regions may connect to the background regions with similar brightness, and the width of the brightness window may become large. They are also corrected based on the motion window as shown in Fig. 5.15(a).

5.6.3.2 Modification of Motion Window by Brightness. When a target object overlaps with another object with similar motion, the motion window becomes wider. The motion window can be modified by the brightness window if it is correctly obtained. Fig. 5.15(b) shows an example of this case.

On the other hand, if the motion window includes some of the uniform brightness regions of the other object, the motion window cannot be modified correctly due to the spurious brightness regions. In order to solve this problem, we introduce a reliability measure: if a region has been tracked in many frames, the region is regarded as reliable. Regions outside of those reliable regions are regarded as belonging to the other object. In Fig. 5.16(a), the foreground person has been tracked in many frames. After overlapping as shown Fig. 5.16(b), the background person appears again in Fig. 5.16(c) where spurious brightness regions also appear. Based on the reliable regions, the spurious regions are removed as shown in Fig. 5.16(d).

5.6.3.3 Determination of Search Area for the Motion and Brightness Windows. If both of the motion and the brightness windows are extracted, their positions in the next frame are predicted from the current mean flow vector. If only the brightness window is lost, the

(a) Modification by motion

(b) Modification by brightness

Figure 5.15 Tracking by modification.

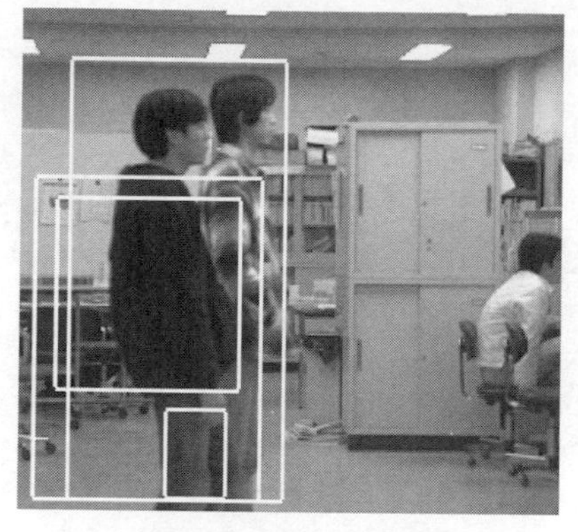

(a) Tracking before overlap

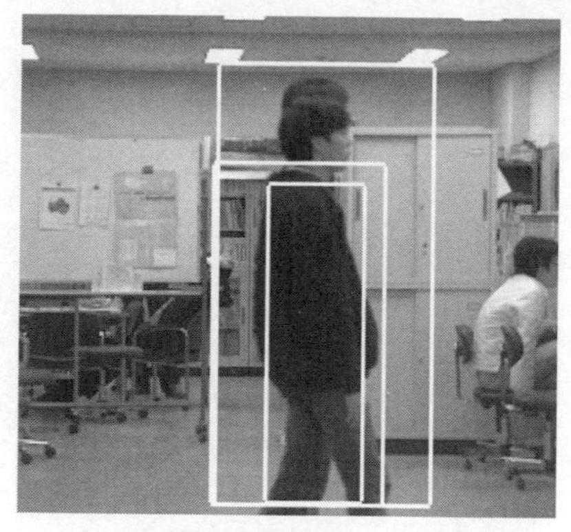

(b) Tracking during overlap

(c) Spurious brightness regions included

(c) Spurious regions removed

Figure 5.16 Motion window modified by reliable brightness regions.

target brightness regions is searched for near the motion window. If only the motion window is lost, the target motion region is searched for near the brightness window. If both windows are lost, they are searched for in the same way as the initial search.

5.7 CONCLUSION

Reliable visual tracking is realized by integrating many cues. One of the important problems to handle more complicated cases is how to realize real-time processing with low cost.

REFERENCES

[1] B. K. P. Horn and B. G. Schunck, "Determining optical flow," *Artificial Intelligence*, Vol. 17, pp. 185–203, 1981.

[2] M. V. Srinivasan, "Generalized gradient schemes for the measurement of two-dimensional image motion," *Biological Cybernetics*, vol. 63, pp. 421–431, 1990.

[3] Y. Mae, S. Yamamoto, Y. Shirai, and J. Miura, "Optical Flow Based Object," *Tracking by Active Camera, Proceedings of the Second Japan-France Congress on Mechatronics*, pp. 545–548, 1994.

[4] S. Yamamoto, Y. Mae, Y. Shirai, and J. Miura, "Realtime Multiple Object Tracking Based on Optical Flows," *Proceedings IEEE International Conference Robot and Automation*, pp. 2328–2333, 1995.

[5] D. Coombs and C. Brown, "Real-time Smooth Pursuit Tracking for a Moving Binocular Robot," *Proceedings CVPR '92*, pp. 23–28, 1992.

[6] N. Kita, S. Rougeaux, Y. Kuniyoshi, and S. Sakane, "Through ZDF-based Localization for Binocular Tracking," *IAPR Workshop on Machine Vision Applications*, pp. 190–195, 1994.
[7] R. Okada, Y. Shirai, and J. Miura, "Object Tracking Based on Optical Flow and Disparity," *Proceedings of IEEE/SICE/RSJ International Conference on Multisensor Fusion and Integration for Intelligent Systems*, pp. 565–571, 1996.
[8] T. Yamane, Y. Shirai, and J. Miura, "Person Tracking by Integrating Optical Flow and Uniform Brightness Regions," *Proceedings IEEE International Conference on Robot and Automation*, pp. 3267–3272, 1998.

Chapter 6 TWO-DIMENSIONAL MODEL-BASED TRACKING OF COMPLEX SHAPES FOR VISUAL SERVOING TASKS

Nathalie Giordana, Patrick Bouthemy,
François Chaumette, and Fabien Spindler
IRISA/INRIA Rennes

Jean-Claude Bordas and Valéry Just
DER-EdF Chatou Cedex

Abstract

Visual servoing needs image data as input to realize robotics tasks such as positioning, docking, or mobile target pursuit. This often requires to track the 2D projection of the object of interest in the image sequence. To increase the versatility of visual servoing, objects cannot be assumed to carry landmarks. We have developed an original method for 2D tracking of complex objects which can be approximately modeled by a polyhedral shape. The proposed method fulfills real-time constraints as well as reliability and robustness requirements. Real experiments and results on a positioning task with respect to different objects are presented.

6.1 INTRODUCTION

The visual servoing approach, which consists of controlling movements of a robot from the estimation of image features, is attractive for industrial use in changing or hostile environments such as a nuclear power plant. In order to follow this approach, the extracted image information must be robust, accurate, and computed in real time. Current techniques exploited in an industrial environment run on marked and simple objects. Our goal is to design a method to extract relevant image features without such constraints. Several authors have proposed ways to solve tracking of features in image sequences with monocular vision [1, 2, 3, 4, 5] or stereo vision [6]. We have developed an original method for 2D tracking of complex objects which can be approximately modeled by a polyhedral shape. The efficiency of this method is demonstrated through a visual servoing homing task which consists in positioning a camera mounted on the end effector of a six d.o.f cartesian robot with respect to objects.

The chapter is organized as follows. In Section 6.2, we briefly recall the visual servoing approach, and we specify the considered task. Section 6.3 describes the initialization step of the tracking algorithm. In Section 6.4, we present the tracking algorithm which relies on 2D global parametric motion model and 2D deformable template. Experimental results are reported in Section 6.5. Section 6.6 contains concluding remarks.

6.2 SPECIFICATION OF THE HOMING TASK

6.2.1 Image-Based Visual Servoing

The image-based visual servoing approach consists in specifying a task as the regulation in the image of a set of visual features [7, 8]. Another approach consists in using a model of 2D image motion [9]. Let us denote p the visual features involved for the task. The task function is defined by

$$e = \hat{L}^{T^+}(p(t) - p^*) \tag{6.1}$$

where
- $p(t)$ is the current value of the considered image features (e.g., coordinates of the particular object points);
- p^* is the desired value of p;
- $\hat{L}^{T^+}$ is the pseudo inverse of a model or an approximation of the interaction matrix L_p^T defined by $\dot{p} = L_p^T T_c$, T_c being the camera velocity.

The goal is to minimize $\|e\|$. In order that e exponentially decreases, the desired evolution of e takes the form:

$$T_c = -\lambda e \tag{6.2}$$

where λ tunes the speed of convergence.

6.2.2 Positioning with Respect to an Object

We have considered a generic homing task that positions an eye-in-hand system with respect to a given object. For this application, we take as p the coordinates of an appropriate set of points on the object silhouette: $p = \{(x_j, y_j), j = 1, \ldots, k\}$, $k \geq 4$. Considering the perspective projection model, a point in the image plane with coordinates (x_j, y_j) corresponds to a 3D point (X_j, Y_j, Z_j) in the camera coordinate system with $x_j = X_j/Z_j$ and $y_j = Y_j/Z_j$. The related interaction matrix is given by

$$L_p^T = \begin{pmatrix} -1/Z_j & 0 & x_j/Z_j & x_j y_j & -1 - x_j^2 & y_j \\ 0 & -1/Z_j & y_j/Z_j & 1 + y_j^2 & -x_j y_j & -x_j \\ & & & \vdots & & \end{pmatrix} \tag{6.3}$$

The model $\hat{L}^T$ of the interaction matrix chosen as $L^{T^+}_{p=p^*}$, where Z_j^*, $j = 1, \ldots, k$ is obtained by a pose computation, as explained in Section 6.3.

6.3 SEMI-AUTOMATIC INITIALIZATION

To initialize the tracking algorithm, we have to determine a number of control points on the contour of the object projection in the first image of the sequence. These points will then form a polygonal shape which is assumed to correctly model the object appearance in the image. To identify this 2D polygonal shape, we estimate the camera pose from the first image of the sequence. To this end, we exploit a CAD polyhedral 3D model of the object. We have to find the 3D rotation and the 3D translation which map the object coordinate system with the camera coordinate system. The 3D CAD model is then projected onto the image by perspective projection in order to match the silhouette of the object projection in the first image. We use the intrinsic camera parameters given by the maker. A number of methods to compute perspective from N points have been proposed [10, 11, 12]. We resort to the method designed by Dementhon and Davis [13]. This method calculates the rigid transformation in an iterative way from the knowledge of the coordinates of at least four non-coplanar points, in the object coordinate system, and of their corresponding projections in the image. Its principle consists in approximating perspective projection by scaled orthographic projection, and then iteratively modifying the scaled orthographic projection to converge to the perspective projection. The initialization step is semi-automatic, because the correspondence of at least four non-coplanar

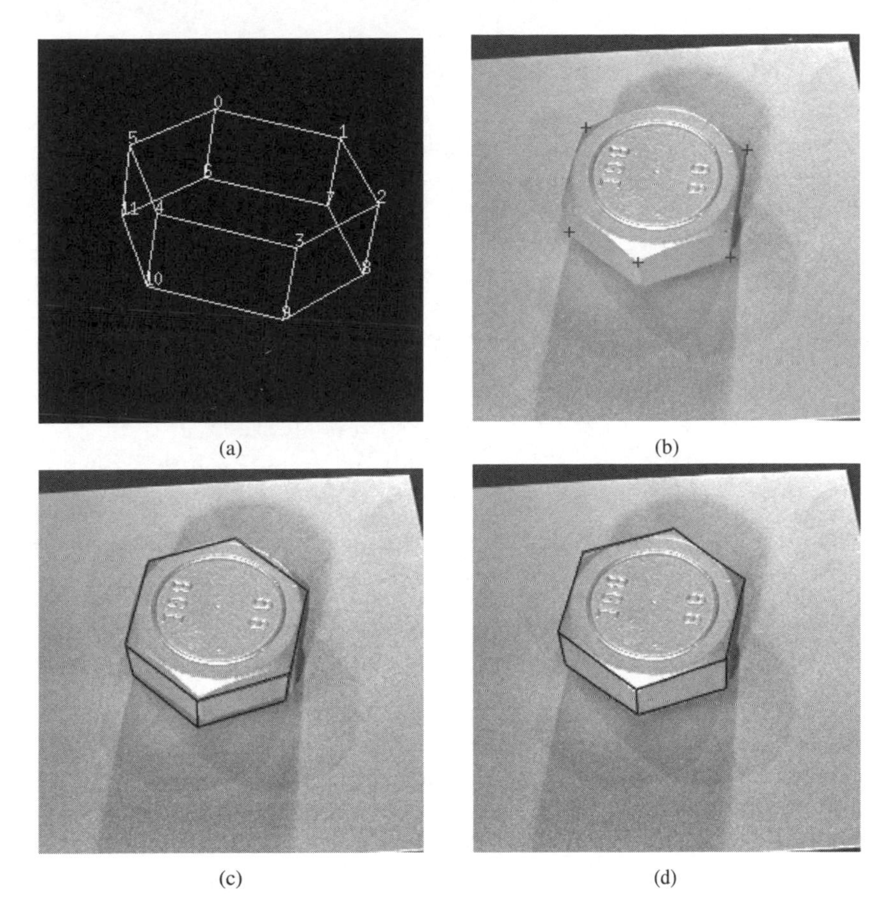

Figure 6.1 An example of initialization step. (a) 3D CAD model of the nut. (b) Crosses represent the points selected to calculate the pose of the object. (c) Projected model superimposed on the image. (d) Projected model superimposed on the same image after the refinement step using the tracking algorithm.

points (typically 4 or 5) of the 3D model with image points is achieved in an interactive manual way. Because the faces of the 3D CAD model are oriented by construction, we can determine the visible parts after projection of the 3D model. In order to refine the projected contour obtained after pose calculation, we apply the tracking algorithm presented in Section 6.4 on the same first image. The points used in p are a subset of points characterizing the projected contour of the object (typically, the corners). An example of initialization step is presented in Fig 6.1, for one of the real objects we have dealt with.

6.4 TWO-DIMENSIONAL TRACKING OF POLYHEDRAL OBJECT

As described in the previous section, the 2D projection of the object to be tracked is characterized by points on the object contour supplied by the initialization step. We consider that the 2D global transformation between two successive projections of the object in the image plane can be represented by a 2D affine displacement model augmented with local deformations. The aim is to estimate the parameters of the 2D global transformation.

6.4.1 Transformation Model

Let $X^t = [X_1^t, \ldots, X_n^t]^T$ a vector composed by the image coordinates X_i^t of points along the contour of the object projection at time t, and Γ_{X^t} the contour associated with the vector X^t. Let us denote ${}^lX^t$ the optimal shape of the object projection estimated at time t, and ${}^fX^t$ a filtered version of ${}^lX^t$ (to be defined in Subsection 4.2). The optimal shape ${}^lX^{t+1}$, at time $t+1$, will be given by

$$ {}^lX^{t+1} = {}^lX^{t+1}(\Theta, \delta) = \Psi_\Theta({}^fX^t) + \delta \tag{6.4} $$

where - Ψ_Θ is a 2D affine transformation given by

$$ \begin{bmatrix} x' \\ y' \end{bmatrix} = \begin{bmatrix} a_1 & a_2 \\ a_3 & a_4 \end{bmatrix} \begin{bmatrix} x \\ y \end{bmatrix} + \begin{bmatrix} T_x \\ T_y \end{bmatrix} \tag{6.5} $$

with $\Theta^T = (a_1, a_2, a_3, a_4, T_x, T_y)$, $X = (x, y)^T$ and $X' = (x', y')^T = \Psi_\Theta(X)$.

- $\delta = (\delta_1, \ldots, \delta_n)$, with $\delta_i = (\delta_{x_i}, \delta_{y_i})$ denotes the local deformation introduced at point X_i. It will be modeled by a centered Gauss-Markov process with σ_i and correlation factor ϵ_{ij}.

6.4.2 Tracking Algorithm

The tracking algorithm is articulated into five steps as outlined in Fig. 6.2. The first two steps are concerned with the estimation $\hat{\Theta}$ of the global affine parameters Θ. The third step computes the optimal shape ${}^lX^{t+1}$ by minimizing an energy function E_Θ with $\Theta = \hat{\Theta}$. In the fourth step, the model shape denoted ${}^mX^t$, undergoing only the global affine deformation, is computed. Finally, the fifth step delivers the final shape ${}^fX^{t+1}$. It is given by

$$ {}^fX^{t+1}(\Theta_f, \delta) = \Psi_{\Theta_f}({}^fX^t) + G\,\delta $$

where Θ_f is the 2D affine deformation obtained at step 4, and G is a validation factor of local deformation.

6.4.3 Estimation of the Model Parameters

To estimate the optimal shape ${}^lX^{t+1}$, i.e., (Θ, δ), we adopt a Bayesian criterion, which turns out to lead to a minimization problem. More precisely, the problem is to estimate $(\hat{\Theta}, \hat{\delta})$

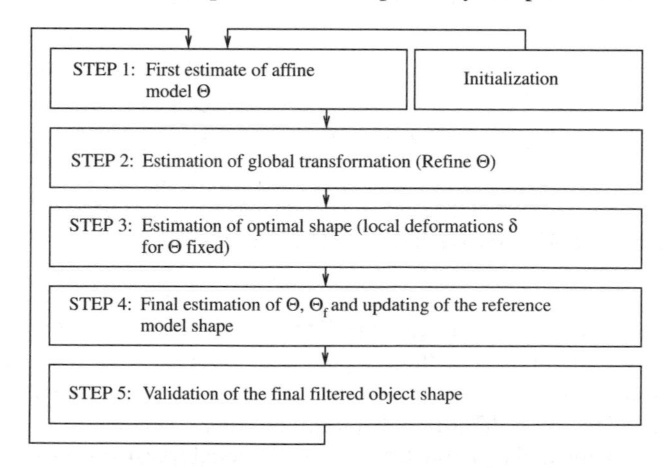

Figure 6.2 Outline of the tracking algorithm.

by minimizing an energy function E_Θ. For more details, the reader is referred to [14, 15]. This implies that $(\delta_i)_{i=1,\ldots,n}$ are supposed to be of low magnitude. This assumption is verified in our application of visual servoing. A first step supplies an initial estimate of Θ using as input normal displacements evaluated along the object shape contour with the ECM algorithm [16, 15]. Then, this estimation will be refined as explained below. E_Θ is decomposed in two terms E_d and E_p:

$$E_\Theta({}^lX^{t+1}, d^{t+1}) = E_d({}^lX^{t+1}, d^{t+1}) + E_p({}^lX^{t+1}) \tag{6.6}$$

E_d expresses the adequacy between the variables to be estimated and the observations $d^{t+1} = \{d_s^{t+1}\}$. This is the "data-driven" energy term. The observations d^{t+1} are given by

$$d_s^{t+1} = -\min(\|\nabla I_s(t+1)\|, Gr_{max}) \tag{6.7}$$

where ∇I_s denotes the spatial gradient of the intensity function at point s along the contour, and Gr_{max} is a threshold which permits to saturate the too high intensity gradient values. E_p represents the *a priori* information on the local deformations δ. This is the regularization term.

The optimal shape of the object projection will be given by ${}^lX^{t+1}(\hat{\Theta}, \hat{\delta})$ where

$$(\hat{\Theta}, \hat{\delta}) = \arg\min_{(\Theta,\delta)}\{E_d({}^lX^{t+1}, d^{t+1}) + E_p({}^lX^{t+1})\} \tag{6.8}$$

Let us define energy terms E_d and E_p. E_d is given by

$$E_d({}^lX^{t+1}, d^{t+1}) = \sum_{s \in \Gamma_{{}^lX^{t+1}}} d_s^{t+1} \tag{6.9}$$

where $\Gamma_{{}^lX^{t+1}}$ represents the contour of the 2D shape ${}^lX^{t+1}$.

Concerning E_p, two deformation processes are in fact introduced. As previously mentioned, we consider the local deformation field δ with variance σ_i^2 and correlation factor ϵ_{ij}. We also take into account the "reference" shape ${}^mX^t$, which is provided at time t by the transformation of the initial 2D object projection model, resulting only from the combination of successive estimated 2D global affine transformations. Then, ${}^m\delta$ is given by ${}^m\delta_i = \Psi_{\hat{\Theta}}({}^mX^t) - {}^lX^{t+1}$ with variance ${}^m\sigma_i^2$ and correlation factor ${}^m\epsilon_{ij}$. The interest of the deformation field ${}^m\delta$ is to avoid undesirable deformation of the shape over time.

The expression of E_p is thus defined as

$$E_p({}^lX^{t+1}) = \sum_i \left(\frac{\rho(\|\delta_i\|)}{\sigma_i^2} + \frac{\|{}^m\delta_i\|^2}{{}^m\sigma_i^2} \right) + \sum_{(i,j)neighbour} \left(\frac{\rho(\|\delta_i - \delta_j\|)}{\epsilon_{ij}^2} + \frac{\|{}^m\delta_i - {}^m\delta_j\|^2}{{}^m\epsilon_{ij}^2} \right) \tag{6.10}$$

where ρ is a quadratic truncated function.

Two points indexed by i and j are called "neighbor" if they are located at two successive positions along the shape contour.

The criterion (6.8) cannot be directly solved. We resort to an alternative iterative procedure. First, we estimate Θ using

$$\hat{\Theta} = \arg\min_{\Theta} E_d(\Psi_\Theta({}^fX^t), d^{t+1}) \tag{6.11}$$

then, for $\hat{\Theta}$ fixed, we estimate δ using

$$\hat{\delta} = \arg\min_{\delta} E_{\hat{\Theta}}({}^lX^{t+1}, d^{t+1}) \tag{6.12}$$

The optimization of E_d is performed by a gradient algorithm, whereas the optimization of E_p is achieved by simulated annealing.

6.5 EXPERIMENTAL RESULTS

The complete experimental implementation and validation of the visual servoing task, including initialization and tracking, have been carried out. We have conducted experiments dealing with a positioning task. Several objects of interest have been considered. This task has been performed on an experimental testbed involving a CCD camera mounted on the end effector of a six DOF cartesian robot.

The experiment comprises the following steps:

- In an off-line step, the camera is first positioned at the final desired position and a number of points (at least 4) on the object image are selected to specify the control law. The 2D model of the object projection is initialized, as explained in Section 6.3.
- The camera is then positioned at the initial position. The 2D model of the object projection is then also initialized, as explained in Section 6.3.
- At every intermediate camera position between the initial and the final ones, the contour of the object projection in the image is updated by the tracking algorithm presented in Section 6.4. Then, the control law is activated to reach the next position.

A first real example involving a nut as object of interest is now reported. The tracking algorithm runs on an Ultra-Sparc-1 Sun workstation, equipped with a Sunvideo image capture board, at the rate of 1Hz for images of 256×256 pixels. This relative low processing rate implies that the positioning task is specified in position, i.e., $\Delta r = T_c \Delta t$, where Δr is the camera displacement. Otherwise, the control could be performed on the velocity.

Figures 6.3 and 6.4 show the temporal evolution of the components of $(p - p^*)$, in pixels, and of T_c, in cm/s and dg/s. These curves show the stability and the convergence of the control law. Indeed, the error on each coordinate of the six points specifying the task and the components of the control vector T_c converge to zero. Figures 6.5(a) and 6.5(b), respectively, contain the images delivered by the camera at its initial position and final reached one. Crosses overprinted in the image indicate the target position of the points used to specify the control law. Figure 6.5(b) depicts the apparent trajectory in the image of these points during the achievement of the task. The initial and the final polygonal shape contours accounting for the tracking of the nut projection in the image are also drawn.

We can point out on this example that the tracking of the object contour in the image must tackle with low intensity contrast, presence of cast shadows, mirror specularities and so

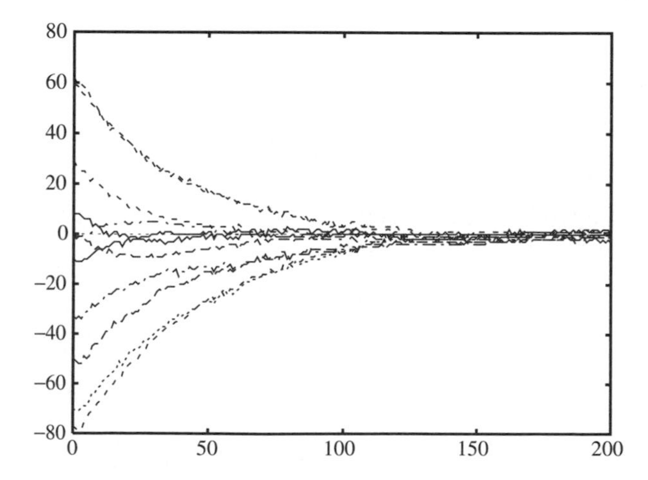

Figure 6.3 Temporal evolution of $(p - p^*)$ for the nut experiment.

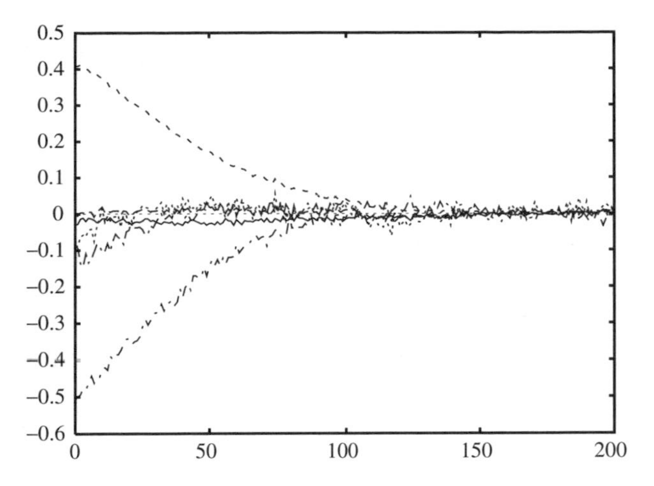

Figure 6.4 Temporal evolution of T_c for the nut experiment.

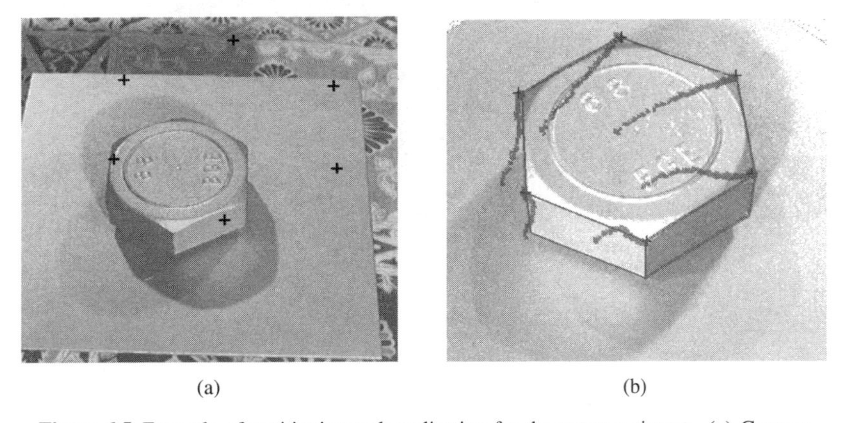

(a) (b)

Figure 6.5 Example of positioning task realization for the nut experiment. (a) Crosses indicating the desired position and plot of the polygonal model contour of the nut projection after initialization. (b) Apparent trajectories of the points used to specify the task, and plot of the contour of the nut projection model at the convergence of the task.

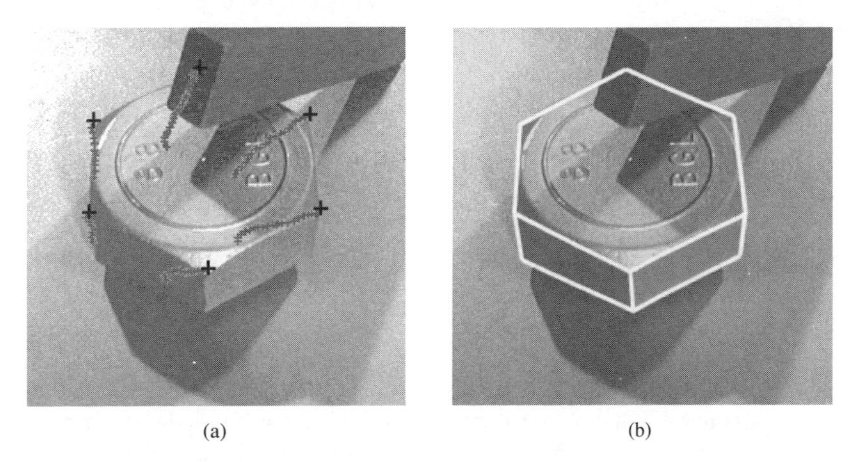

(a) (b)

Figure 6.6 Example of positioning task with partial occlusion. (a) Apparent trajectories of the points used to specify the task. (b) Contour of the nut projection model at the convergence of the task.

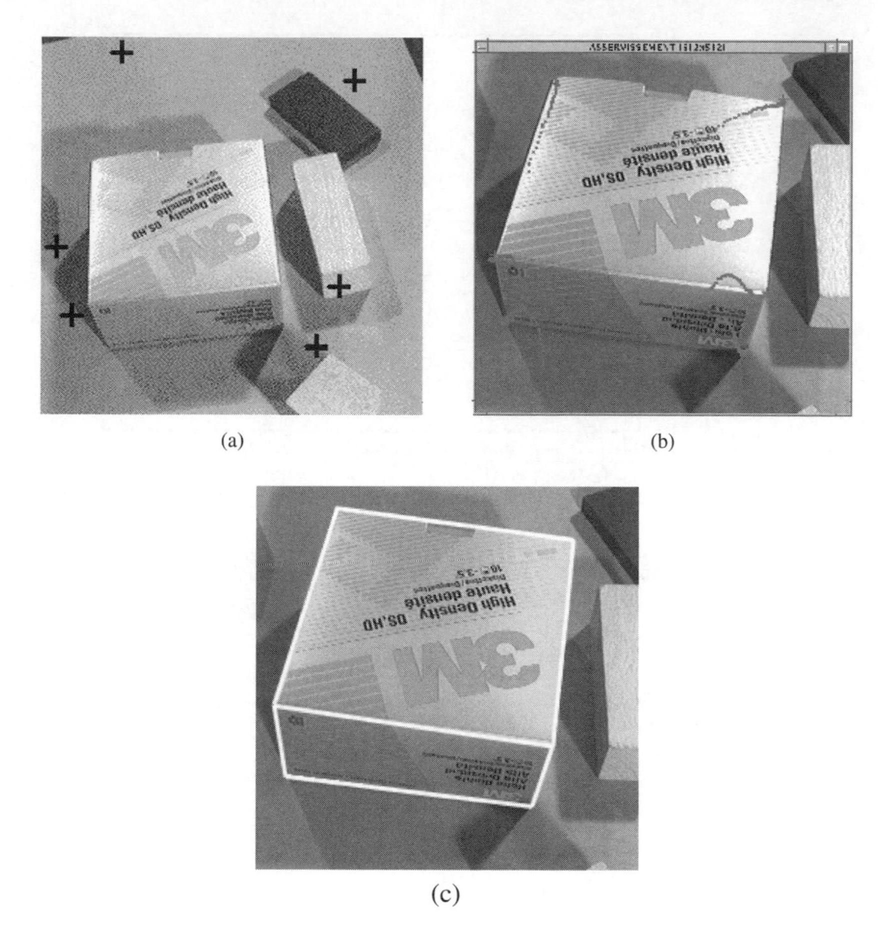

Figure 6.7 Example of positioning task on a box with possible false matches. (a) Crosses indicating the desired position. (b) Apparent trajectories of the points used to specify the task. (c) Contour of the box projection model at the convergence of the task.

on. Moreover, the object is not exactly polyhedral, and the object edges cannot be physically precisely defined. Despite these difficulties, the proposed method have proven its efficiency on different classes of objects such as box or nut. Experiments in the presence of partial occlusion (Fig. 6.6) or possible false matches (Fig. 6.7) have been performed with success. However, the tracking method contains some limitations. The method cannot take into account important changes of appearance of the projection of the object in case of large displacements of the camera.

6.6 CONCLUSIONS

We have presented an original method for tracking complex objects in an image sequence. It allows us to carry out visual servoing task of positioning with respect to real objects (without any landmarks). Initialization of the algorithm is based on pose computation while exploiting the 3D CAD model of the object of interest. The tracking is based on the estimation, between two successive images, of a global affine transformation augmented with local deformations. It is formulated within a Bayesian framework. A real practical implementation has been realized. Results on different examples of positioning task have demonstrated the robustness and the reliability of the proposed method. In order to increase the processing rate, several improvements are under investigation. For instance, the tracking stage could exploit the 3D model of the object of interest. This could avoid to estimate any local deformations, which represents the main part of the computational load. The handling of the appearance in the image of previously hidden object parts will also be considered.

6.7 ACKNOWLEDGMENTS

This study has been supported by DER-EDF.

REFERENCES

[1] A. Blake, R. Curwen, and A. Zisserman. "A framework for spatiotemporal control in the tracking of visual contours." *International Journal of Computer Vision*, vol. 11, no. 2, pp. 127–145, 1993.

[2] D. B. Gennery. "Visual tracking of known three-dimensional objects." *International Journal of Computer Vision*, vol. 7, no. 3, pp. 243–270, 1992.

[3] G. D. Hager and K. Toyama. "X Vision: A portable substrate for real-time vision applications." *Computer Vision and Image Understanding*, vol. 69, no. 1, pp. 23–37, 1998.

[4] M. Isard and A. Blake. "Contour tracking by stochastic propagation of conditional density." *Proceedings of European Conference on Computer Vision*, vol. 1, pp. 343–356, Cambridge, England, April 1996.

[5] C. Meilhac and C. Nastar. "Robust fitting of 3D CAD models to videostreams." *Proceedings of International Conference on Image Analysis and Processing*, vol. 1, pp. 661–668, Florence, Italy, September 1997.

[6] P. Braud, M. Dhome, J. -T. Lapresté, and N. Daucher. "Modelled object pose estimation and tracking by a multi-cameras system." *Proceedings of IEEE International Conference on Computer Vision and Pattern Recognition*, pp. 976–979, Seattle, Washington, June 1994.

[7] B. Espiau, F. Chaumette, and P. Rives. "A new appraoch to visual servoing in robotics." *IEEE Transactions on Robotics and Automation*, vol. 8, no. 3, pp. 313–326, 1992.

[8] S. Hutchinson, G. D. Hager, and P. I. Corke. "A tutorial on visual servo control." *IEEE Transactions on Robotics and Automation*, vol. 12, no. 5, pp. 651–670, 1996.

[9] V. Sundareswaran, P. Bouthemy, and F. Chaumette. "Exploiting image motion for active vision in a visual servoing framework." *International Journal of Robotics Research*, vol. 15, no. 6, pp. 629–645, 1996.

[10] M. Dhome, M. Richetin, J. -T. Lapresté, and G. Rives. "Determination of the attitude of 3-D objects from a single perspective view." *IEEE Transactions on Pattern Analysis and Machine Intelligence*, vol. 11, no. 12, pp. 1265–1278, 1989.

[11] R. Kumar and A. R. Hanson. "Robust methods for estimating pose and a sensitivity analysis." *Computer Vision and Image Understanding*, vol. 60, no. 3, pp. 313–342, 1994.

[12] D. G. Lowe. "Three-dimensional object recognition from single two-dimensional images." *Artificial Intelligence*, vol. 31, no. 63, pp. 355–394, 1987.

[13] D. Dementhon and L. Davis. "Model-based object pose in 25 lines of codes." *International Journal of Computer Vision*, vol. 15, no. 1, pp. 123–141, 1995.

[14] C. Kervrann and F. Heitz. "A hierarchical statistical framework for the segmentation of deformable objects in image sequences." *Proceedings of IEEE International Conference on Computer Vision and Pattern Recognition*, pp. 724–728, Seattle, WA, June 1994.

[15] J-M. Odobez, P. Bouthemy, and E. Fleuet. "Suivi 2D de pièces métalliques en vue d'un asservissement visuel." *Proceedings of French Congress RFIA'98*, vol. 2, pp. 173–182, Clermont-Ferrand, France, January 1998.

[16] P. Bouthemy. "A maximum-likelihood framework for determining moving edges." *IEEE Transactions on Pattern Analysis and Machine Intelligence*, vol. 11, no. 5, pp. 499–511, 1989.

Chapter 7 INTERACTION OF PERCEPTION AND CONTROL FOR INDOOR EXPLORATION

D. Burschka, C. Eberst, C. Robl, and G. Färber
Technische Universität München

Abstract

In this chapter we present our approach to improve vision-based exploration by cooperative interaction of the perception and control modules. We describe the way the sensor data from a video camera are processed and how this information is transformed into a 3D environmental model of the world. We introduce a data-stabilizing module that operates in a closed loop with the sensor systems, improving their performance by verifying the explored information from different positions and by predicting reliable sensor features. An important part of our approach is the interaction between the verification module and the interpretation module that predicts missing sensor features. This interaction helps to generate a more accurate model containing poorly detectable features that are impossible to extract from a single sensor view. The control of the sensors and perceptors based on the acquired knowledge is also described.

7.1 INTRODUCTION

An autonomous mobile robot (AMR) requires a dependable representation of the environment for a proper operation. Permanent changes in the area of operation require frequent updates of the internal representation to maintain an exact and valid model of the environment. This representation is then used for sensor-based localization of the AMR to allow a long-term operation with sufficient position accuracy. The position error must be corrected at fixed time intervals depending on the speed of the vehicle and the convergence of the pose estimation algorithm used to keep the resulting error within tolerable limits.

In this chapter we present our approach to fulfill these requirements without usage of specialized hardware. This approach is based on the interaction between the modules compensating the flaws of the fast simplified algorithms used in them. The sensor data is processed in four subsequent stages extracting the environmental description at increasing levels of abstraction: sensor data preprocessing, 3D reconstruction, filtering and stabilization, and interpretation. In common systems [1]–[3], the data processing uses only the information available at the given stage. The data flow is unidirectional to the higher level. In contrast, we use the information from the next higher level to improve the performance in each stage (Fig. 7.1). A related approach applying the information of higher stages in the lower stages on a restricted class of features is described in [4], [5].

Section 7.2 describes the concept and gives an overview of our system structure. The feature extraction based on fast contour tracing and the 3D reconstruction is described in Section 7.3. The interpretation and integration, the final stage of our data processing, is introduced in Section 7.4. Section 7.5 describes the control of the perception modules, the platform and sensors. Section 7.6 presents our results.

7.2 CONCEPT

Our exploration system is designed to operate with a minimum of sensors. Robustness is reached based on the following ideas:

- The sensor data processing is tuned to low cost to allow a short cycle time without specialized hardware. The inadequacies of the sensor data processing are compensated by verification of the information from different positions.
- Hypothetical features are generated to complete the sensor information based on statistics and known geometrical primitives. These features are used to improve the sensor data extraction and the 3D reconstruction. This additional information helps to resolve difficult conditions such as poor illumination or multiple corresponding lines in a pair of images.
- Complementary and competitive interpretation modules are integrated to improve the reliability and to accelerate the sensor data processing.

7.2.1 System Architecture

The concept of hierarchical data processing with feedback from higher stages finds its equivalent in system architectures that excel by tightly coupling of all data acquisition modules. Figure 7.1 illustrates the information flow within our multi-staged architecture. The primary sensor of our system is a multipurpose binocular video system, supported by a panoramic laser range finder (LRF) designed for localization [6].

The line extraction module is based on a contour tracer that is described in Section 7.3.1. It generates two-dimensional line segments from the sensor images. The 3D reconstruction is done in a binocular stereo module, where the line segments are matched to reconstruct their position in space. These 3D lines are stored and fused in the Dynamic Local Map (DLM), which stores the lines at their geometrical position in a 3D model. The lines extracted by the sensor system are stabilized and filtered to remove false interpretations. The stabilized information is aggregated by the Predictive Spatial Completion (PSC) module to reach the required level

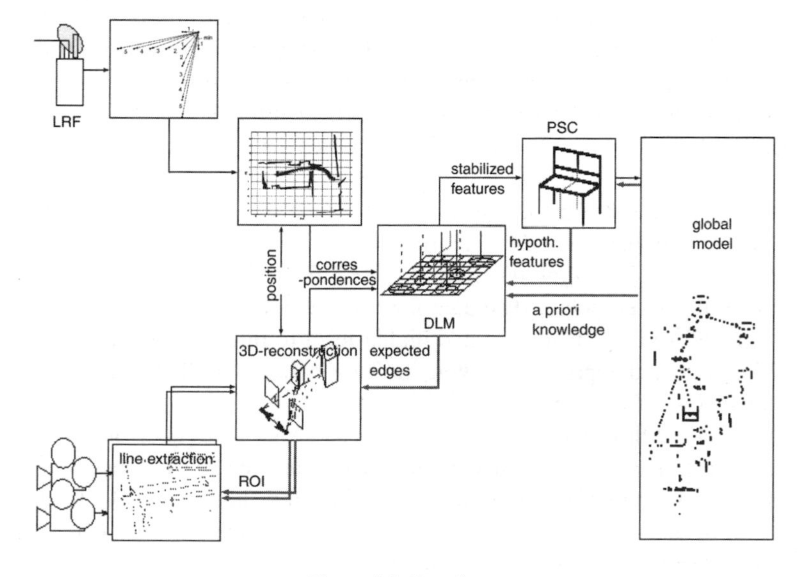

Figure 7.1 Data flow.

of abstraction for structures and objects that are stored in a global multipurpose model of the environment [7]. Hypothetical completions of the recognized objects and structures are fed back into the DLM to be verified by the sensor system.

7.3 SENSOR DATA PREPROCESSING

The main sensor used for localization and exploration is a CCD camera system for both binocular and monocular stereo, based on line features.

7.3.1 Image Processing

Sensor data preprocessing has to fulfill four requirements: data reduction, transformation to a symbolic description, noise reduction and real-time capability. First, the raw data, coming with a high rate (video field rate of 50Hz) from the CCD-sensor, must be reduced, because the high-level interpretation is computational expensive. Second, sensor data have to be transformed into a symbolic form that is suitable for higher level interpretation. Third, the influence of noise has to be limited by averaging over time and space. Fourth, the task has to keep up with the video rate for better stability of robot motion control is gained.

Because the sensor data preprocessing is computationally expensive and must work under real-time conditions, the method used to extract features must be adapted as close as possible to the application. Within the scope of this discussion there are four different application-dependent tasks possible. The first task is the exploration, which uses full- or half-frame processing. The second task is the object or state identification, processing only parts of a frame. The third task is the full-frame processing with prioritized parts around known or hypothetical regions for localization, and the fourth task is surveying parts of a frame for obstacle detection. The method we use that can keep up with all these demands is a line feature extraction with a contour tracer.

In classical edge line extraction schemes like Canny or Laplacian-of-Gaussian all pixels undergo processing with sometimes huge convolution matrices (up to 20×20). Therefore, the real-time processing requirements can only be achieved using expensive massively parallel computers. Another technique is contour tracing, which is normally used for binary images, in which a large amount of information is lost because of the application of a global threshold.

We apply a different approach and use a contour tracing algorithm on a single processor system, combining most steps needed for extracting contour segments into three small operators, built by 5×5 matrices. This algorithm is self-steering as soon as a valid starting point is found. The next contour spots are determined by the convolution results of these operators applied to the prior contour spot. A tremendous gain in efficiency is reached by computing only about 15% convolutions compared to classical gradient matrices like Canny operator. Compared with these methods we process only those parts of the image that contain contours. For example, only about 38% of the pixels of the sample image in Fig. 7.12 have to be processed. The matrices have been originally derived by an electrodynamical analogy, considering an electron or an positron that is driven along a contour by Coulomb- and Lorentz Forces caused by electrostatic and magnetic fields (Fig. 7.2). We found, however, that using classical gradient and Laplacian-of-Gaussian matrices for contour tracing produced similar results. Nevertheless, the analogy helps to understand how contour tracing works, and therefore, it will be explained in principle in the following paragraph.

Each pixel forms a horizontal and a vertical electric dipole. The dipole charges are derived from the horizontal and vertical local gradients of the grey level image. The electric field is composed of the distributions of the dipoles in the 5×5 vicinity. The magnetic field is

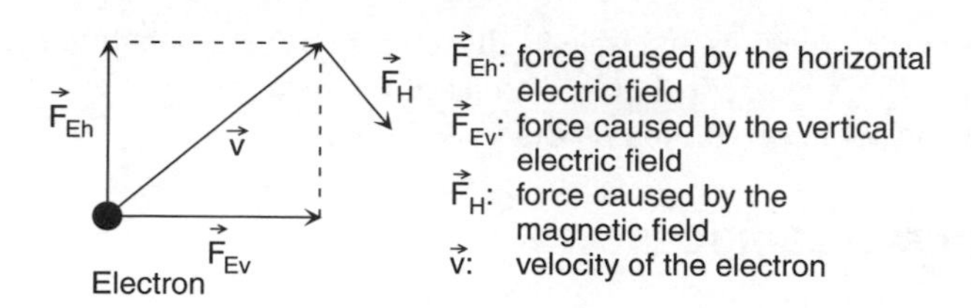

Figure 7.2 Electric and magnetic forces influencing a moving electron.

formed by a current flow between the inexhaustible charges of each dipole. The electric field causes the electron to be driven along the contour, but the electron cannot be hold on the ridge of the contour. The trace of the electron is stabilized on the ridge by adding the Lorentz Force. Thereby, strong divergences are avoided even if the contour is strongly curved. The electron can, however, overshoot a corner due to its inertia. This problem can be efficiently solved by changing the ratio between the electric and magnetic field when a corner is expected. The derived matrices and physical equations, presented in [8], are used to compute the successive contour spot. The tracing direction can easily be changed, substituting the virtual electron by a virtual positron.

To take advantage of the contour tracer's performance for each single application, an efficient and adaptable search for a starting point of each isolated contour is necessary. Therefore, we use an adaptive grid-based search scheme, which, on one hand, minimizes the number of pixels to be examined to find a contour and, on the other hand, maximizes the contour detection rate. A contour is assumed to be found if a local maximum of the electric field is by a factor greater than the local averaged electric field in search direction. The starting point is valid only if it does not belong to a previously extracted contour.

In this search scheme, an image is divided into small commensurate windows with application-dependent side length, called *grid elements*. A given grid element is scanned vertically and horizontally through the center. With this approach, contours with a length greater than twice the grid element's side length are always detected and shorter ones, for example, due to noise or texture, may be detected. The electric field along the search direction is computed with only a 5×1 vector for the vertical direction and a 1×5 vector for the horizontal direction. The vector consists of the middle row of the Eh-matrix and of the middle column of the Ev-matrix, respectively. After finding starting points for a grid element, the single contours are traced in both directions to their ends as described above, even if a contour crosses other grid elements. Segmentation into line segments is done with a fast scan-along algorithm [8] and least square regression analysis[1] in conjunction with contour tracing. In order to suppress outliers, the extracted line segment is merged with a neighboring one if their angle difference is smaller than a threshold. Each successive line segment is corrected using the equation of Tsai for camera calibration [9] and in a second merge stage, concatenated with the prior one if possible.

The scan-along algorithm used detects a corner if the pixel deviation to the averaged direction is greater than a threshold value. Recursive algorithms that may produce better results cannot be used, because the segmentation is done in parallel to the contour tracing. In addition, recursive algorithms have no deterministic processing time, which is necessary for real-time applications.

The variable access to the single grid elements is controlled using a translation table. This table contains either the ID of a grid element or the mark "READY," which means that no starting point is required. The ID of a grid element is substituted by 'READY' as soon as a

[1]The segmentation is not restricted to least square regression analysis.

certain number of contour spots have been found within that grid element, in order to increase performance. Although no starting point is searched, a contour crossing this grid element is of course extracted.

It is possible to apply contour tracing to real-time image processing using this translation table. Therefore, after each processed grid element the time left in a given real-time interval is checked, and the contour extraction of the current image is aborted, when the limit is reached. In this case the processing order of the grid elements is prioritized in order to achieve a more optimal result. The content of the translation table additionally depends on the application and determines the processing sequence of the grid elements. Within the scope of this paper there are four different applications determining how the table is to be filled:

Localization The grid elements touched by a priori known edges are registered first. The other grid elements are added for new edges or left out for less interesting parts of the image (Fig. 7.3, upper left).

Exploration The whole image should be processed to gain as much information as possible. The grid elements are registered according to their priority, e.g., like a spiral or even controlled by hypothetically predicted edges (Section 7.4, Fig. 7.3, upper right).

Obstacle Detection The part of the frame representing the path of the robot has to be surveyed. Therefore only the touched grid elements are processed in a prioritized order (Fig. 7.3, lower left).

Object and State Identification Only the grid elements encompassed by significant windows (e.g., when identifying a door wing angle) are registered in a prioritized order; all the others are left out (Fig. 7.3, lower right).

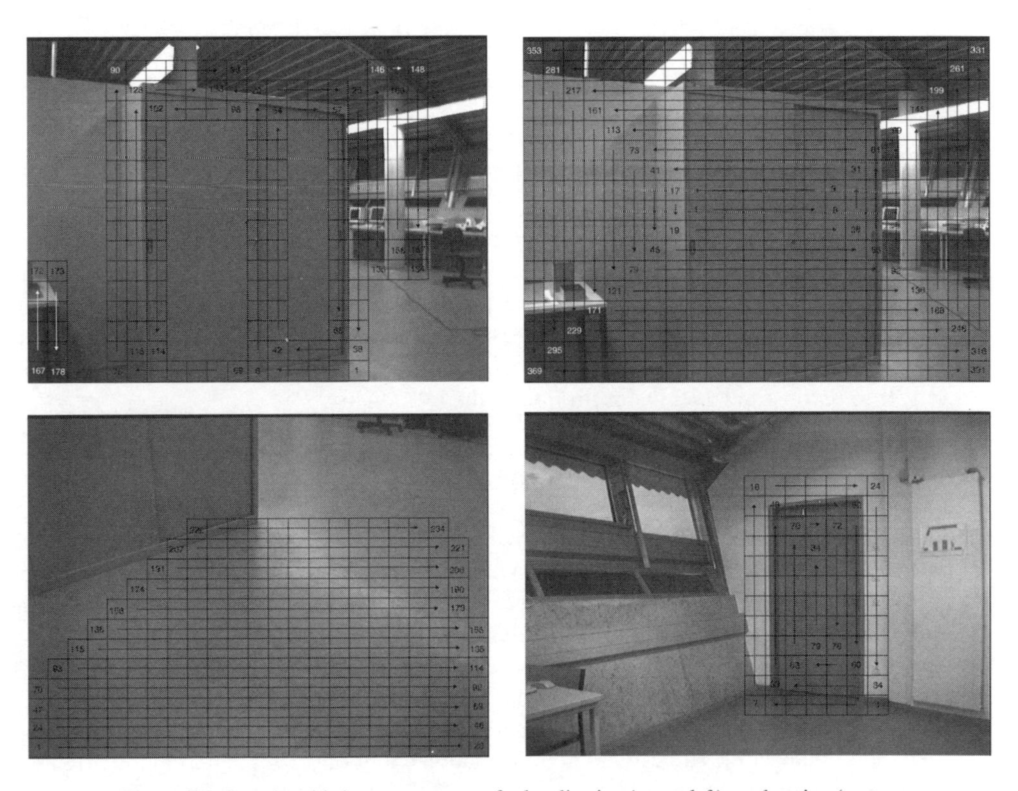

Figure 7.3 Sample grid element sequences for localization (upper left), exploration (upper right), obstacle detection (lower left), and state identification (lower right).

7.3.2 Three-Dimensional Reconstruction

In our stereo system we use the boundary lines of the objects to reconstruct the environmental description of a local area. The line segments extracted in an image pair are matched to reconstruct their 3D description. The established matches are based on two different constraints:

- DLM predictions—for each sensor view the expected lines are predicted from the information stored in the DLM. These lines are matched to the extracted line segments to verify the prediction. All verified lines are stored back into the DLM, where their confidence and accuracy is modified (Subsection 7.3.4).
- Geometric constraints—the corresponding features must meet some constraints to be matched. The epipolar constraint and a limited disparity range help to filter out most of the false correspondences in a current sensor reading. The remaining lines are passed on to the DLM were they are verified from different viewpoints.

The 3D reconstruction is based on the interaction with the DLM. The information stored in the DLM can be used for two different purposes. The first problem to be solved is the recalibration of the camera system. Due to vibrations and repositioning errors of the camera head, the actual orientation of the single camera or of the entire camera system can differ from the assumed orientation. These orientation errors cause two different errors in the reconstructed data. The orientation error of a single camera changes the disparity of the matched line segments, deteriorating the accuracy of the extracted features. The resulting error depends on the distance of the extracted feature from the camera system. The orientation error of the entire system preserves the correct distances between the extracted features, but rotates the coordinate system of the camera with respect to the global coordinate system. This error prevents correct verification of the stored features. Therefore, information with high confidence stored in the DLM is used to recalibrate the external parameters of the sensor system. The predicted 3D lines are projected on the image planes of both cameras and matched to the extracted line segments. The orientation errors are computed from the divergence of the matched lines.

The other use of the information predicted from the DLM is to establish correct correspondences between the extracted line segments without the exhaustive search of the stereo match algorithm. As already explained for the recalibration, the remaining predictions are also projected on the image planes of the cameras and matched to the extracted line segments. The lines found are removed from the pool of the unknown lines, reducing the computational effort to reconstruct the 3D description of the local view.

7.3.3 Obstacle Detection

The 3D reconstruction of the environment is performed from discrete points chosen by the navigator (Subsection 7.5.1). While traveling between these points the vehicle uses its stereo system for obstacle avoidance. In this mode the cameras are pointed at the floor to observe the space in front of the vehicle. The knowledge about the expected extent of the ground plane is used for detection of obstacles obstructing the planned path. The sensor feature extraction is restricted to the path in front of the robot (Fig. 7.4).

All objects detected within the darker area (Fig. 7.4, left) cause an immediate stop of the vehicle to inspect the situation. The information about obstacles found in the brighter area is used to slow down the vehicle.

Figure 7.4 Obstacle detection (left) and geometrical arrangement (right).

The system with a baseline distance B has to distinguish between texture or lines on the floor and obstacles obstructing the way. From the knowledge about the extent of the ground plane the disparity d for each scan line can be estimated as

$$d(z) = \frac{B}{z \cdot \cos\alpha - H \cdot \sin\alpha} \cdot \frac{f}{p_{size}} = \frac{K}{z \cdot \cos\alpha - r} \tag{7.1}$$

It is constant for a given image line defined by the image coordinate y_p. The value z can be computed from the image coordinate y_p (7.2) for a given camera with a focal length f and pixel size p_{size} (Fig. 7.4, right).

$$z = \frac{H}{\tan(90^\circ - (\alpha + \arctan \frac{y_p \cdot p_{size}}{f}))} \tag{7.2}$$

7.3.4 The Dynamic Local Map

The DLM decouples the sensor system from the other modules utilizing the sensor information. There are two information cycles within the system (Fig. 7.5) with different kinds of data processed in them.

On the one side there are the sensor systems with a short cycle time producing large quantities of data with a poor accuracy and confidence. In this cycle, the false information caused by misinterpretations of the sensor readings appears. This information must be filtered before it can be used for other purposes, such as object recognition or path planning. In this

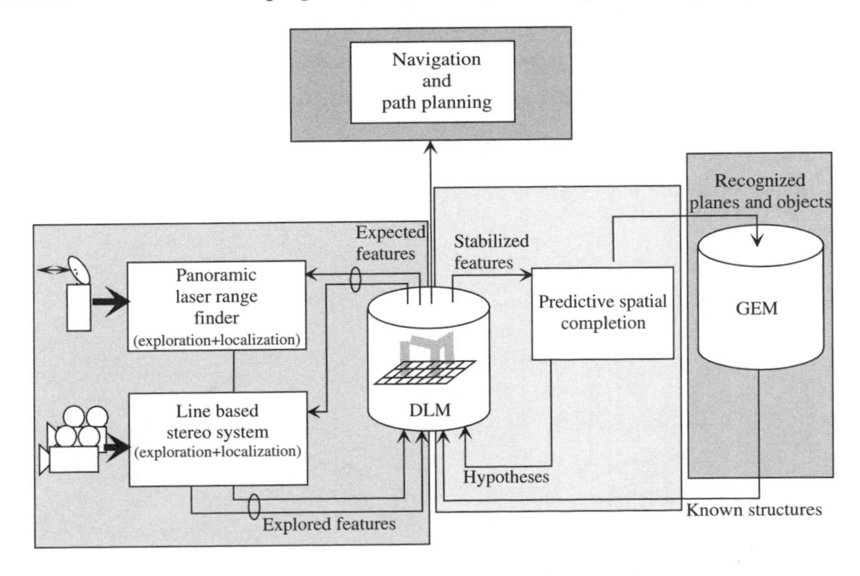

Figure 7.5 Communication structure with the DLM.

cycle the DLM predicts the expected features for the current sensor reading. This additional information simplifies the data processing in the sensor system. Previously known features in the world are only verified in the current sensor reading. The time-consuming match of the features in case of stereo vision is not necessary.

On the other side, there are modules utilizing the stored information for other purposes, such as object recognition or path planning. In this path, the processing time is not dependent on the cycle time of the sensor system. The cycle time instead depends on the current task. The information in this path has a higher accuracy and confidence, because only information verified in several sensor readings is passed on to this cycle.

The DLM must operate in a closed loop with various sensor systems of an autonomous mobile robot (see Fig. 7.1). Because of this requirement, the abstraction level of the data is similar to that used by the corresponding sensor system. This supports fast access by the sensor system to the data stored in the DLM. The interaction with different modules requires a fast access to the map, limiting the available time for the analysis of the stored information. Therefore, the time-consuming grouping of the features to reconstruct known reference structures is done by the PSC module (Section 7.4). The feedback between the DLM and the PSC refining the stored information distinguishes our approach from similar ones using local maps of the environment [1]. The interaction with the PSC results in additional hypothetical features giving clues to still undetected features (Subsection 7.3.4).

With a CCD camera, the environmental representation consists of 3D line segments representing the boundaries of the objects (sensor specific features). The DLM stores the recent sensor information after comparing it to previous sensor information to improve and stabilize the sensor readings (Fig. 7.6). Each feature line is represented by its geometrical position as well as the expected *accuracy* and *confidence* in its existence (Subsection 7.3.4).

The presented local map stores only a local region of the environment adapted to the sensor range and the size of the operating area. This local area can be a room or a part of a big hall in which the robot operates. Frequent changes in the environment due to object relocations or moving objects require a permanent adaptation of the stored information. Therefore, the information becomes uncertain if it cannot be verified for a given period of time. The reason for the limitation to a local area is the possible use for topological navigation, in which the local map is not organized relative to a global coordinate system but relative to some significant structure within the local area. This structure reduces the requirements of knowing the exact relations between distant objects. The distances are measured only within the local area. The local maps can be stored in a global map and regained in case of re-entry into this area.

7.3.4.1 Update of the Map. The DLM was tested with a line based stereo system, but it is able to cooperate with other sensor types, such as a laser range finder. The stereo system

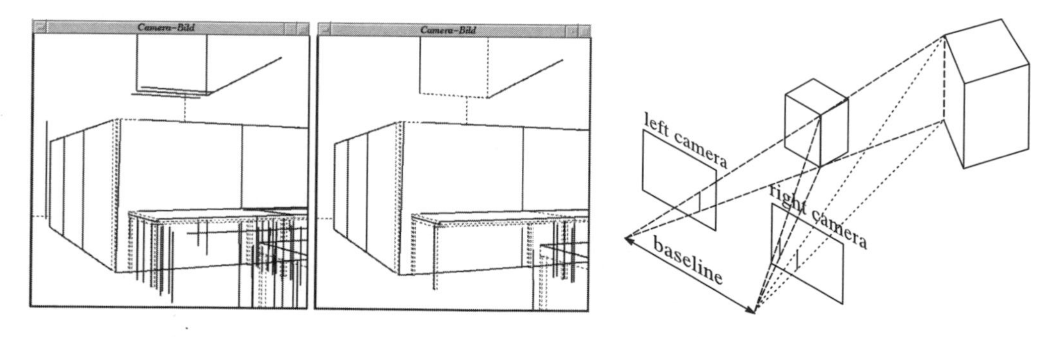

Figure 7.6 Left: The features are filtered from different points of view to reduce misinterpretations. Right: The correspondence ambiguity in the stereo vision.

used delivers 3D line segments reconstructed from image lines. The image lines themselves correspond to object boundaries and illumination changes such as patterns and shadows. The resulting 3D features are described by their endpoints and by three additional values: the *accuracy*, *confidence,* and *age*.

The accuracy describes the assumed maximum divergence of the endpoints depending on the current distance from the vehicle and the length of the extracted line segments [10]. New features inserted into the map are matched with the previously stored information to improve its accuracy. The shift of the stored feature depends on its accuracy compared to the accuracy of the newly detected one.

The confidence value is used to distinguish between real features and misinterpretations due to the correspondence ambiguity in the stereo vision (Fig. 7.6, right). The value of the confidence depends on the number of the match candidates after the epipolar constraints are applied. The internal confidence within the DLM is increased if the feature could be matched in several updates. The idea is to benefit from the motion of the vehicle to filter out false correspondences, because the artifacts caused by misinterpretations change their position depending on the current position of the mobile robot.

The age value is necessary to remove false features that could not be verified, or features for objects that have been removed. This value is decremented each time the feature appears in the sensor view but cannot be verified.

7.3.4.2 Interaction with other Modules. The features stored in the DLM come from different sources, such as the sensor system, the global map or the PSC. All input sources to the DLM are treated as different "virtual sensors." The features from all these sources are matched in the 3D space to obtain a uniform description of the environment. Recent changes in the environment are taken into account and the initial feature detection in a-priori known environments[2] is simplified. The features are tagged with additional information about their origin, which can be a combination of these three sources.

This additional information is important for a proper handling of the stored features. For example, the hypothetical features stored by the PSC are important to plan the exploration course. On the one hand, the regions containing these kind of features are very interesting for the exploration, since additional information in this area is expected. On the other hand, it is unwise to plan a path with these hypothetical features, because they are not verified and often several alternatives for the appearance of an object are inserted. The stored information is also used for the sensor-based localization of the vehicle.

The interaction with the sensor system reduces the correspondence ambiguities and accelerates the sensor data processing. The DLM is a substitute for a third camera used in some approaches to verify the correspondences found [10]. It allows to store multiple matching candidates to be verified in following sensor readings from other positions, exploiting the mobility of the AMR.

The hypotheses generated by the PCS are used as additional hints for correct correspondences. They are predicted in the respective local areas, and the sensor system tries to verify them in the current view. Each time the hypothetical feature is verified its confidence value is increased and after it reaches a defined threshold, a successful verification is reported to the PSC. On the other hand, if a hypothetical feature cannot be verified, its confidence value is decreased and the feature is deleted if the value falls below a minimum confidence threshold. This mechanism is important for deleting false information, because it is possible to store several hypotheses for a given structure. The wrong hypotheses must be removed. All removed hypotheses are also reported to the PSC. Occlusions may also cause deletion of hypotheses. In this case, they can be generated to be verified later.

[2]Stored in the global model as a-priori or already explored information.

Each feature stored in the DLM has its own unique ID. The IDs are generated by the DLM for the new explored features or adopted from the external sources. The PSC requires its own IDs to assign the verified features. Therefore, in case hypotheses are matched to other features, such as model lines, the IDs of the hypothetical features are kept.

7.4 INTERPRETATION

The *Predictive Spatial Completion* (PSC) integrates the sensor data, extracted with no relation to each other, to a joint representation on a higher level, such as planes, clusters and objects, essential for a proper treatment of objects in the global model. The PSC also transfers uncertain information as hypothetical features to the sensor system via the DLM.

7.4.1 Recognition and Completion

The sensor features are interpreted and completed, based on object recognition, structure recognition, and structure conversion. The incoming features are exposed to these different, redundant and complementary strategies to increase the ratio of sensor features that can be interpreted. All strategies have no known time-consuming sensor data processing but operate on the stream of features, stabilized in the DLM. This implies that just newly explored or altered features are utilized. This is cost efficient but poses temporal effects on the recognition that must be considered. Because the sensors used are complementary and have also strongly differing temporal properties, this is not an additional drawback.

7.4.1.1 Determination of Structures. To contribute to the environmental modeling, the PSC checks the unrelated incoming features against typical structures as faces and convex, concave, parallel, or symmetrical feature groupings and eventually clusters them. The established higher abstraction levels improve the usability of the extracted information, for example, to facilitate the assignment of features to object hypotheses. The delays in extracting features and the demand for a fast contribution of information are taken into account by a bottom-up promotion of features to structures.

As a first step, relationships among features are determined that are candidates for being assigned to parts of typical structures in the environment (see Fig. 7.7). Relationships searched reflect combinations of features that are likely to be extracted with little or no delay and that can be found in a wide variety of environmental structures. The determination of relationships is partially processed redundantly to compensate for imperfections because of optimization for low computational costs and the unpredictable feature appearance. Costs are kept moderate by stopping processing if previously found relationships are sufficient, reliable, and complete.

All features that overlap by a minimum amount the inspected feature are selected and checked for typical groupings as vertices, T-intersections, and parallel or aligned arrangements. For each grouping, different guidelines are applied, e.g., the visibility, the enclosed azimuth and elevation angle and the orientation, the distance between close endpoints or between the features, aligned and normal to the current feature. The plausibility of a relationship is updated according to the deviation from the optimum by a combination of a ramp function and a function of $1/x$ with upper and lower saturation, restricting the influence of a single guideline. Occlusion, noise and modern asymmetric architectural design, are taken into account for all guidelines. Relationships, with plausibility above threshold are marked as valid and are candidates for being assigned to joint substructures. Since the best-fitting feature for a standard relationship can be misleading with the restricted information, we select the best relationship for each feature and use all combinations of features and relationships that are above the plausibility threshold for determining substructures. This allows to recover from incorrect assignments.

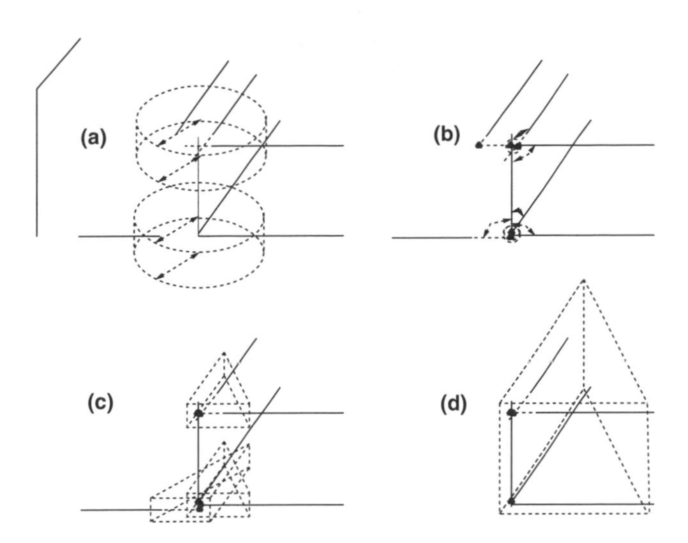

Figure 7.7 Stepwise structure recognition. (a) Selection of neighboring features, (b) relationships and related features, (c) local substructure, one side, (d) structure.

Next, substructures are determined by combining non-contradictory relationships and the involved features. Since relationships are based only on a locally restricted geometrical arrangement, they can be contradictory or wrong and are therefore tested for mutual supporting or contradictory relationships in semantic and geometrical fit, separately and reciprocal. If a contradiction between two relationships can be resolved in this stage, the offending relationship is rejected; otherwise the plausibility of both relationships is reduced. The plausibility for mutually supporting relationships and the resulting substructure is increased.

Third, all plausible substructures, assigned to the feature, are combined to form a structure, after a test for contradiction is performed. If more than one substructure was determined that occupies the same space and does not resemble a more complex structure, and the contradiction cannot be resolved, or if substructures from different sides of the feature are contradictory, the substructures are not combined to form a structure. The recognition process will resume interpreting them, based on the maintained relationships, as soon as more features in the vicinity are found or changed in their description.

Structures and substructures, built up of features with plausible relationships, are clustered. If relationships do not include a connection (e.g., a parallel arrangement) or if they are of medium certainty, the assignments of the features are marked, but features are not clustered. This makes it more unlikely that the features belonging to different neighboring objects are clustered and the cluster must be resolved when contradictions arise later. With more subsequent processing, more complex and meaningful structures are built up.

Incomplete structures, substructures, or relationships are used as clues for missing sensor features. These missing features are generated as hypothetical by (a) projecting features from other subparts of the structure, (b) connecting parallel features, and (c) building vertices. Statistics learned from the environment facilitate the assignment of related features and improve the quality and hit rate of the generated hypothetical features.

7.4.1.2 Object Recognition. Recognized objects represent a valuable source of information for improving perception and tasks such as localization or obstacle avoidance. To identify these objects as soon as possible, recognition is performed continuously. To cope with the restricted computational power, the recognition is optimized for low cost, rather than to recognize objects under all conditions. Time spent for the object recognition should be partially regained by reducing the costs of sensor feature extraction. Furthermore, the recognition is restricted to objects and groups of objects that are relevant for localization, obstacle avoidance, and map construction. Recognition is facilitated by focusing only on objects that stay in the pose of their natural use. The identification applied during the exploration presumes an approach with low computational costs that does not require separate, time-consuming sensor preprocessing. The features supplied by the sensors for the purpose of localization, exploration, and obstacle avoidance are stabilized in the DLM.

Structure recognition works on a stream of 3D edges, which are stabilized in the DLM, and groupings of of 3D edges. The identification is based on 3D geometrical models.

The models are analyzed off-line for symmetries. Symmetries found are marked for the on-line recognition and triggers for object recognition are reduced according to the results. Increasing modularity in the construction of, for example, furniture and the uniformity in their style increases demands on distinction between objects that show a similar appearance or that resemble a combination of other objects. The distinction must be made as soon as possible to avoid the handling and promoting of multiple hypotheses. Therefore, we compare objects off-line with the same recognition process as on-line and modify objects that lead to a misleadingly high plausibility in a wrong interpretation. To these objects, features are added that belong to the description of the other object as "contradictive features." If an equivalent feature is assigned in the on-line recognition, the plausibility of the hypothesis is reduced, while the plausibility of the correct hypothesis is increased, speeding up the distinction.

Object hypotheses are evolved when triggering features are detected and promoted by a 3D match of their representation against features previously received from the DLM and verified in a variation of a "prediction and verification" approach: for each hypothesis that exceeds a threshold, the features of the object description that have not been assigned are generated as hypothetical ones and prepared for being sent to the DLM. Hypothetical features and assigned features are marked to belong to an object. The applied object recognition is described in more detail in [11].

7.4.2 Transfer of Predictions

Hypothetical features generated from all strategies are kept in a buffer before being inserted into the DLM. Similar hypothetical features are merged, reducing the communication requirements. The hypothetical features are stored inside the PSC with their entire evolution history for reevaluation and to be taken into account in further recognition processes, suppressing the evolution of new ones. A reduced description of the features is then transfered to the DLM, independently of their visibility from the current position of the robot. This allows the verification of the hypothetical features, equivalent to the verification of the object and structure hypotheses, from different points of views—exploiting the advantages of the robot's motion. The asynchronous coupling separates the generation and verification of hypothetical features.

7.4.3 Evaluation

Newly sensed features and validated hypothetical features are received equally by the PSC from the DLM. Therefore, the features received are evaluated to retrieve their possible origin from a recognition strategy and eventually to regain their entire description. This also suppresses possible contradictions that can arise from delays in the asynchronous communication between DLM and PSC. The search of the received feature inside the PSC is based on its minimum evolution history, including ID and origin, and on its geometrically significant properties. With this complete evolution history, the access on all structures and objects that led to their generation or to which they are linked, is sped up.

Three main cases can be distinguished. First, newly explored features are tested against the set of generated hypothetical features that have not yet been sent to the DLM. If a hypothetical one is found similar to the received one, all or some of its relationships and assignments to structures are adopted after a test for contradictions, depending on the geometrical similarity of the features. Their histories are also merged. The hypothetical feature is then deleted. By this mechanism, hypothetical features that anticipated their detection by the sensor system do not increase communication. Furthermore, they are already assigned to a structure and an object and are not presented to further recognition strategies.

Second, a feature is received that is already known to the PSC and therefore already stored in the map. This case includes features that are modified, deleted, or hypothetical as indicated by the history flag. Modified or confirmed hypothetical features are updated by replacing the geometrical description of the equivalent feature stored in the map and objects with the new one and by merging their history and relationships after a test for fit, preserving the latest changes due to fusion inside the DLM and PSC. Hypothetical features are directed to all strategies that contributed to their evolution such as to the object recognition, for recalculating plausibility, pose, or shape. Valid relationships to directly linked features lead, in case of structure recognition, to a clustering, further converting the representation to higher levels. Structures or objects that are confirmed are inserted in a global model.

Third, features to delete are removed from the map and objects. Relationships, in which these features are involved in are left untouched. Resulting inconsistencies must be removed afterwards by the ruling out mechanisms of structure recognition.

7.5 SENSOR AND PLATFORM CONTROL

The processing modules of an autonomous mobile robot can be subdivided in at least two different layers: data processing and control layer. In our system we use three layers, as shown in Fig. 7.8.

The data processing layer is responsible for sensor data processing and 3D reconstruction. The modules in this layer offer different capabilities and require specific system resources. The requirements of some modules are often contradictory and must be coordinated. For example

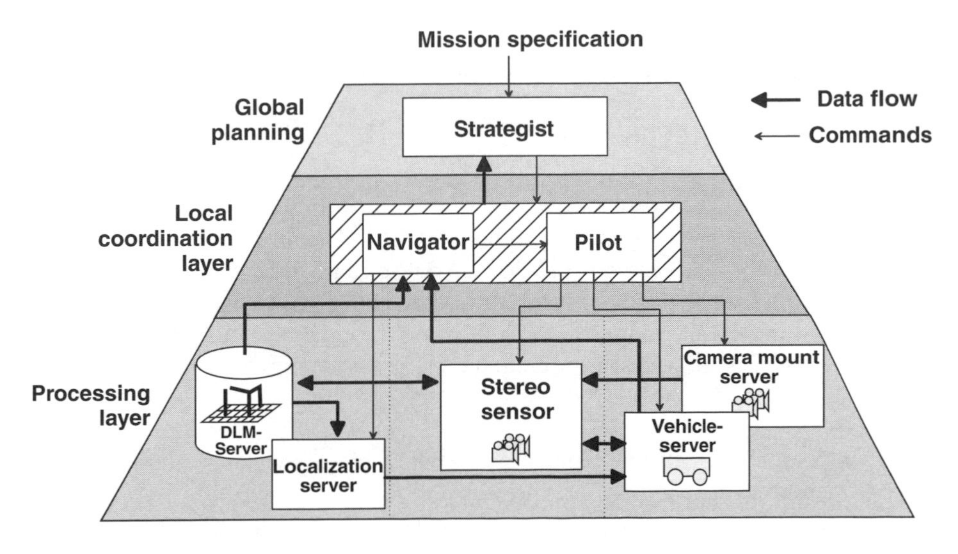

Figure 7.8 System structure.

a physical sensor can be used for exploration tasks and for obstacle avoidance. These two tasks often require different camera orientations and cannot be done simultaneously. Therefore, a coordinating layer is added to control the vehicle and the camera head and to initiate the sensor data acquisition and processing. All actions within the processing layer are supervised by the coordinating layer.

The actions are planned depending on the current mission goal and the abilities of the system. The mission is specified in the layer responsible for global planning. This layer does not know anything about the hardware configuration of the system. It plans the actions on an abstract level using topological information about the environment. It knows the global goal, the degree of familiarity with the environment, and the priority of the current task. Based on this information it subdivides the global task in partial goals and specifies the amount of time available to reach them.

The local coordinating layer knows the hardware configuration and the abilities of the applied systems. To reduce the complexity of this layer, it gets only local goals planned by the global planning and the amount of time to reach the goal. It does not know anything about the global mission. Depending on the local goal and the remaining time, the coordinating layer plans the actions in the local area. It uses the informations stored in the DLM to determine the free space and the obstacles surrounding the vehicle.

7.5.1 Exploration Strategy

The goal of the system is a sensor-based generation of the environmental description. The generated descriptions are stored as 3D geometrical representations in a cartesian coordinate system. All objects stored in the model are referred to an origin defined by a reference structure. The limited sensor range restricts the area which can be explored by the sensor system. Therefore, the whole environment is subdivided into local areas represented by exact geometrical 3D maps. Each map is referred to a reference structure that must be identified to localize in this area [12]. The exact geometrical relations between the local areas are not necessarily known.

The goal of the coordinating instance is an efficient exploration of the local map of the environment in a given period of time. The information stored in the DLM is clustered and rated. Each cluster is a possible subgoal to be inspected next depending on the accuracy of the contained information. Other possible goals are unknown areas and clusters containing hypothetical features generated as a completion of the already explored information [13].

7.5.2 Free Space Estimation

The problem is to estimate the free space from the information stored in the DLM. There is no surface description which could be used for hidden line removal. All lines in the local area are visible. An access to the DLM also returns lines, which are invisible from the current position, but are known from prior readings from other positions in the environment. We use a 2D grid for path planning, in which all obstacles are projected. The information accessed from the DLM is analyzed to generate the grid. We subdivide the stored line segments into three layers:

- *Roof layer*—all line segments above the vehicle, which should not be considered as an obstacle.
- *Obstacle layer*—all line segments that cause a collision with the vehicle. These lines are higher then several centimeters above the floor and lower than the height of the vehicle.

- *Ground layer*—all line segments with a maximum height within up to several centimeters above the floor that can be passed by the vehicle.

All lines from the obstacle layer are directly projected on the grid. The problem is to decide the passability of the lines in the ground layer. These can be lines or textures on the floor or they belong to objects standing on the floor, e.g., walls. In the first case the corresponding line should not be inserted into the grid as an obstacle. To decide it the surface description of the objects is required, but this information is not available from the DLM. The only possibility to check the visibility from the current position is to use the sensor, but the information returned by the sensor system sometimes contains false lines caused by misinterpretations. The idea is to check the region behind a line in question for other visible features. In case the sensor can see behind the line it should be passable (Fig. 7.9).

False correspondences hinder the use of unfiltered sensor information. The solution is to use the information stored in the DLM to filter the current sensor reading. All line segments which could not be found in the DLM are deleted, but no additional information about the local area is added.

7.5.3 Path Planning

The resulting grid used for path planning is shown in Fig. 7.10. All lines are projected on the grid and each line is surrounded by a safety margin of some centimeters to force a minimum distance to the obstacles.

The availability of the possible gloals (see Subsection 7.5.1) is checked in this grid. The unsteady diffusion equation strategy is used to plan the path in the extracted free space [14].

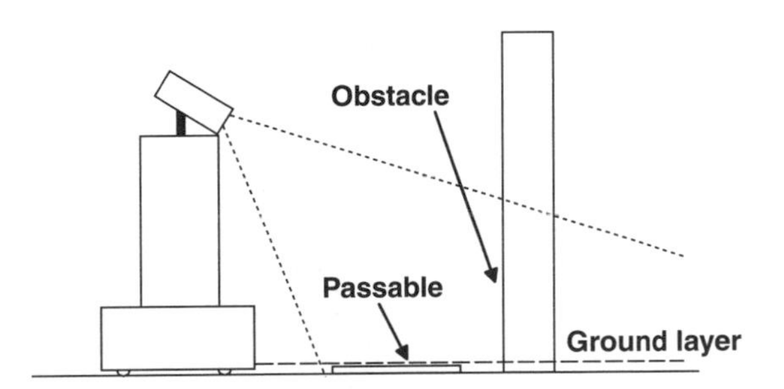

Figure 7.9 Free-space estimation.

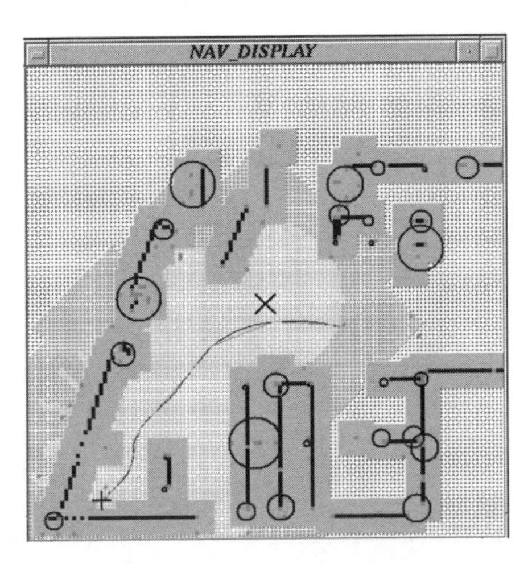

Figure 7.10 Path planning.

Equation (7.3) describes the diffusion from the starting point x_0 to the goal x_G.

$$\frac{\partial u}{\partial t} = a^2 \cdot \nabla^2 u - g \cdot u, \qquad (7.3)$$
$$u(t; \underline{x}_G) = 1, \qquad \underline{x}_G \in \Omega \subset R^n$$

We use a nonoscillating equation 7.5 to compute the diffusion.

$$\tilde{u}_{k+1;r} = \frac{1}{1+M} \cdot \left(\sum_{m=1}^{M} u_{k;m} + u_{k;r} \right) - \tau \cdot g \cdot u_{k;r} \qquad (7.4)$$

$$u_{k+1;r} = \begin{cases} \text{eqn. 7.4} & \text{for} \quad r \in \Omega' \\ 0, & \text{for} \quad r \in \delta\Omega' \\ 1, & \text{for} \quad r = r_G \end{cases} \qquad (7.5)$$

The value M describes the number of the neighbors of a grid element, Ω' is the free space, $\delta\Omega'$ are the grid elements containing obstacles and r_G is the goal.

7.6 RESULTS

7.6.1 Image Processing

The contour tracer is implemented on different target systems, e.g., PC with Linux or Windows NT or Alpha Workstation with Digital Unix. The following processing times were determined for a DEC 21164 at 266MHz. Processing times are plotted in Fig. 7.11.

The processing times of the contour tracer for non-real-time full-frame feature extraction using the presented algorithm on 768x576 8 bit grey level images of different indoor scenes (Fig. 7.12, left) compared to those of the image processing package VISTA [15] can be seen in Fig. 7.11, right.

As a result we cut processing time by a factor of about 37. The feature extraction with VISTA, using the Canny-operator [16] shows comparable results. However, the Canny-operator is more sensitive to short edges and the contour tracer merges as many line segments as possible. In addition, the grid-based search scheme prefers long edges. This is the reason for the different number of edges found in the same scenes (Fig. 7.11, left).

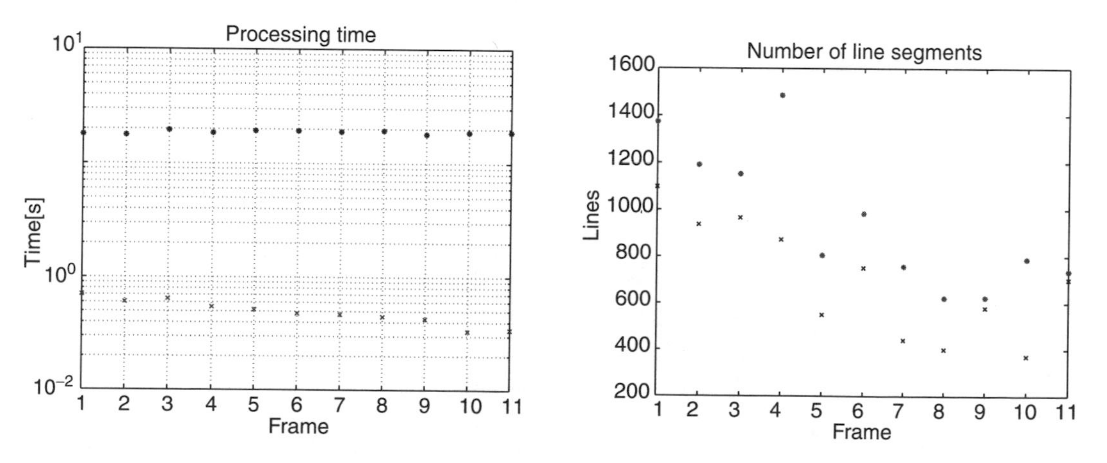

Figure 7.11 Processing time for sequence of full frames without using knowledge from higher stages for acceleration (*:vista, x:contour tracer) (left). Number of resulting edges for the same frame sequence (*:vista, x:contour tracer) (right).

Figure 7.12 Sample frame with extracted line segments (left). Zoom of the top left corner of the door with edge line segments and contour spots (right).

The accuracy of the extracted contours depends on the length of the line segment due to least square regression analysis. The divergences to real edges are less than 0.4 pixel for line segments with a length that is sufficient for higher level interpretation (Fig. 7.12 right). Because of this, our contour tracer can also be used in image processing applications with a demanded high accuracy, such as a micropositioning system [17] that mounts parts with an accuracy of about 10 microns.

7.6.2 Three-Dimensional Reconstruction

The 3D reconstruction is very sensitive to orientation errors of the camera system (Fig. 7.13). Small deviations in the orientation of the cameras can cause large errors in the reconstructed data. Therefore, we use the data stored in the DLM to correct the orientation errors of the camera system. Line segments with a high confidence are matched to the extracted lines in both images. The computed orientation error is used to correct the external camera parameters. An example for the improvement in the reconstructed data is shown in Fig. 7.14.

7.6.3 DLM

The data inserted into the DLM are matched to the previously known information to improve their accuracy and confidence. The original data reconstructed from a single sensor view have a poor accuracy and contain some false features caused by misinterpretations. An example is shown in left image Fig. 7.15 where, due to limitations of the feature extraction,

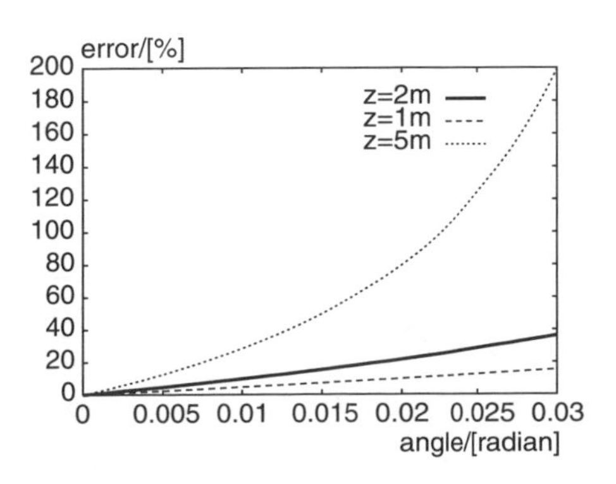

Figure 7.13 Effect of the orientation error for different distances z from the camera.

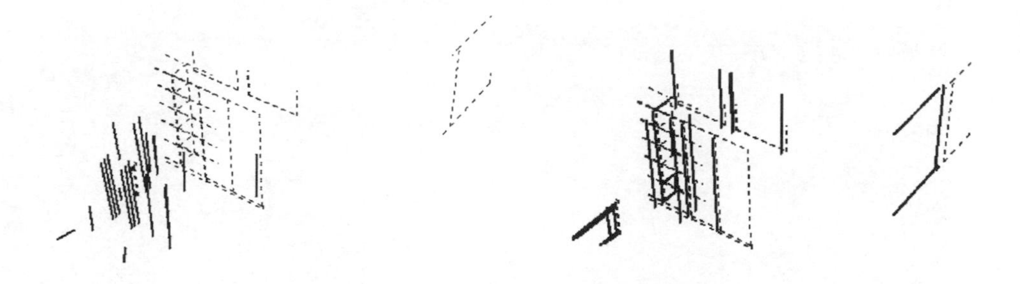

Figure 7.14 (Left) Reconstructed data without recalibration. (Right) recalibrated result.

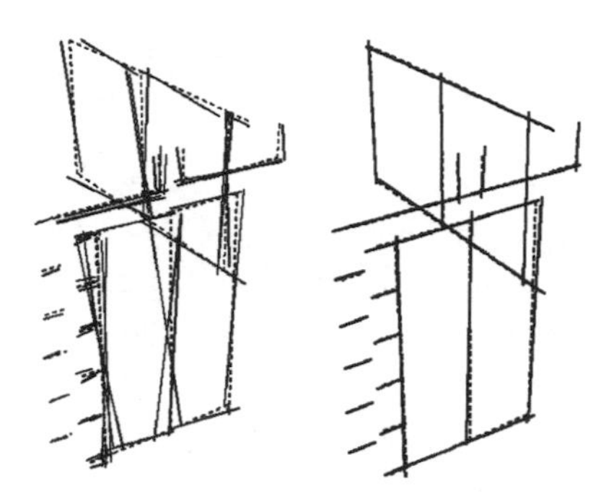

Figure 7.15 Geometric fusion of data.

the reconstructed data deviate from the true position. In the right image the result of the fusion of the reconstruction results from several viewpoints is shown.

7.6.4 Sensor and Platform Control

Our approach allows the planning of the exploration course depending on the explored information. The three types of goals generated by the pilot, uncertain clusters, hypothetical features, and unknown areas can have different priorities to achieve specific behaviors of the vehicle. The prioritization of uncertain clusters results in extensive verification of the perceived information. In this case, the vehicle inspects individual clusters from different viewpoints without moving to other unknown regions. On the other hand, the prioritization of unknown regions results in a fast superficial exploration of the area.

The quality of the generated path depends on the number of iterations computed for a given situation (Fig. 7.16). The paths chosen for a given situation improve with the increasing number of iterations.

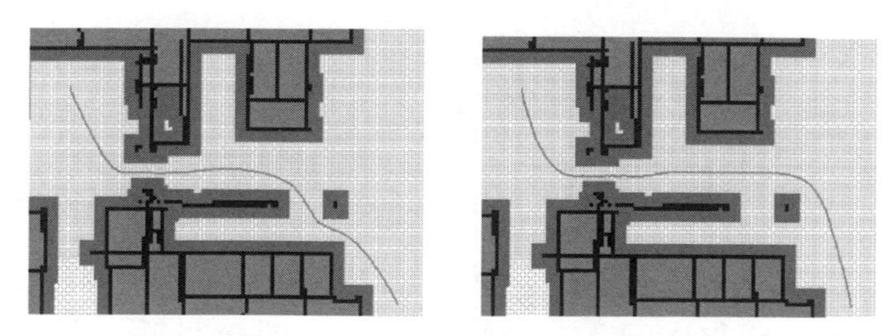

Figure 7.16 Generated paths after 100 and 400 iterations.

7.7 CONCLUSION

This chapter describes our approach to couple perception and control for video-based exploration of indoor environments. The generation of a dependable 3D model of the environment by interaction of contour tracing, 3D reconstruction, filtering, and interpretation is described. Usage of specialized hardware is avoided by reducing the complexity of the individual steps of the perception and exploration task. A multistage extraction of the environmental description is combined with the feedback of acquired information from the higher to the lower stages. The coordination of the perception modules according to resources, sensor information, and mission is described, focusing on data extraction and collision-free motion in the environment.

7.8 ACKNOWLEDGMENT

The work presented in this paper was supported by the *Deutsche Forschungsgemeinschaft* as a part of the research project SFB331 and by the *Bayerische Forschungsstiftung* as a part of the research project FORMIKROSYS. Communication with *no wait rpc* [18] was provided by Thomas Hopfner.

REFERENCES

[1] H. Bulata and M. Devy. "Incremental construction of a landmark-based and topological model of indoor environments by a mobile robot." In *Proceedings of IEEE International Conference on Robotics and Automation*, vol. 2, pp. 1054–1060, Minneapolis, MI, April 1996.

[2] P. Skrzypczynski. "Building geometricals map of environment using IR range finder data." In *Proceedings of International Conference on Intelligent Autonomous Systems*, vol. 1, pp. 408–411, Karlsruhe, Germany, March 1995.

[3] P. Weckesser, F. Wallner, and R. Dillman. "Position correction of a mobile robot using predictive vision." In *Proceedings of International Conference on Intelligent Autonomous Systems*, vol. 1, pp. 78–85, Karlsruhe, Germany, March 1995.

[4] X. Lebegue and J. K. Aggarwal. "Extraction and interpretation of semantically significant line segments for a mobile robot." In *Proceedings of IEEE International Conference on Robotics and Automation*, vol. 3, pp. 1778–1785, Nice, France, May 1992.

[5] X. Lebegue and J. K. Aggarwal. "Generation of architectural CAD models using a mobile robot." In *Proceedings of of the IEEE International Conference on Robotics and Automation*, vol. 1, pp. 711–717, San Diego, CA, May 1994.

[6] T. Einsele. "Real-time self-localization in unknown indoor environments using a panorama laser range finder." In *Proceedings of IEEE/RSJ International Conference on Intelligent Robots and Systems*, vol. 2, pp. 697–703, Grenoble, France, September 1997.

[7] A. Hauck and N. O. Stöffler. "A hierarchic world model supporting video-based localisation, exploration and object identification." In *Second Asian Conference on Computer Vision*, vol. 3, pp. 176–180, Singapore, Singapore, December 1995.

[8] G. Magin and C. Robl. "A single processor realtime edge-line extraction system for feature tracking." In *Proceedings of MVA'96 IAPR Workshop on Machine Vision Applications*, vol. 1, pp. 422–425, Tokyo, Japan, November 1996.

[9] R. Y. Tsai. "A versatile camera calibration technique for high accuracy 3D machine vision metrology using off-the-shelf TV cameras and lenses." *IEEE Transactions on Robotics and Automation*, vol. 3, no. 4, pp. 323–344, IEEE Robotics and Automation Society, August 1987.

[10] O. Faugeras. *Three-Dimensional Computer Vision*. Massachusetts Institute of Technology, Cambridge, MA, London, England, The MIT Press, 1993.

[11] C. Eberst and J. Sicheneder. "Generation of hypothetical landmarks supporting fast object recognition with autonomous mobile robots." In *Proceedings of IEEE/RSJ International Conference on Intelligent Robotic Systems*, vol. 2, pp. 813–820, Osaka, Japan, November 1996.

[12] D. Burschka and S. Blum. "Identification of 3d reference structures for video-based localization." In *Proceedings of third Asian Conference on Computer Vision*, vol. 1, pp. 128–135, Hong Kong, China, January 1998.

[13] D. Burschka and G. Färber. "Active controlled exploration of 3d environmental models based on a binocular stereo system." In *Proceedings of Eighth International Conference on Advanced Robotics*, vol. 1, pp. 971–977, Monterey, CA, July 1997.

[14] G. Schmidt and K. Azarm. "Mobile robot navigation in a dynamic world using an unsteady diffusion equation strategy." *Proceedings of IEEE/RSJ International Conference on Intelligent Robots and Systems*, vol. 1, pp. 642–647, Raleigh, North Carolina, July 1992.

[15] A. R. Pope and D. G. Lowe. "Vista: A software environment for computer vision research." In *Proceedings of International Conference on Computer Vision and Pattern Recognition*, vol. 1, pp. 768–772, Seattle, WA, June 1994.

[16] J. Canny. "A computational approach to edge detection." In *IEEE Transactions on Pattern Analysis and Machine Intelligence*, vol. 8 no. 6, IEEE Computer Society, 1986.

[17] C. Robl, S. Petters, B. Schäfer, U. Reiländer, and A. Widl. "Micro positioning system with 3dof for a dynamic compensation of standard robots." In *Proceedings of IEEE/RSJ International Conference on Intelligent Robots and Systems*, vol. 2, pp. 1105–1110, Grenoble, France, September 1997.

[18] T. Hopfner, F. Fischer, and G. Färber. "NoWait–RPC: Extending ONC RPC to a fully compatible message passing system." In *Proceedings of the First Merged International Parallel Processing Symposium & Symposium on Parallel and Distributed Processing*, pp. 250–254, Orlando, FL, March 1998.

Chapter **8**

REAL-TIME IMAGE PROCESSING FOR IMAGE-BASED VISUAL SERVOING

Patrick Rives and Jean-Jacques Borrelly
INRIA Sophia Antipolis Cedex

Abstract

In this chapter we point out some specific characteristics of image-based visual servoing schemes and we discuss how to account for such characteristics at the image processing level. We present a hardware and software architecture specially designed for implementing image-based visual servoing applications. We also present the results obtained using such machine vision for an underwater robotics application.

8.1 INTRODUCTION

With regard to the general class of vision-based control schemes, image-based visual servoing techniques [1, 2] are characterized by the fact that the signals used in the feedback loop are directly extracted from the image without explicit 3D reconstruction step. Thus, the control goal is specified as a certain configuration of the features in the image to be reached when the task is well performed. This approach is particularly sense-full when the robotic task can be specified, in a natural way, as a relative positioning of the camera frame handled by the robot with regard to a peculiar part of the environment. In practice, many applications can be stated like this. For example, in a road following application, the desired configuration (position and orientation) of the vehicle with respect to the road can be specified as a desired position and orientation of the lines, in the image, resulting from the projection of the white lines lying on the road.

In a more general way, let us now consider a set of 2D features resulting from the projection onto the image frame of geometric 3D primitives in the scene. We denote $s = (s_1 \dots s_n)^T$ a vector of parameters characterizing these 2D features. For example, a 3D line in the scene will project onto the image as a 2D line which can be represented by its slope θ and its distance to the origin ρ, thus we shall have $s = [\theta, \rho]^T$ as vector of parameters. The projective model of the camera defines a local differential mapping between the configuration space (position and orientation of the camera frame parameterized by an X with respect to the scene reference frame) and the sensor output space (the image plane). Referring to previous papers [1, 3, 4], we shall denote $L^T = \frac{\partial s}{\partial X}$ the Jacobian (*interaction matrix*) of this mapping which relates the variation of the parameters of the 2D features to the relative motion (expressed by the velocity screw $T_{ST} = \frac{dX}{dt}$) between the vehicle and the 3D primitives. In a general case involving a 3D geometric primitive moving in the scene with its own motion represented as a function of the time t, we get

$$\dot{s}\ (X, t) = (\frac{\partial s}{\partial X}\ .\ T_{ST} + \frac{\partial s}{\partial t}) \tag{8.1}$$

So, using the *task function approach* introduced by C. Samson [9], we can express an image-based visual servoing scheme in terms of regulating to zero a certain *output function*

$e(X, t) = (s(X, t) - s^*(X^*, t))$ directly expressed in the image. In this equation, $s(X, t)$ is the current value of the 2D features in the image corresponding to the current configuration X of the robot (or the vehicle) carrying the camera, and $s^*(X^*, t)$ is the desired value of the 2D features in the image corresponding to the desired configuration of the robot X^* at the time t. We consider the task is perfectly achieved during the time interval $[0, T]$ iff : $e(X, t) = 0, \ \forall t \in [0, T]$.

Using a very simple *gradient based approach* embedded in a velocity control scheme is sufficient to ensure an exponential convergence to 0 of the task function $e(X, t)$ by using the following desired velocity screw T_{ST} as control input

$$\begin{cases} \dot{e} &= -\lambda.e \quad \text{with} \quad \lambda > 0 \\ U &= T_{ST} \end{cases} \tag{8.2}$$

Using (8.1) and (8.2), we obtain as control input

$$U = -L^{T^+}. \left(\lambda \left(s(X, t) - s^*(X^*, t) \right) + \frac{\partial s}{\partial t} \right) \tag{8.3}$$

where λ is a positive gain and L^{T^+} is the pseudo inverse of the interaction matrix computed at the desired value $s(X, t) = s^*(X^*, t)$.

In terms of image processing, the computation of the control input requires the measurement, at each sampling period of the control loop, of the parameters of the 2D features $s(X, t)$ (located around a desired known value $s^*(X^*, t)$) and their derivatives $\frac{\partial s}{\partial t}$.

To do that, we have developed, in collaboration with the *LASMEA* in Clermont Ferrand [6] a dedicated machine vision based on a parallel architecture concept and able to tackle such requirements efficiently.

8.2 IMAGE-BASED VISUAL SERVOING REQUIREMENTS

Implementing efficiently a vision based control approach needs to solve several problems at the image processing level. The major difficulties are induced by real-time constraints and the processing of large data flow from the sequence of images. These problems can be summarized as follows.

- *Timing aspect.* In an image-based visual servoing scheme, the sampling rate of the closed loop control is directly given by the image processing time. Thus, to ensure the performances of the control, this sampling rate has to be sufficiently high. For the most current applications in visual servoing, a rate of 10 images per second can be considered as the lower limit.
- *Image processing aspect.* In a *scene analysis approach*, the main goal is to get high level description of a scene by means of sophisticated time consuming algorithms. By opposite, in an image-based visual servoing scheme approach, we only want to extract from the image the necessary and sufficient informations able to constitute a vector for the closed loop control scheme. As established above, for image-based visual servoing schemes, the feedback loop requires the measurement of the 2D features and their derivative in the image plane at each sampling period. Consequently, we only request low or intermediate levels processings providing contour based or region based features like points, lines, area, center of gravity and so on. Moreover, due to the fact that we have an a priori knowledge on the desired value of the feature $s^*(X^*, t)$, in the most cases, we can reduce the research area in the image plane at some regions of interest and then, the processing of the whole image is not necessary.

- *Tracking aspect.* The features extracted from the image have to be tracked along the sequence. We have also to take into account appearing and disappearing features due to the motions of the camera and objects in the scene.

From these requirements, we have developed an original architecture characterized by its modularity and its real-time capabilities (Fig. 8.1). This architecture implements the concept of active window. An active window is attached to a particular region of interest in the image and has in charge to extract a desired feature in this region of interest. This concept is quite similar to the *logical sensors* introduced by Henderson [5]. At each active window is associated a temporal filter able to perform the tracking of the feature along the sequence. Several active windows can be defined, at the same time, in the image and different processings can be done in each window. The functionalities of such an architecture are presented at the Fig. 8.2.

As above mentioned, we deal with image processing algorithms which only request low or intermediate levels processings. This aspect is clearly taken into account at the hardware level by using a mixed architecture:

- *At the low level*—We have to deal with repetitive processings through the window like convolution or filtering. *Pipeline Architecture* has been chosen to perform efficiently such processings. In practice, that is done by means of VLSI convoluors which admit until 5×5 mask size. This step can be performed in real time with a delay directly function of the size of the mask. The input of the algorithm is the active window and the values of the mask parameters corresponding to the desired processing. The output will be constituted by a list of pixels which potentially belong to the geometric features present in the window.
- *At the intermediate level*—We find merging-like algorithms which take in input the list of pixels of interest provided by the low level stage and merge them to compute a more structured representation of the geometric features present in the window (e.g., slope and distance to the origin for a line or coordinates of center and radius for a circle). In our hardware, this part is done by using a *MIMD Architecture* and it is practically implemented by means of multi-DSP's boards.

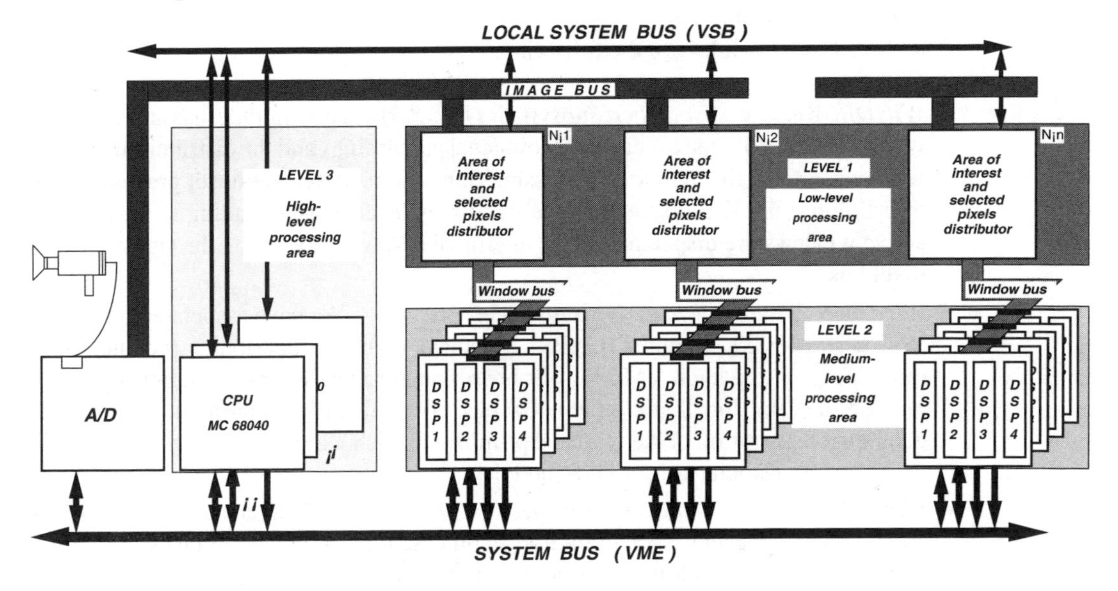

Figure 8.1 Overview of the general system.

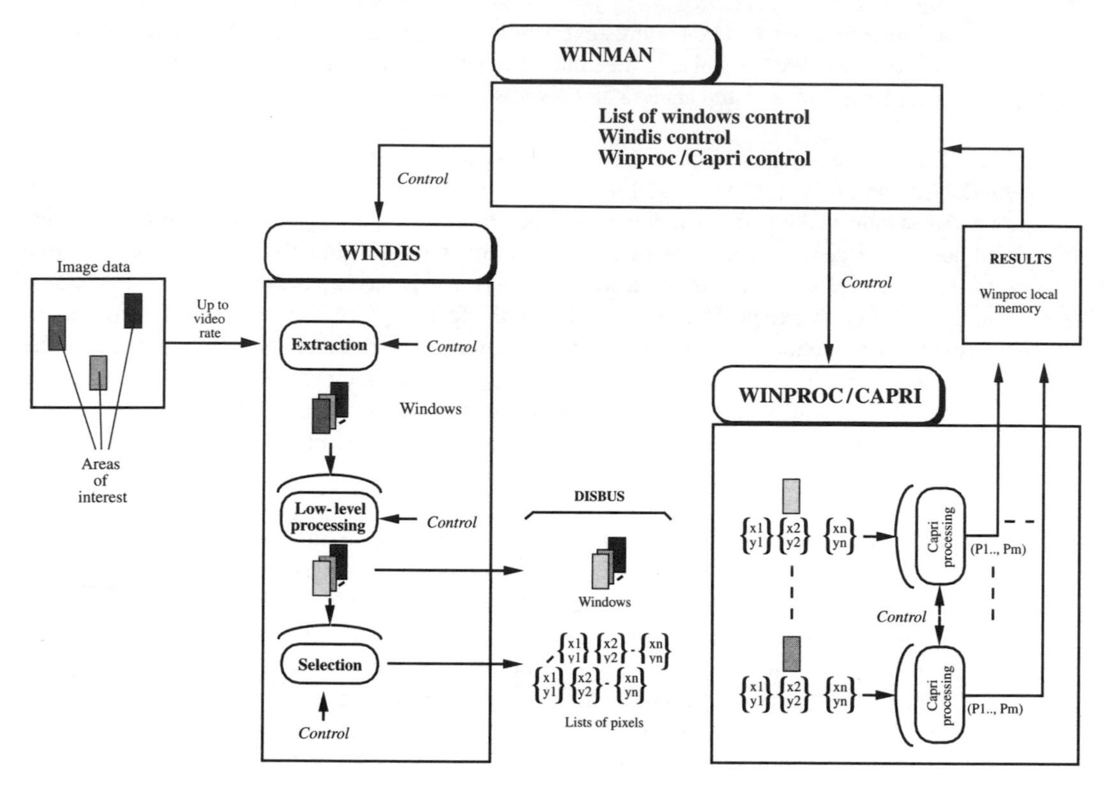

Figure 8.2 Data flow through Windis/Winproc subsystem.

The details of the hardware has been already described in a previous paper [6], [8], and we just recall the principles. Three basic modules have been defined corresponding, in practice to three different VME boards. Depending on the complexity of the application, the system will be built around one or more of these basic modules in such a way that we will be able to attempt video rate performances.

The three basic modules are the following:

- *WINDIS Window Dispatcher Subsystem (Fig. 8.3)*—This module has in charge the window extraction, the execution of low-level processings and the distribution of active windows toward the Window Processing Subsystem. After low-level processing and thresholding, the list of selected pixels and grey levels corresponding to the different active windows are dispatched to the intermediate-level processing through a 20 MHz pixel bus.
- *WINPROC Window Processing Subsystem (Fig. 8.4)*—We have associated one to sixteen DSP 96002's modules (CAPRI modules) with one distributor module. DSP modules are plugged on VME mother boards and execute intermediate-level processings on the data flow from the window bus. Window processing modules provide a geometric description of the searched primitive in each window. At each VME mother board can be associated four DSP modules.
- *WINMAN Window Manager Subsystem*—The window manager controls distributor and DSP modules, and executes high-level processing of applications tasks. Moreover, it has in charge the tracking of the active windows along the sequence. A 68040-based cpu board implements this module.

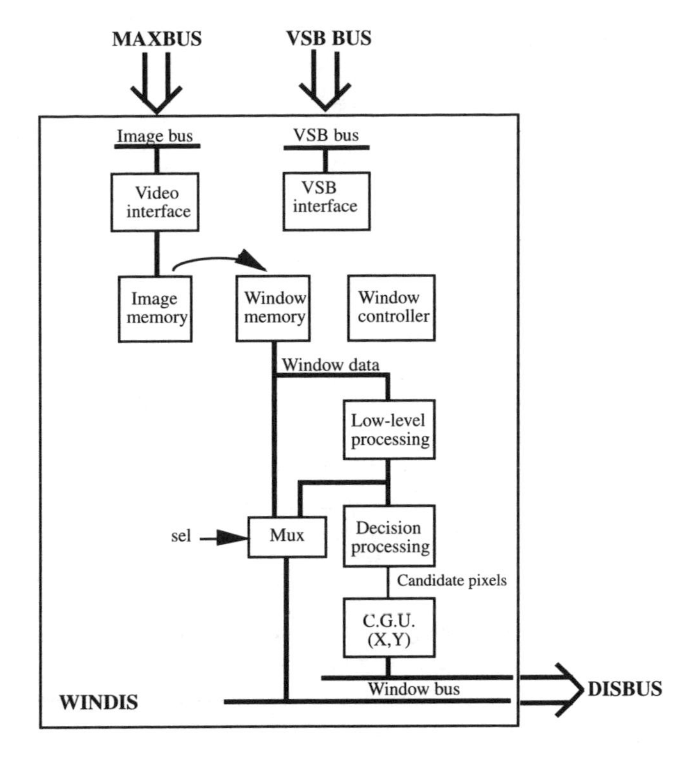

Figure 8.3 WINDIS subsystem.

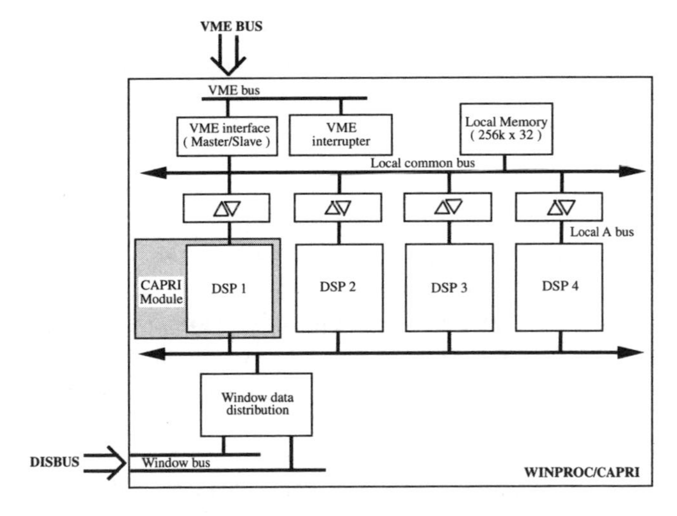

Figure 8.4 WINPROC/CAPRI subsystem.

For each level, we have introduced parallelism allowing us to reach video rate for most of applications tasks. All the modules satisfy to VME requirements and accept in input MAXBUS video bus from Datacube. The management of such a system is a tricky job and requests the use of a real-time operating system. For facility reasons, we have chosen to develop all the system level under VxWorks O. S. We present in the next section the different layers of the software from the application level until the implementation aspects using the real-time tasks scheduled by VxWorks.

8.2.1 Building an Application

To build an application the user must first define the DSP processing functions and assign them an unique identifier as well as the data structures for the function parameters and results. These functions will be called by the *Windows* driver and must respect the following prototype:

```
int      process_func(x,y,a,b,p,r)
int      x,y,a,b; /* window geometry */
param*   p;       /* pointer to parameters */
result*  r;       /*pointer to results */
```

The DSP programs can then be compiled and linked with the *Windows* driver. The processing functions need not exist on all the CAPRI modules provided they have a unique identifier.

The parameters and results of each DSP function will be located in the WINPROC shared memory, and will be used by the DSP and the application (*Results*) through pointers. The *winDrv* driver provides functions to allocate double banks of parameters or results for a window and the associated DSP. Then, one must fill the window descriptor of each window, including :

- *Geometry*—The window geometry is defined by the window position and size in the source image, windows are rectangular areas.
- *Low-level processing*—The low-level processing concerns the WINDIS attribute such as the filter size and coefficients and the selection of the pixels of interest.
- *Intermediate-level processing*—The intermediate level processing takes its input data from the low level and is done by up to four CAPRI modules. Each of the DSP module will run the appropriate user defined function with standard parameters and results.
- *Global processing*—This level is in charge to get the DSP results for each window, update the window geometry and compute the control vector to send to the robot.

An application supports three types of processing:

- *Multiple-dsp* is used when the *same* window is sent to several CAPRI modules after the low-level processing. A maximum of four modules is allowed provided they lie on the same WINPROC device.
- *Multiple-window* is used when a list of windows is send to a CAPRI module before starting the processing. The multiple-window mode can be used together with the multiple-dsp mode provided that all the windows of the list are sent to the same set of CAPRI modules.
- *Multiple-rate* is used when some windows are activated less frequently than other. The window rate is derived from the 40 ms video rate. All windows running at a selected rate must be sent to the same set of CAPRI modules.

The user also has to define the window distribution policy by building the lists of windows and configure the *digDrv* to generate appropriate events from the video rate. One or more lists of windows will be attached to each generated event.

8.3 APPLICATION TO A PIPE INSPECTION TASK

In this section we present some experimental results using the machine vision described above customized with one WINDIS board and one WINPROC with four CAPRIs modules. The application consists to follow a pipeline with an autonomous underwater vehicle using an image-based visual servoing technique. The synthesis of the control loop is detailed in [7], and we focus here on the image processing aspects. In terms of modeling, the pipe is represented as a cylinder. The projection of the pipe onto the image plane is (for non-degenerated cases) a set of two straight lines which can be parameterized by the vector of 2D features $s = [\theta_1, \rho_1, \theta_2, \rho_2]^T$. Using the *Task function framework* previously described, the control objective will be to keep into the image, the vector s as closer as possible to the desired known reference value s^* corresponding to the desired position of the vehicle with respect to the pipe.

8.3.1 Image Processing Aspects

In this section we present the techniques used to extract the sensor signals $s = [\rho_1, \theta_1, \rho_2, \theta_2]^T$ from the sequence of images provided by the vision sensor. Due to the bad conditions of lightning and the lack of dynamic in the video signal which are characteristic of the underwater environment (Fig. 8.5), the image processing techniques used have to be particularly robust. Moreover, in order to ensure good performances at the control level, real-time performances are also required. To satisfy these two aspects, the image processing mixes both data-based and model-based approaches. The basic idea is to use local regions of interest for detecting the edges of the pipe and to use global knowledge on the pipe to check the spatial and temporal consistency of the local detections and if needed, correct them.

Local processing (data-based) We use at this level the concept of active window previously presented. The aim of this local processing is to detect a piece of edge which belongs to the pipe into a region of interest and to track it along the sequence. To perform that, we have developed a token tracker with an edge operator based on Zernike's momentums. The tracking along the sequence is performed by a Kalman filter. A matching algorithm based on a geometrical distance between the predicted segment at time $(k-1)$ and the segments into the active window detected at time (k) allows to select the best candidate for updating the estimate. Such a processing provides a local estimate of (ρ, θ) at video rate. Moreover, at each active window is associated two logical states:

- **seg:** This flag is "TRUE" (T) if the matching process selects one segment among the candidates. If this flag is "FALSE" (F) the Kalman filter propagates its prediction during a given number k_{max} of images.
- **win:** This flag is "TRUE" if *(seg = TRUE) OR ((seg = FALSE) AND (* $k \leq k_{max}$ *))*.

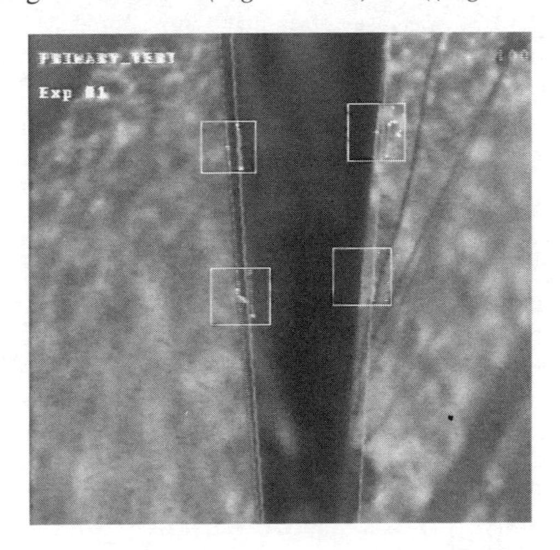

Figure 8.5 Example of image.

Global processing (model-based) Theoretically, using only two active windows located on each limb of the pipe, it is possible to build the complete sensor signals vector used in the pipe inspection task. Unfortunately, due to the bad lighting conditions and to the dirtiness around the pipe, only local processing is not sufficient in terms of robustness. To increase performances, we added over the local processing a more global one checking the consistency of the data provided by the local step. For each limb, the global processing uses two active windows providing two local estimates (ρ, θ). These two local estimates are used to build a global estimate according to control rules based on their logical states. Moreover the logical state of the window is also used to positioning the window at the next image $(k + 1)$. In practice, these rules allow us to compute the best estimate with regard to the quality of the local signals and to reinit a dead window using the data provided by the window still alive. Table 8.1 summarizes the logical behavior of the global processing.

The whole image processing algorithm was implemented on our real-time machine vision. Four active windows were used (two for each limb), and the complete processing is performed in 80 ms. Figure 8.6 illustrates the reinitialization of a dead window by using data provided by the window which remain alive.

TABLE 8.1 Global control rules.

WIN1		**WIN2**			**Window position at (k+1)**	
seg	**win**	**seg**	**win**	**Computation of the feedback vector**	**WIN1**	**WIN2**
T	T	T	T	WIN1 + WIN2	GLOBAL	GLOBAL
F	T	T	T	WIN1 + WIN2	WIN1	WIN2
F	F	T	T	WIN2	WIN2	WIN2
T	T	F	T	WIN1 + WIN2	WIN1	WIN2
T	T	F	F	WIN1	WIN1	WIN1
F	F	F	F	IDLE	REINIT	REINIT

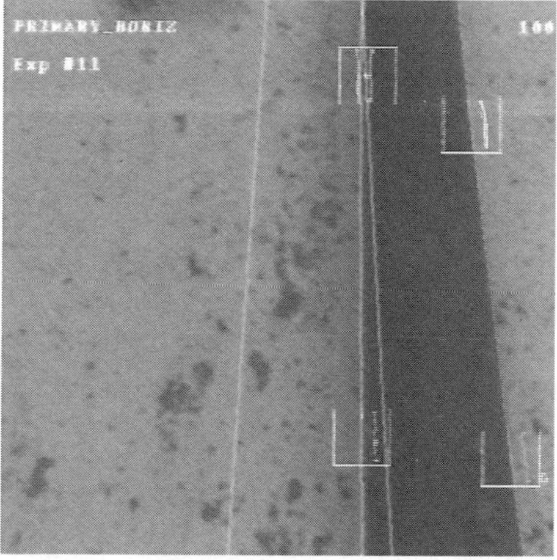

Figure 8.6 Reinitialization of the windows.

8.4 CONCLUSION

The pipe inspection task was validated in a real experiment in *Ifremer's* center of Toulon. Many problems were encountered both at the image processing level and at the vehicle level. The first one was mainly due to the bad condition of lightning (reflected light on the waves) and dirtiness in the bottom of the pool. However, the algorithms used at the image processing level were robust enough to overcome these difficulties. Major problems occurred with the control of the *Vortex*. Due to important geometric bias on the locations and orientations of the propellers, strong couplings appear on the behavior of the vehicle, even in tele-operated mode. For example, when the vehicle dives, it turns around its vertical axis; when it goes ahead, it turns on the left and so on. These defaults are largely increased by the presence of the umbilical. Moreover, the response and the dead zones of the propellers are strongly nonlinear and assymetric mainly due to the lack of velocity loop at the low-level control. Despite these imperfections, we succeeded in doing a complete pipe inspection task confirming the robustness both at the image processing level and at the control level.

REFERENCES

[1] F. Chaumette. "La relation vision-commande: theorie et applications a des taches robotiques." Master's thesis, Ph.D. thesis, University of Rennes I, July 1990.

[2] F. Chaumette, P. Rives, and B. Espiau. "Classification and realization of the different vision-based tasks." *Visual Servoing - Automatic Control of Mechanical Systems with Visual Sensors; World Scientific Series in Robotics and Automated Systems*, Vol. 7, K. Hashimoto (ed.), London, 1993.

[3] B. Espiau, F. Chaumette, and P. Rives. "A new approach to visual servoing in robotics." *IEEE Transactions on Robotics and Automation*, vol. 8, no. 3, pp. 313–326, 1992.

[4] P. Rives, R. Pissard-Gibollet, and L. Pelletier. "Sensor-based tasks: From the specification to the control aspects." *Sixth International Symposium on Robotics and Manufacturing*, Montpellier, France, May 28–30, 1996.

[5] T. Henderson and C. Hansen. "Multisensor knowledge systems: interpreting a 3D structure." *International Journal of Robotic Research*, vol. 7, no. 6, pp. 114–137, 1988.

[6] P. Martinet, P. Rives, P. Fickinger, and J. J. Borrelly. "Parallel architecture for visual servoing applications." *Workshop on Computer Architecture for Machine Perception*, Paris, France, December 1991.

[7] P. Rives and J. J Borrelly. "Visual servoing techniques applied to an underwater vehicle." *IEEE International Conference on Robotics and Automation*, Albuquerque, April 20–25, 1997.

[8] P. Rives, J. J. Borrelly, J. Gallice, and P. Martinet. "A versatil parallel architecture for vision based control applications." *Workshop on Computer Architectures for Machine Perception*, New Orleans, December 15–17, 1993.

[9] C. Samson, B. Espiau, and M. Leborgne. "Robot control: The task function approach." *Oxford Engineering Sciences Series 22*. Oxford University, 1991.

Chapter
9

PROVEN TECHNIQUES FOR ROBUST VISUAL SERVO CONTROL

K. Arbter, G. Hirzinger, J. Langwald, G.-Q. Wei, and P. Wunsch
German Aerospace Center - DLR

Abstract

This chapter discusses a variety of techniques that can significantly improve the robustness of visual servoing systems. Unlike other disciplines that address robustness issues such as robust control, we do not adopt a formal notion of robustness but rather concentrate on two phenomena which are typically the reason for a failure of vision systems: unexpected *occlusion and infavorable illumination conditions*. We discuss a variety of image processing techniques that ensure stable measurements of image feature parameters despite these difficulties. In particular, we will cover feature extraction by Hough-Transforms, color segmentation, model-based occlusion prediction, and multisensory servoing. Most of the techniques proposed are well known, but have so far been rarely employed for visual servoing tasks because their computational complexity seemed prohibitive for real-time systems. Therefore an important focus of the chapter is on efficient implementation. The robustness of the techniques is highlighted by a variety of examples taken from space robotics and medical robot applications.

9.1 INTRODUCTION

During the last few years we have observed a tremendous growth in research of closed-loop control of robotic systems by visual information, which is commonly referred to as visual servoing (see [1] or [2] for survey of recent research). So far a major part of research has been mainly devoted to the kinematic aspects of visual servoing, that is techniques for computing robot actuation commands from 2D visual measurements (e.g., [3]). In addition, there has been some work that addresses the dynamic aspects of closed-loop visual control (e.g., [4]), methods for the automatic learning of control algorithms (e.g., [5]), or techniques for specification of vision-based tasks [6, 7].

However, in order to achieve real-time operation at video frame rate, most of the visual servoing systems that have been demonstrated so far rely on rather simple methods for deriving measurements from the raw image data. Image processing is usually restricted to small regions of interest which are often only a few pixels wide and feature extraction is rarely more than binary centroid computation, correlation matching or 1D gradient detection. Such simple image processing techniques allow for frame-rate tracking of several image features even on standard workstations without any specialized hardware [8]. However, it is intuitively clear that such simple image processing may fail, if imaging conditions differ from those for which the image operators were designed. The consequence is a well-known lack of robustness that many visual servoing systems clearly exhibit.

The reason why simple image operators, which are, in addition, restricted small image regions, may fail lies in the complexity of the image formation process. Small changes in the environment that is imaged by a camera can result in significant variations of the image data. The most common effects are abrupt *changes in contrast* and *occlusion of expected features*.

Thus, in a sense, the image formation process poses a discontinuity problem against which image operators need to be robust. An example may illustrate this point.

The left side of Fig. 9.1 shows the laboratory setup of the German national space robotics project *Experimental Servicing Satellite* (ESS). The goal of this project is to build a satellite that is equipped with a servicing robot designed to do repair and maintenance work on failed geostationary satellites by teleoperation from ground. A key step in this process is the automatic, vision-based capturing of the satellite, which is simulated in the lab by a two robot system. One of the robots carries a simplified satellite mockup and performs a tumbling motion which is typical of a rigid body floating under zero gravity. The other robot is equipped with a stereo camera system and a capture tool that is supposed to be inserted into the apogee motor in order to rigidly attach the satellite to the service robot.

The right side of Fig. 9.1 shows a gradient-filtered image of the left-hand camera. First of all, you can observe that a significant portion of the features of the satellite (edges of the satellite body, rim of the apogee motor) are occluded by the capture tool. But, what is more, a small change of the relative position or attitude of gripper and satellite may result in completely different occlusion, which can amount in a complete disappearance of features as well as in an appearance of totally new features. In addition, the image clearly demonstrates the influence of illumination on image features, which is usually unknown when image operators are designed. As we can see in this example, we cannot detect any contrast on left part of the apogee motor's contours. But, again, a slight change of the orientation may result in a different reflection, such that we will encounter contrast on the left part of the contour, but will probably fail to find edges elsewhere. Thus, in the context of this chapter, robustness means that *stable measurements of local feature attributes, despite significant changes in the image data, that result from small changes in the 3D environment.*

In the following we will propose some image processing techniques for visual servoing that have proven successful in enhancing robustness in the sense defined above. Some of these techniques are well known, but need adaptation in order to allow for real-time tracking, which is prerequisite for visual servoing. In particular we will discuss

- the Hough-Transform
- color segmentation operators
- a model-based approach to occlusion prediction
- multisensory visual servoing

For all these techniques we will show successful results from both space robotics and medical robot applications that have been developed at the Institute of Robotics and System Dynamics of the German Aerospace Center (DLR).

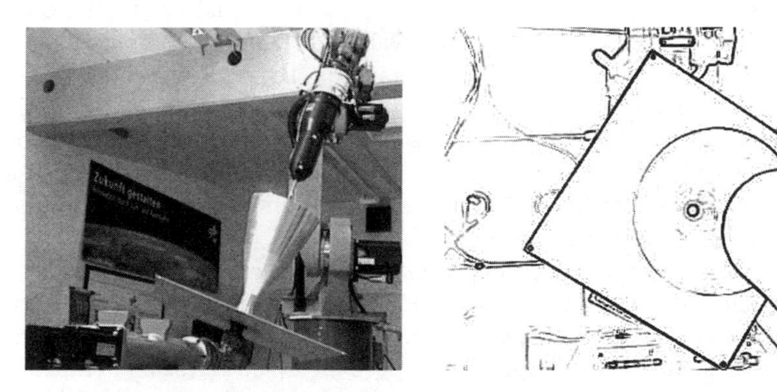

Figure 9.1 Left: Laboratory setup of the space robotics experiment "Experimental Service Satellite"; right: Image from the "Capture Tool Camera."

9.2 ROBUST FEATURE EXTRACTION

First we will address two distinct methods to extract measurements of object feature properties from digital image data, which are particularly robust with respect to the difficulties described above.

9.2.1 The Hough-Transform

The Hough-transform (HT) [9] is a well-known robust technique to extract a large variety of geometrical object features from digital images such as straight lines, parametric curves, or shapes defined by digital patterns [10, 11].

The Hough-transform

$$H(\mathbf{p}) = \iint_D f(x, y)\, \delta(\, g(x, y, \mathbf{p})\,)\, dx\, dy \tag{9.1}$$

is an integral transformation, which maps all information of an edge filtered image $\mathbf{F}(x, y)$ to the parameter space of a searched pattern. The elements of $\mathbf{F}(x, y)$ are the gradient magnitude $f(x, y)$ and the gradient direction $\varphi(x, y)$. The HT operates in a selective way; the adaption is performed through the description $g(x, y, \mathbf{p}) = 0$ of the searched pattern, where $\mathbf{p}$ is a parameter vector that is to be determined by the feature extraction process. Local maxima in the Hough accumulator $H(\mathbf{p})$ point to corresponding features in the image; thus feature extraction reduces to a simple maximum detection in the accumulator space.

As an example, the following equations describe geometric features that are often encountered in visual servoing tasks in technical environments such as straight lines $g(x, y, d, \vartheta)$, circles $g(x, y, x_0, y_0, r)$ or ellipses $g(x, y, x_0, y_0, a, b, \psi)$. The description of ellipses by the parameters x_0, y_0, a, b, ψ was chosen to clarify the analogy to the circle description. A parameter space without singularities should be selected for HT implementations.

Lines:

$$\mathbf{p} = (d, \vartheta) \tag{9.2}$$

$$g(x, y, \mathbf{p}) = x \cos\vartheta + y \sin\vartheta - d \tag{9.3}$$

Circles:

$$\mathbf{p} = (x_0, y_0, r) \tag{9.4}$$

$$g(x, y, \mathbf{p}) = (x - x_0)^2 + (y - y_0)^2 - r^2 \tag{9.5}$$

Ellipses:

$$\mathbf{p} = (x_0, y_0, a, b, \psi) \tag{9.6}$$

$$g(x, y, \mathbf{p}) = \frac{[(x - x_0)\cos\psi + (y - y_0)\sin\psi]^2}{a^2} + \frac{[(y - y_0)\cos\psi - (x - x_0)\sin\psi]^2}{b^2} - 1 \tag{9.7}$$

Unlike more traditional ways of computing feature properties from images, this type of feature detection exhibits two significant advantages:

- Noise and background clutter do not contribute to the accumulator in a systematic way and thus do not impair a detection of local maxima.
- As each image point is transformed independently of the others, feature parameters can already be determined if parts of the pattern are transformed. Thus we obtain robustness with respect to partial occlusion and varying contrast.

However, the HT requires time and storage space that increases exponentially with the dimensionality of the parameter space. The accumulator $H(\mathbf{p})$ has a size of $\mathcal{O}(n^p)$, where p is the number of parameters and n the number of discrete values per parameter. The size of $H(\mathbf{p})$ is $\mathcal{O}(n^2)$ for straight lines, $\mathcal{O}(n^3)$ for circles and $\mathcal{O}(n^5)$ for ellipses. For every point (x, y) of the image with a magnitude $f(x, y) > 0$, all valid parameter combinations $\mathbf{p}$ must be computed and the corresponding cells of H updated. The computational load per pixel (x, y) is $\mathcal{O}(n^{p-1})$, thus $\mathcal{O}(n)$ for lines, $\mathcal{O}(n^2)$ for circles, and $\mathcal{O}(n^4)$ for ellipses.

For this reason, real-time tracking applications of complex features based on HT are very rare, although quite a few approaches to speed up the computation have been suggested in the literature. But often, the proposed simplifications eliminate the advantages of HT. Examples are the exploitation of contour chains [12] or of pairs of edge points [13], which require clearly detectable object contours that must not be interrupted. With such assumptions, the algorithms cannot be robust against partial occlusions, low contrast, variable lighting, or overlapping patterns.

9.2.1.1 Efficient Computation of the HT. The computational load per pixel can be reduced from $\mathcal{O}(n^{p-1})$ to $\mathcal{O}(n^{p-2})$ by incorporating the gradient constraint $\varphi(x, y)$. It is $\mathcal{O}(1)$ for straight lines, $\mathcal{O}(n)$ for circles, and $\mathcal{O}(n^3)$ for ellipses [14].

Using the information about gradients direction computing the HT for straight lines, there is only one accumulator cell $H(d, \vartheta)$ per pixel to update. With $\vartheta = \varphi(x, y)$ the distance d from the origin is

$$d = x \cos \varphi(x, y) + y \sin \varphi(x, y) \tag{9.8}$$

For the HT of circles, the accumulator cells $H(x_0, y_0, r)$ to be increased dependent on $\varphi(x, y)$ are determined by the following relations:

$$\begin{aligned} x_0 &= x \pm r \cos \varphi(x, y) \\ y_0 &= y \pm r \sin \varphi(x, y) \end{aligned} \tag{9.9}$$

Equation (9.9) demonstrates the linear relation between the number n of discrete values per parameter, here dependent on the radius r, and the number of accumulator cells $H(\mathbf{p})$ to take into account. The reduction of effort taking advantage of the gradient constraint is shown in Fig. 9.2.

A real-time application requires both a fast image preprocessing step and an efficient implementation of the HT. A circle tracking algorithm based on HT is implemented in a real-time visual servoing system.

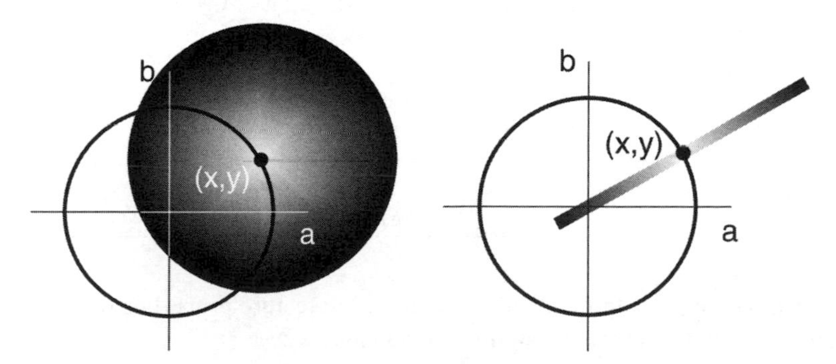

Figure 9.2 Shown highlighted is the space of all accumulator cells $H(a, b, r)$ to update accordingly to the marked point of the image (x, y) (left side without, right side with consideration of the gradient direction information φ). Intensity corresponds to parameter r, larger radii are dark.

The steps of image acquisition, image rectification to eliminate lens distortions, and horizontal and vertical gradient filtering with a 7×7 kernel run on a Datacube MV250 image processing system, a VMEbus based pipeline processor specialized in image preprocessing tasks. Also, it computes the gradient magnitude and the gradient direction combined with a data reduction and packing. Because the HT computation needs the Hough accumulator in randomly accessible memory space, the HT is computed on the host processor board also controlling the image processor. The host board is a VMEbus based PowerPC CPU board produced by Motorola. Preprocessed data are transferred between the image processing unit and the host computer over the VMEbus. Because of the bandwidth limitations of the VMEbus and the involved boards, there is the need for a data reduction in real-time applications. This is done by the reduction of gradient magnitude $f(x, y)$ and direction $\varphi(x, y)$ to a four-bit value per pixel each and packing in one byte. Besides, only the transfer of regions of interest is possible.

For tracking tasks, the feature extraction step only needs to detect a small deviation from an expected feature position within a small region of interest, feature parameters are determined by HT in an extremely limited parameter space. A practicable number n of discrete values per parameter is in the range of $(13 \leq n \leq 25)$. For the HT of circles, there is a linear relation between n and the computational load per pixel.

9.2.1.2 Discretization Effects. The restricted gradient direction φ coding using only four bits per pixel yields to a large quantization error. Using an n bit coding, the number of intervals is 2^n, the size per interval is $\frac{\pi}{2^n}$ and the maximum discretization error per interval is $\pm\frac{\pi}{2^{n+1}}$. This quantization error is reduced by invalidating different ranges of gradient directions to improve the correctness of valid ranges. Using only $\frac{1}{2^k}$ values of all possible gradient directions, the size per interval shrinks to $\frac{\pi}{2^{n+k}}$ and the maximum discretization error is reduced to $\pm\frac{\pi}{2^{n+k+1}}$ conforming a $(n + k)$ bit coding. The disadvantage of this solution is taking only $\frac{1}{2^k}$ image points of the tracked circle into account. This invalidation and the packing can be done simultaneously in an efficient way using a look up table on the pipeline processor. A grey level image, taken with $k = 2$, is shown in Fig. 9.3.

9.2.1.3 Results. A circular structure within an unfavorable environment being robustly tracked in real time is shown in Fig. 9.4. The hardware used for this implementation permits the processes of image acquisition, preprocessing, and HT to run in parallel. Based on the

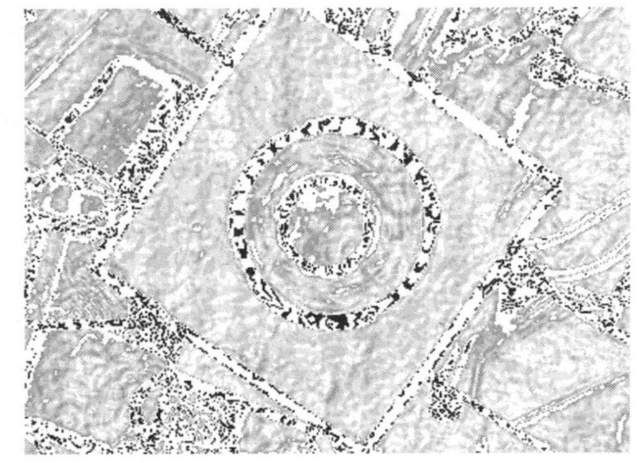

Figure 9.3 (Left) Rectified grey level image of a tracked satellite model carried by a robot, seen from the ESS manipulator's hand camera on a second robot [15]. (Right) Image of gradient *magnitude* after the image preprocessing step and invalidating of uncertain ranges of gradient direction.

Figure 9.4 In unfavorable environmental conditions tracked circle.

field rate of a CCIR camera and an image resolution of 384×287 pixels the acquisition takes 20 ms, the preprocessing step also requires approximately 20 ms, and the HT for circles has also a time frame of about 20 ms on the specified system. The time needed for the HT linearly depends on the number of discrete values per parameter and the size of the regions of interest. Accordingly, circular objects are robust trackable at a rate of 50 Hz with a period of about 60 ms between camera shuttering and the computed result.

Because the algorithm needs for a reliable detection of the best fitting parameters a distinct peak in the Hough accumulator $H(\mathbf{p})$, effects concerning the discretization of the gradient direction influence the necessary size of the parameter space. The relation between the discretization error and the necessary minimal size of the parameter space is presented in Table 9.1.

Restriction of quantization errors by invalidating of uncertain ranges of gradient direction improves the correctness of valid ranges, reduces noise, achieves a distinct peak in Hough space, and permits a smaller size of the parameter space. The disadvantage of using only $\frac{1}{2^k}$ image points belonging to the tracked circle is acceptable with regard to the stability of the algorithm.

9.2.2 Robust Color Classification

Because color has more disambiguity power than intensity values, using color information to obtain image features for visual servoing has some specific robustness properties. Because of the real-time requirement of visual servoing, we choose colors which are known a priori to ease this. In this sense we are doing supervised color segmentation. We analyze the color distribution of the scenes we are working with and choose colors that do not appear in them as marker colors [16]. These markers are brought onto the scene for real-time segmen-

TABLE 9.1 Influence of the discretization error of the gradient direction $(0 \leq \varphi < \pi)$ on the minimal possible dimension of the search area, four-bit coding of the gradient direction.

k	Used Values 2^{-k}	Equally to Coding	Minimum Search Area
0	1	4 bit	13^3
1	1/2	5 bit	9^3
2	1/4	6 bit	7^3
3	1/8	7 bit	5^3

tation. We implemented it on a DataCube image processing system. The positions of these markers are used as the input to our visual servoing system.

A color can be represented by the RGB (red-green-blue) components. In digital images, the RGB values are between 0 and 255. Thus all possible colors can be represented by the points within the RGB cube of size $256 \times 256 \times 256$. Another representation of color is the HSV (hue-saturation-value) representation, where H and S are directly related to the intrinsic color and V to the brightness. Because we would like the segmentation results to be insensitive to the strength of illumination, the HSV representation is more suitable for our purpose: the color signature of a color image can be directly analyzed in the H-S plane. Figure 9.5(a) shows an H-S plane filled with the corresponding color, where the brightness is set to 255. In the H-S coordinate system, the H value is defined as the angle from the axis of red color, and the S value (normalized to the range of zero to one) is the length from the origin (the white color).

9.2.2.1 Color Training and Color Segmentation. By analyzing the HS histograms of typical images of our robotic environment, we choose four colors as our marker colors. The corresponding color markers are then brought onto the object we want to track, This is shown in Fig. 9.5(b). The colors of the markers are analyzed again in the HS histogram to determine their positions in the color space. We use an interactive procedure to do so.

We first manually outline the markers in the image with polygonal boundaries. Then, the pixels within each polygon are used to compute the color distribution in the H-S plane. Figure 9.5(c) shows the color cluster of the markers. To represent the color cluster, we again use a polygonal approximation of the cluster boundary. (All the polygonal approximations can be done interactively with a computer mouse.) The above process is called color training.

At the segmentation stage, all pixels whose hue and saturation values fall within the cluster polygons and whose brightness value is higher than a threshold are initially labeled as the markers. The reason to use brightness thresholding is that low intensity pixels do not provide robust HS values and are themselves of no interest. The classification of color pixels can be realized by a look-up table. Due to image noises and artifacts of color cameras on image edges, pixels classified by the look-up table are not all marker pixles, that is, there are

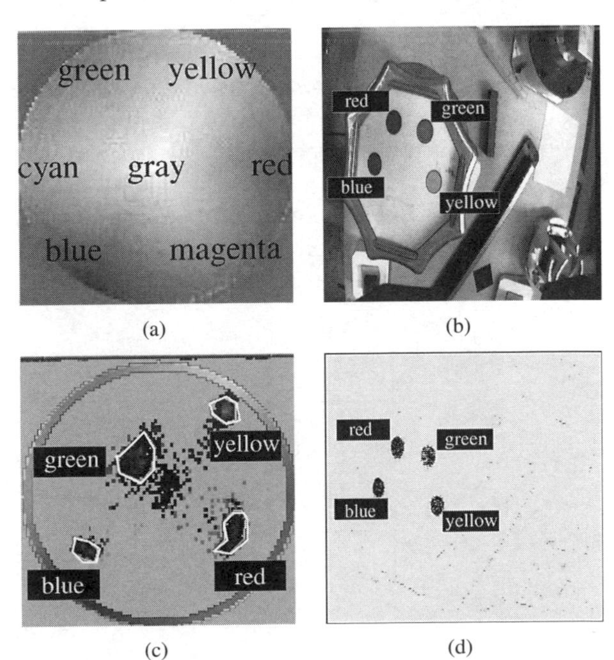

Figure 9.5 Using color markers for robust feature extraction. (a) HS plane; (b) four-color markers in our robotic environment; (c) color cluster of the markers; (d) initial segmentation.

false classifications. Figure 9.5(d) shows the initial segmentation results. To remove these false segmentations we postprocess the segmented, binary image in two stages. In the first stage, we apply spatial filtering. Since the classification errors of background pixels tend to be spatially scattered (they are not connected to form a regular object due to our choice of unique colors), we can remove some of the false classifications by a low-pass filter. We convolve the segmented image with a 7×7 kernel of uniform values, which is equivalent to a 7×7 local averaging. By thresholding the low-pass filtered image, we have reduced the rate of false classification. The second stage of postprocessing is based on temporal filtering. This is done by first adding two segmented images in successive frames and then thresholding the results. This removes segmentation errors due to image noises which tend to be temporally random; that is, they do not tend to appear at the same image position in successive frames.

9.2.2.2 Real-Time Implementation. All the above were done first interactively in a software environment, to check whether the chosen cluster-polygon and the thresholds were appropriate for correct segmentation. Then, the parameters obtained are transferred to the DataCube hardware environment.

Figure 9.6 shows the system diagram of the DataCube implementation, where the multiplexer switches between two stereo cameras, so that only one hardware system is needed to process stereo images. A statistic processor provides the centroid of each color blob. The coordinates of the centroids are used as image features for visual servoing. The current positions of the markers also provide *Region of Interest* (ROI) for the processing of the next frame. This has significantly speeded up the segmentation procedure. The system can run at about 15 Hz for stereo images during tracking, i.e., when ROIs are already available. When the ROIs are the whole image, the system operates at about 5 Hz. The advantage of using colors for robust feature extraction is that no initialization for the marker position is needed. When any marker falls out of any camera view, it can be immediately captured as soon as it re-enters into the camera view (at the rate of 5 Hz). Furthermore, since each marker has a different color, the identification of each feature is always known; that is, when one feature goes out of view, we know which one it is. This allows us to switch to the specific control mode prepared for a subset of the features. This greatly broadens the validity area of the sensory patterns for visual servoing.

9.2.2.3 Applications in Medical Visual Servoing. We applied the above techniques of supervised color segmentation to automatic visual guidance of laparascopic cameras in minimally invasive surgery. We brought a color marker on the instrument tip so that the position of the marker in the image can be found in real time by color segmentation and used to control a medical robot (AESOP). Since safety is of the highest priority in surgery, correct segmentation of instruments is crucial for correct visual guidance. A problem particular to laparoscopic images is that the received light by the narrow lens system is usually very weak,

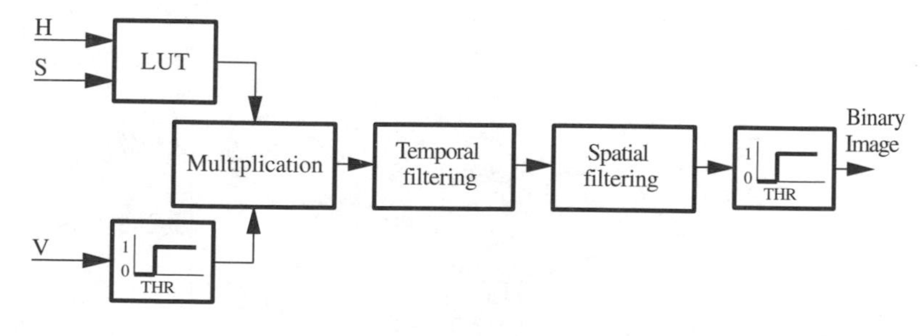

Figure 9.6 The block diagram for the classifier.

so that the CCD signal (including noise) must be highly amplified. For this reason, the signal-to-noise ratio is considerably lower than that of a standard CCD camera. In our system, the high rate of correct color segmentation is attributed to the use of low-pass filtering. Noises which have the same color as the color code and are not correlated in the spatial and temporal domain will be removed.

Another characteristic of laparoscope images is that saturation may occur due to too strong illumination. A highly saturated part of the image is white in color. When saturation occurs on the instrument, e.g., when the instrument is placed too near to the laparoscope, the saturated part loses its original color of the color-code. But since the instrument is cylinder in shape, i.e., the surface normals vary in a wide range, there are always pixels which can be correctly segmented. This enables the position of the marker still usable in lateral guidance of the laparascope, since our goal is to center the instrument in the image and the precision of placement is of no importance. A similar situation occurs when there are partial occlusions of the marker, due to either another instrument or blood contamination.

Figure 9.7 shows experimental set-up. Figure 9.7(b) shows a stereo image superimposed with the bounding box obtained from color segmentation.

The system works very robustly. It is now at the Klinikum rechts der Isar of the Technical University of Munich for clinical evaluation. Figure 9.8 shows a minimally invasive surgery using robot camera assistance.

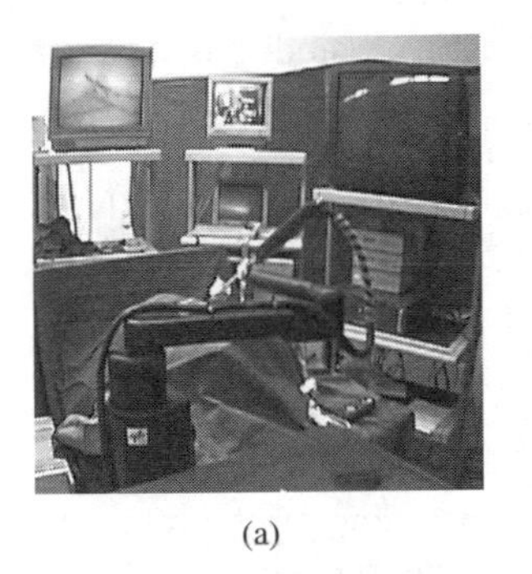

(a)

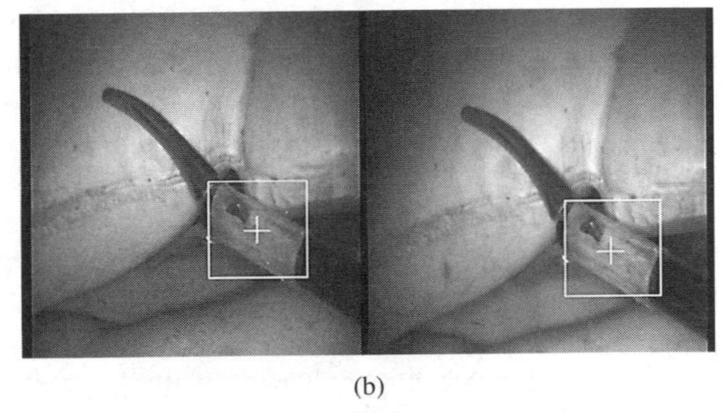

(b)

Figure 9.7 Tests in a laboratory environment. (a) Experimental setup. (b) A stereo laparoscopic image superimposed by the bounding box of the segmented mark.

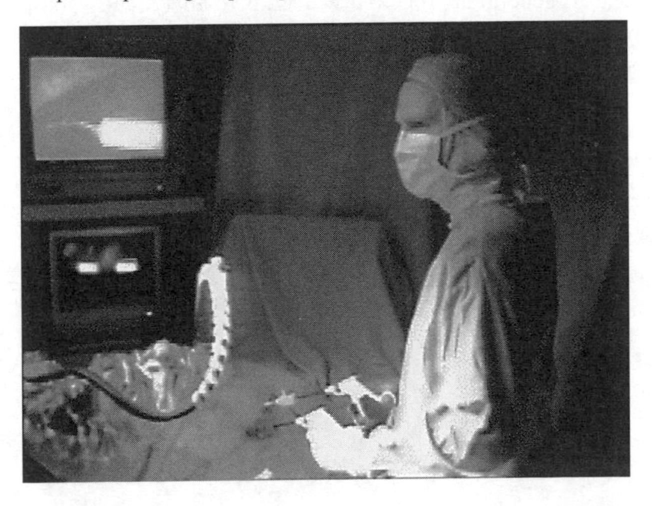

Figure 9.8 Minimally invasive solosurgery based on autonomous robot camera assistance.

9.3 MODEL-BASED HANDLING OF OCCLUSION

Robust feature extraction methods such as described in the previous section can very well cope with both unfavorable illumination and partial occlusion. However *aspect changes*, that is, complete occlusion of object features due to a change of the spatial attitude of an object with respect to the camera cannot be handled by these techniques. If in a particular visual servoing application the motion of a target object cannot be restricted, aspect changes must be explicitly considered.

9.3.1 Approach

The visibility conditions in a 2D camera image depend on the 3D geometry of the imaged object. So the key to a general technique for dealing with occlusion during tracking is to build and maintain a 3D model of the observed objects. A pragmatic approach is to base such a model on a 3D CAD-like description of the scene, especially because such models are often available in high-level robot programming systems (e.g., [6, 7]). The structure of the resulting tracking system is depicted in Fig. 9.9. From the raw sensor data we can extract measurements of object features from which we estimate the spatial position and orientation of the target object. Based on the history of estimated poses and assumptions about the object motion you can predict an object pose expected in the next sampling interval. With this predicted pose and the 3D geometric object model, we can determine feature visibility in advance, select an appropriate subset of the visible features, and thus guide the feature extraction process for the next frame without a risk of searching for occluded features.

As we need to compute object pose relative to the camera we can also use this information for the robot controller, so this approach naturally leads to position-based visual servoing.

This structure is of course not new and has been proposed by many authors. However, this approach is very computation intensive. Due to the computational complexity we have so far only seen real-time implementations for rather restricted cases: Usually occlusion are only considered for single convex objects, for which visibility determination is trivial [17, 18]. A few authors also precompute feature visibility off-line depending on the relative orientation of the target, such that during run time it only needs to be checked whether and which aspect change has occurred [19, 20]. Unfortunately this implies the restriction to a fixed object distance. The most general approach, however, is to analytically compute a complete analytic hidden line removal at each tracking cycle.

So, basically two major challenges have to be met, when implementing a visual servoing system according to the structure above:

- efficient 3D pose estimation
- efficient hidden line removal

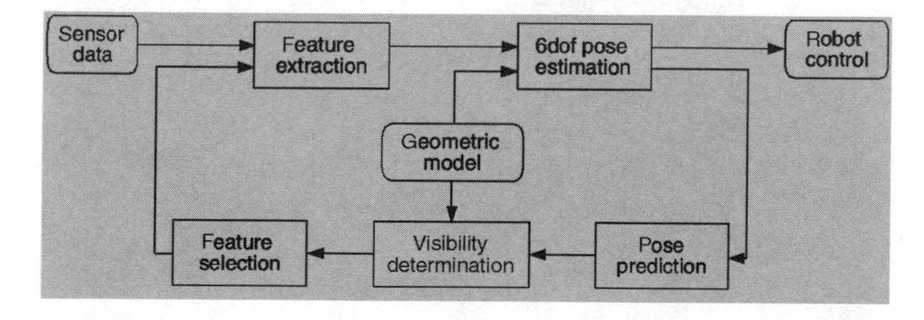

Figure 9.9 Structure of a tracking system designed to handle aspect changes online.

As a variety of efficient methods for 3D pose estimation have already been proposed [21, 22, 23], we will focus on hidden line removal in the following.

9.3.1.1 Efficient Hidden Line Removal. In order to guide a feature extraction process visibility determination must provide exact location information of object features such as lines or curves. Therefore we cannot make use of the well-known z-buffering algorithm [24] that is often supported by 3D graphics hardware of standard workstations, because it only yields a digitized image and thus no information on object features.

Therefore, we have developed an analytic hidden line removal algorithm that combines a variety of techniques to reduce the computational complexity [25]. The algorithm is based on a polyhedral object model that has been extended such that circular features are explicitly modeled [2]. In particular, high performance is achieved by

- an extension of the face-priority concept [26] that allows to determine during run time without any additional computation which faces of the polyhedron can be excluded a priori when computing occlusions of a particular face
- an efficient method to compute visibility values locally, such that in contrast to standard algorithms it is no longer required to presort faces according to their distance from the camera
- highly optimized geometric basis routines to compute intersections of lines and faces
- the implementation of optimistic strategies, that yield correct results with little computation in frequent cases but that may require more sophisticated postprocessing in some rare cases.

In addition, the explicit modeling of curved object structures allows us to eliminate *virtual lines*, i.e., line segments that are included into a polyhedral object model in order to approximate curved face, but that do not have a physical correspondence in a camera image. Figure 9.10 shows an example.

The cycle times for object of medium complexity such as the one in Fig. 9.10 (282 vertices, 366 edges, and 139 faces) ranges, depending on the aspect, between 6 msec and 12 msec on an SGI MIPS R4000 workstation. Obviously, the algorithms is not capable of real-time operation in the strict sense that computation time is fixed independently of the data processed. If such deviations of processing time are not acceptable in a particular real-time system, the algorithm may be structured such that it exhibits an *any-time* behavior. If the hidden line procedure is interrupted after a predefined time interval, it can be guaranteed that visibility has been completed determined for at least subset of image features onto which subsequent feature extraction can focus.

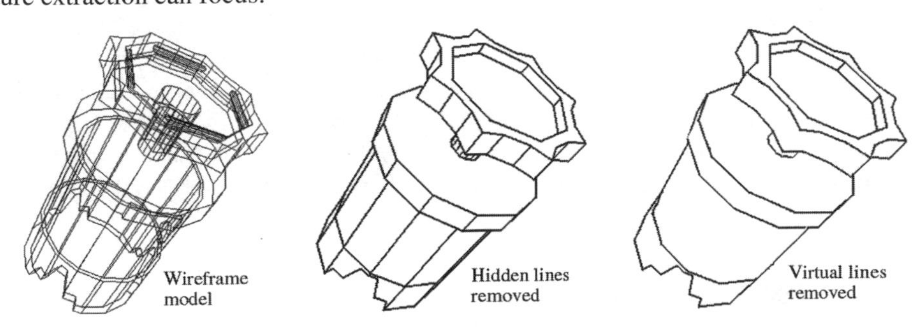

Figure 9.10 If curved structures that are approximated by polyhedral patches are explicitly modeled, virtual edge segments can be easily eliminated.

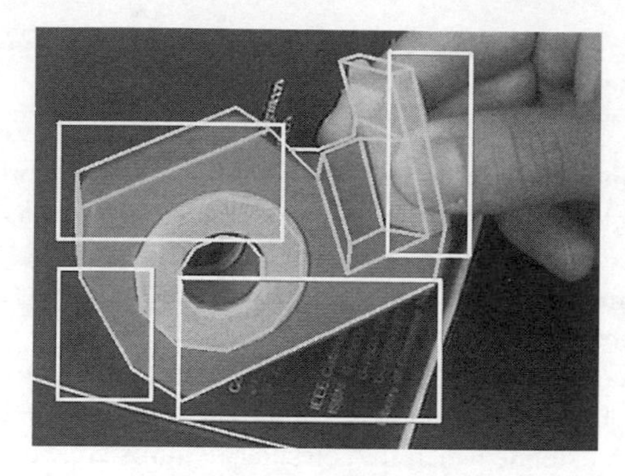
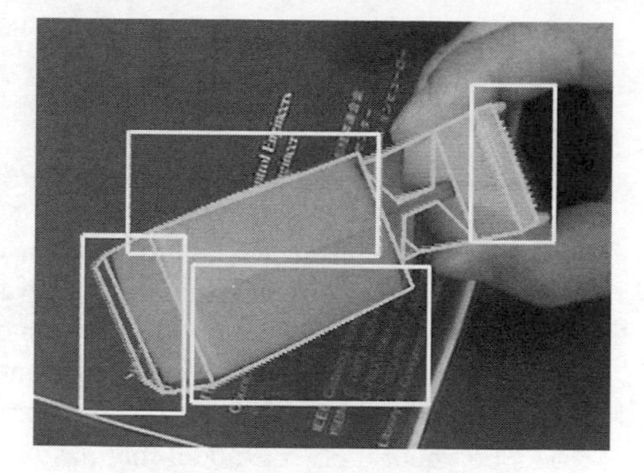

Figure 9.11 Real-time model-based tracking with dynamic handling of occlusion. Online visibility determination enables us to automatically select new features when tracked features become occluded. In addition, due to robust feature extraction by the HT minor unmodeled occlusions, such as by the author's hand, do not affect stability.

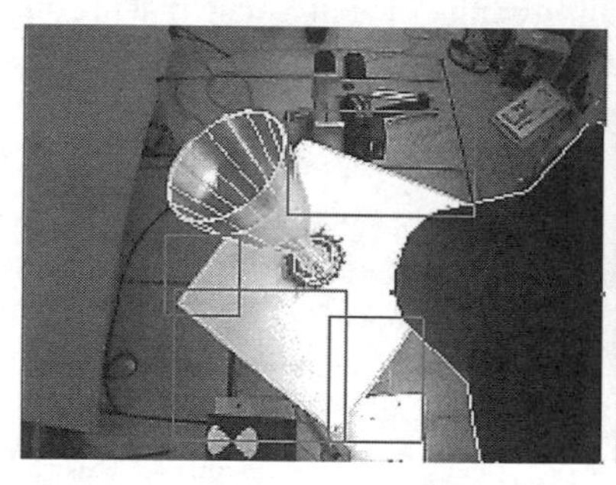
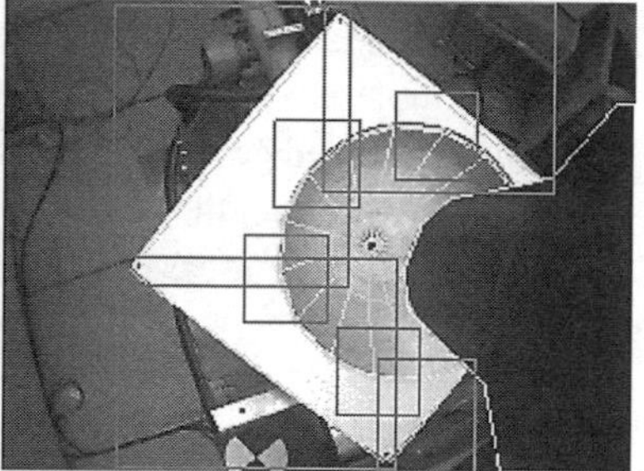

Figure 9.12 Experimental servicing satellite. Visual tracking of edges and circular features despite mutual occlusion.

9.3.1.2 Results. The hidden line algorithm described above along with an efficient pose estimation procedure have been integrated into DLR's model-based 3D tracking system, that has been applied in a variety of sensor-controlled robotics projects. Pose estimation and visibility determination are executed on a standard workstation, while HT-based feature extraction run on a Datacube image processing system. Up to 60 kbyte of image pixels distributed among up to five image features along with full visibility determination and 3D pose estimation may be processed at frame rate (25 Hz). Figure 9.11 shows a typical example for tracking across distinct aspects. The wireframe model of the target object has been projected into the scene according to the currently estimated pose. The bright rectangles highlight the regions of interest where the image processing subsystem is set up to detect object features.

Figure 9.12 is taken from space robotics project "experimental servicing satellite" mentioned in the introduction. Here, in addition to aspect changes, we also need to consider mutual occlusion of distinct objects, such as the robot gripper and satellite model. Also note that our extended polyhedral object model allows us to exactly incorporate the circular rim of the apogee motor into the tracking process.

9.4 MULTISENSORY SERVOING

Servoing by means of multisensory information potentially can solve problems which cannot be solved by the use of only one sensor type. In particular, redundant information may be used to increase the performance of the servoing system as well as the robustness against failing sensors, which include, for instance, the failure to detect expected features in camera images.

A typical configuration, where one sensor type completes the information of the other one, are cameras and laser rangefinders.

Unlike the method described in Section 9.3 which requires camera calibration and hand-eye calibration, we will discuss here an approach which involves no such calibrations, but that is based on *learning*.

We have investigated two different multisensory servoing methods. One is based on classical *estimation theory* [27], the other one on a *neural net* approach [28]. In both cases the control law may be written as

$$v_c = -\alpha \mathbf{C}(m - m^*)$$

where $(m - m^*) \in \mathbf{R}^N$ is the vector-valued deviation between the actual and the nominal sensory pattern indicating the displacement of the actual robot pose $x \in \mathbf{R}^M$ from the nominal pose $x^* \in \mathbb{R}^M$, $v_c \in \mathbf{R}^M$ is the velocity command, α represents a scalar dynamic expression, at least a real constant, determining the closed loop dynamics, and $\mathbf{C}$ represents a projection operator used for mapping the sensor space onto the control space.

Both approaches differ in the type of the projection operator, but they are similar in the way how to get the projection operator. They make use of a learning-by-showing approach having the advantage that no analytical model is needed, neither of the robot kinematics nor of the sensors. But, on the other hand, if one has an simulation environment (as we have it) the learning may easily be performed by simulation instead of in the real environment. We have tested both methods with a setup where two markers on an objects surface have been tracked by two uncalibrated robot-hand-cameras, and four laser range finders. In doing so we got 12 sensor values at each cycle, eight from the camera-system (the two image coordinates of the centers of gravity of two markers in two camera images) and four range values. Using this values, or subsets of them, a robot was servoed in up to six degrees of freedom. Figure 9.13 shows our multisensory robot gripper having integrated the 2 CCD-cameras which have been used (left), 9 laser range finders (1 long-range forward-looking, 4 short-range forward-looking which have been used, 4 inside-looking), 2 tactile arrays, 2 build-in force-torque-sensors, 1 laser scanner (right).

9.4.1 Linear Estimation Approach

A generalized projection operator

$$\mathbf{C} = (\mathbf{J}^T \mathbf{W}_m \mathbf{J} + \mathbf{W}_v^{-1})^{-1} \mathbf{J}^T \mathbf{W}_m, \quad \mathbf{C} \in \mathbf{R}^{M \times N} \tag{9.10}$$

known from linear estimation theory, was used. Here $\mathbf{J} \in \mathbf{R}^{N \times M}$ is the sensor-to-control Jacobian, $\mathbf{W}_m$ the input weight matrix, and $\mathbf{W}_v$ the output weight matrix. We propose to find the elements of the Jacobian experimentally by moving the robot along and around all the degrees of freedom near the nominal pose in order to approximate the partial derivatives by the difference quotients

$$j_{i,j} = \left.\frac{\Delta m_i}{\Delta x_j}\right|_{x^*} \approx \left.\frac{\partial m_i}{\partial x_j}\right|_{x^*}$$

This process may be interpreted as *learning by showing*. In this way the burden of deriving partial derivatives out of complex equations is avoided. The potential of the weight matrices may be shown at simplified cases. Suppose that $\mathbf{W}_m$ and $\mathbf{W}_v$ are originally unit matrices. Then failing sensors can easily be taken into account by setting their corresponding diagonal elements of the input weight matrix to zero. Generally this will reduce the condition number of $(\mathbf{J}^T \mathbf{W}_m \mathbf{J})$. If this matrix is ill conditioned or has a rank deficit or if the control shall be

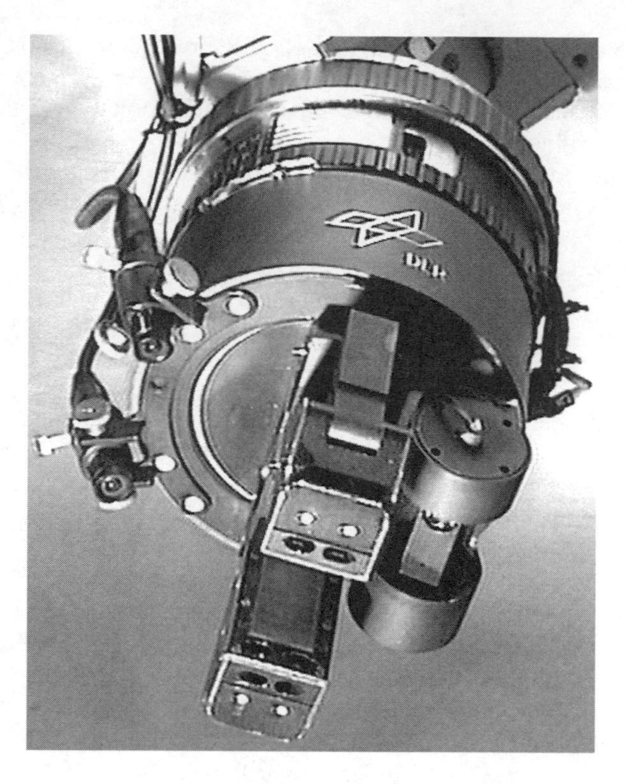

Figure 9.13 Multisensory robot gripper.

restricted to a motion subspace the output weight matrix may be used to improve the condition or to select a preferred control subspace. If we rewrite (9.10) [29]

$$\mathbf{C} = \mathbf{W}_v \mathbf{J}^T (\mathbf{J} \mathbf{W}_v \mathbf{J}^T + \mathbf{W}_m^{-1}) \tag{9.11}$$

then it can be seen that motion degrees of freedom can be taken away by setting the corresponding diagonal elements of the output weight matrix to zero. With this unified approach for sensordata-fusion lots of different situations can be treated on the basis of only one Jacobian without any need for relearning.

9.4.2 Neural Network Approach

The neural network approach learns the direct mapping from the sensory pattern to the motion commands from a set of examples [28, 30]. Unlike the linear method above, here the mapping function $\mathbf{C}$ is position dependent, that is, it varies from position to position. The method works by recalling from and generalization of the learned examples. Unlike previous use of a neural network for learning visual servoing, our method is more flexiable. Usually, learning and recall can be only carried out with respect to one predefined nominal position. Here we are able to modify the trained network such that other desired homing positions can be achieved without the need for retraining. The idea underlying these extensions is best illustrated as follows.

Suppose $\mathbf{C}()$ is the exact mapping, $\mathbf{C}'()$ is the learned mapping. It was proven [30] that in terms of a recursive motion, the robot motion recalled by using the learned mapping $\mathbf{C}'()$ converges to a position having sensory pattern m'^*, instead of the desired reference pattern $m*$. Suppose we would like to move the robot to a position having sensory pattern m_g by using the learned mapping $\mathbf{C}'()$. Now we can modify the learned network according to

$$v_c = \mathbf{C}''(m) = \mathbf{C}'(m) - \mathbf{C}'(m_g). \tag{9.12}$$

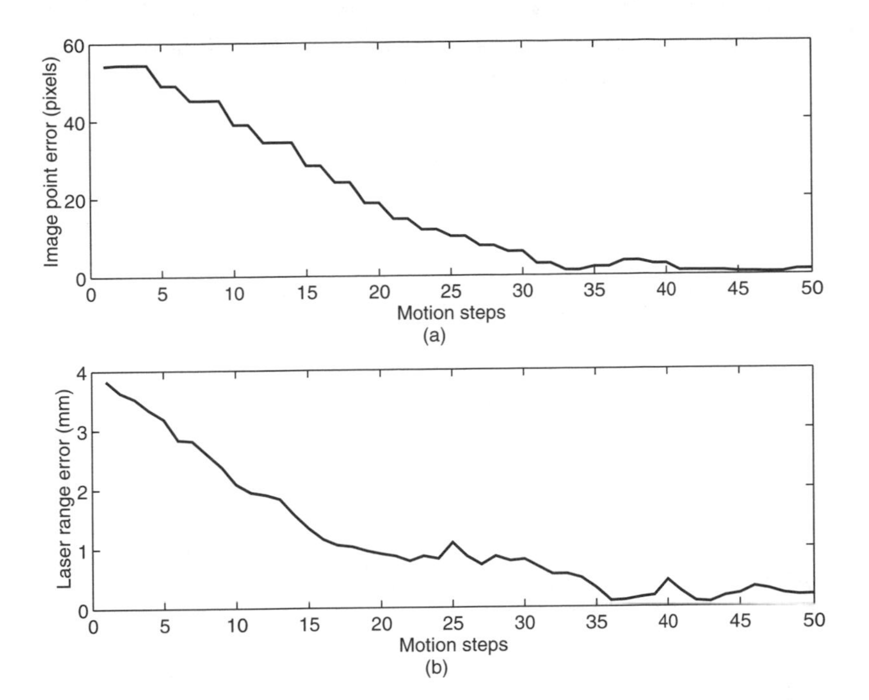

Figure 9.14 The difference of the sensory data from that at the homing position during motion. (a) The image feature coordinate error. (b) The laser range error.

With the modified mapping as (9.12), we have the position converging now at the new reference position $m = m_g$ since $\mathbf{C}''(m_g) = \mathbf{0}$. A special case of the above modification is $m_g = m^*$, which means that the desired reference position is the one used in training.

Figure 9.14 shows the errors of the sensory data (image coordinates and laser ranges) as functions of the step of motion in the homing procedure from a starting position of a robot hand.

9.5 CONCLUSION

In this chapter we have discussed a variety of image processing techniques that can substantially improve the robustness of visual servoing systems and that have been successfully applied in a quite a few research projects involving sensor-controlled robotics. All techniques try to cope with two phenomena that typically cause failures of vision systems, namely occlusion and unfavorable illumination. The methods proposed, although well known, have so far been rarely used in visual servoing systems.

REFERENCES

[1] P. I. Corke. "Visual control of robot manipulators" - A review. In K. Hashimoto, editor, *Visual Servoing*, pp. 1–31. World Scientific, 1993.

[2] P. Wunsch. *Modellbasierte 3-D Objektlageschätzung für visuell geregelte Greifvorgänge in der Robotik*. Shaker-Verlag, 1998. Dissertation. University of Technology, Munich, 1997.

[3] G. D. Hager, S. Hutchinson, and P. I. Corke. "A tutorial on visual servo control." *IEEE Trans. Robotics and Automation*, vol. 12, no. 5, pp. 651–670, 1996.

[4] P. I. Corke. "Dynamic effects in visual closed-loop systems." *IEEE Trans. Robotics and Automation*, vol. 12, no. 5, pp. 671–683, 1996.

[5] P. van der Smagt and B. J. A. Kröse. "A real-time learning neural robot controller." In *Proceedings of the 1991 International Conference on Artificial Neural Networks*, pp. 351–356, 1991.

[6] B. Brunner, K. Arbter, and G. Hirzinger. "Graphical robot simulation within the framework of an intelligent telesensorprogramming system." In W. Straßer and F. M. Wahl (eds.), *Graphics and Robotics*. Springer-Verlag, Berlin, Heidelberg, 1995.

[7] B. J. Neslon and P. K. Khosla. "Task oriented model-driven visually servoed agents." In O. Khatib and K. Salisbury (eds.), *Experimental Robotics IV. Proc of the 4th International Symposium 1995*, pp. 80–85, 1996.

[8] K. Toyama, G. D. Hager, and J. Wang. "Servomatic: A modular system for robust positioning using stereo visual servoing." In *Proceedings IEEE International Conference on Robotics and Automation*, pp. 2626–2642, Minneapolis, MN, April 1996.

[9] P. V. C. Hough. "A method and means for recognizing complex patterns." U.S. Patent 3,069,654, 1962.

[10] R. O. Duda and P. E. Hart. "Use of the Hough transformation to detect lines and curves in pictures." *Comm. Assoc. Comput. Mach.*, vol. 15, no. 1, pp. 11–15, 1972.

[11] D. H. Ballard. "Generalizing the Hough transform to detect arbitrary shapes." *Pattern Recognition*, vol. 13, no. 2, pp. 111–122, 1981.

[12] R. K. K. Yip, P. K. S. Tam, and D. N. K. Leung. "Modification of Hough transform for circles and ellipses detection using a 2-dimensional array." *Pattern Recognition*, vol. 25, pp. 1007–1022, 1992.

[13] H. K. Yuen, J. Illingworth, and J. Kittler. "Detecting partially occluded ellipses using the Hough transform." *Image and Vision Computing*, vol. 7, pp. 31–37, 1989.

[14] D. H. Ballard and C. M. Brown. *Computer Vision*. Englewood Cliffs, NJ, Prentice-Hall, 1982.

[15] G. Hirzinger, K. Landzettel, B. Brunner, and B. M. Steinmetz. "Steuerungskonzepte für internes und externes Roboter-Servicing in der Raumfahrt." In *Tagungsband Deutscher Luft- und Raumfahrtkongreß*, Dresden, 1996.

[16] G.-Q. Wei, K. Arbter, and G. Hirzinger. "A real-time visual servoing system for laparoscopic surgery." *IEEE Engineering in Medicine and Biology*, vol. 16, no. 1, pp. 40–45, 1997.

[17] C. Fagerer, D. Dickmanns, and E. D. Dickmanns. "Visual grasping with long delay time of a free floating object in orbit." *Autonomous Robots*, vol. 1, no. 1, pp. 53–68, 1994.

[18] V. Gengenbach. *Einsatz von Rückkopplungen in der Bildauswertung bei einem Hand-Auge-System zur automatischen Demontage*. Dissertationen zur Künstlichen Intelligenz DISKI-72. Infix-Verlag, 1994.

[19] S. Ravela, B. Draper, J. Lim, and R. Weiss. "Adaptive tracking and model registration across distinct aspects." In *Proceedings International Conference on Intelligent Robots and Systems*, pp. 174–180, Pittsburgh, PA, 1995.

[20] K. Stark and S. Fuchs. "A method for tracking the pose of known 3-D objects based on an active contour model." In *Proceedings International Conference on Pattern Recognition*, pp. 905–909, 1996.

[21] P. Wunsch and G. Hirzinger. "Real-time visual tracking of 3-D objects with dynamic handling of occlusion." In *Proceedings IEEE International Conference on Robotics and Automation*, Albuquerque, NM, 1997.

[22] R. Kumar and A. R. Hanson. "Robust methods for estimating pose and a sensitivity analysis." *Computer Vision, Graphics and Image Processing*, vol. 60, no. 3, pp. 313–342, November 1994.

[23] D. G. Lowe. "Robust model-based motion tracking through the integration of search and estimation." *International Journal of Computer Vision*, vol. 8, no. 2, pp. 441–450, 1992.

[24] J. D. Foley, A. van Dam, S. K. Feiner, and J. F. Hughes. *Computer Graphics: Principles and Practice*. New York, Addison-Wesley, 2nd edition, 1990.

[25] T. Peter. "Effizientes Eliminieren verdeckter Kanten in polyedrischen Szenen zur Simulation von Kamerabildern." Technical Report IB-515-96-5, Institut für Robotik und Systemdynamik. Deutsche Forschungsanstalt für Luft- und Raumfahrt, 1996. Diplomarbeit. Fachbereich Mathematik, Fachhochschule Regensburg.

[26] R. A. Schumacker, B. Brand, M. Gilliland, and W. Sharp. "Study for applying computer-generated images to visual simulation." Technical Report AFHRL-TR-69-14, U.S. Air Force Human Resources Laboratory, 1969.

[27] B. Brunner, K. Arbter, and G. Hirzinger. "Task-directed programming of sensor-based robots." In *Proceedings International Conference on Intelligent Robots and Systems*, pp. 1080–1087, 1994.

[28] G.-Q. Wei and G. Hirzinger. "Learning motion from images." In *Proceedings the 11th International Conference Pattern Recognition*, vol. I, pp. 189–192, 1992.

[29] D. G. Luenberger. *Optimization by Vector Space Methods*. New York, John Wiley & Sons, Inc., 1969.

[30] G.-Q. Wei, G. Hirzinger, and B. Brunner. "Sensorimotion coordination and sensor fusion by neural networks." In *IEEE International Conference on Neural Networks*, pp. 150–155, San Francisco, CA, 1993.

Chapter 10

GLOBAL SIGNATURES FOR ROBOT CONTROL AND RECONSTRUCTION

R. A. Hicks, D. J. Pettey, K. S. Daniilidis, and R. Bajcsy
GRASP Laboratory, University of Pennsylvania

Abstract

We address the problem of control-based recovery of robot pose and the environmental layout. Panoramic sensors provide us with a 1D projection of characteristic features of a 2D operation map. Trajectories of these projections contain the information about the position of a priori unknown landmarks in the environment. We introduce here the notion of spatiotemporal signatures of projection trajectories. These signatures are global measures, characterized by considerably higher robustness with respect to noise and outliers than the commonly applied point correspondence. By modeling the 2D motion plane as the complex plane, we show that by means of complex analysis the reconstruction problem is reduced to a system of two quadratic equations in two variables.

10.1 INTRODUCTION

Consider the following situation. We are given three landmarks in the plane, which we represent as the complex numbers z_1, z_2, and z_3, and a sensor that can measure the angle between any pair of these points, from any position in the plane at which it is placed. Let θ denote the angle between z_1 and z_2 and let ϕ denote the angle between z_2 and z_3. If the sensor undergoes a circular motion (the left of Fig. 10.1) we may record the angles and then plot ϕ versus θ. For example, in the right of Fig. 10.1 we see such a graph for the points $z_1 = 2i$, $z_2 = 2 + 2i$ and $z_3 = 5$.

The primary motivation for this chapter is to investigate the extent to which the curve displayed in the right of Fig. 10.1 characterizes the three points. In other words, we want to solve the inverse problem: given the data depicted in the right of Fig. 10.1, how can the scene be reconstructed? (Generally, a Cartesian frame is not available, so what we seek are the magnitudes of z_1, z_2, and z_3, and the angles between them with respect to the origin, which we take to be the center of the circular motion.) Note that while we require the sensor motion to be constrained to a circle, there is no reference made to where on the circle the sensor is at any time. Rather, we take as our hypothesis that we are densely sampling the angles, although not necessarily at a uniform rate.

10.1.1 Problem Statement

We begin with the general form of our problem. Suppose that $z_1, \ldots, z_n$ are complex numbers representing n landmarks in the plane, and a sensor is available with the ability to measure the angles between the landmarks with respect to any given position of the sensor, w (i.e., it can measure the angle $\angle z_k w z_l,\ k, l = 1, \ldots, n$). Then for m positions $w_1, \ldots, w_m$ of the sensor,[1] can one determine $z_1, \ldots, z_n$ and $w_1, \ldots, w_m$ given only the measurements $\angle z_l w_j z_k, l, k = 1, \ldots, n, j = 1, \ldots, m$?

[1] In our model of the panoramic sensor, the pose of the sensor means only its position, and not orientation. For real applications it is possible to define and compute and orientation with our method if it is needed.

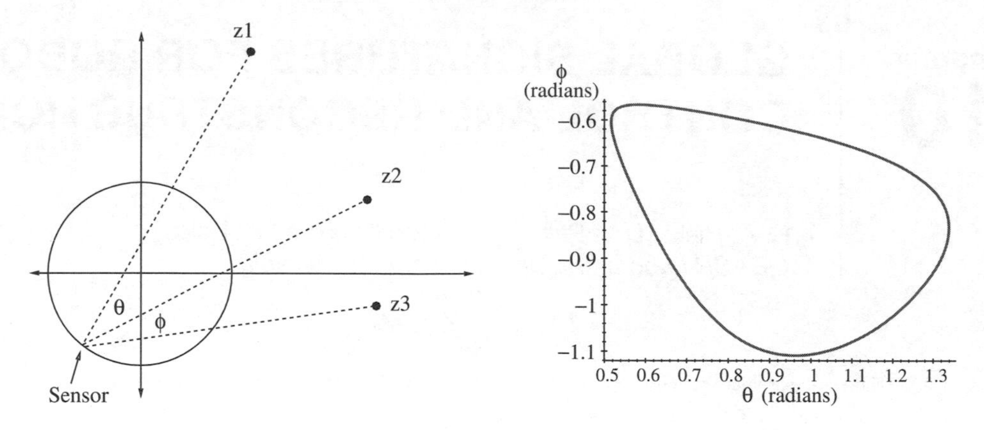

Figure 10.1 The problem: find z_1, z_2, and z_3 by moving on a circle and measuring the angles θ and ϕ during the motion. On the left we see the actual motion of the sensor, while on the right we see a plot of ϕ vs. θ for $z_1 = 2i$, $z_2 = 2 + 2i$, $z_3 = 5$.

One approach to the above is to write down all of the trigonometric equations associated with the configuration. Suppose that $\chi_{k,j}$ is the angle between the segment connecting $z_k = x_k + iy_k$ and the horizontal line through $w_j = a_j + ib_j$ (see Fig. 10.2). What may be measured by the sensor, i.e., the known values in the problem, are the differences $m_{k,j} = \chi_{k,j} - \chi_{1,j}$. From Fig 10.2 we see that we have the equation

$$\tan(\chi_{k,j}) = \frac{y_k - b_j}{x_k - a_j},$$

so that

$$m_{k,j} = \arctan\left[\frac{y_k - b_j}{x_k - a_j}\right] - \chi_{1,j}, \tag{10.1}$$

where $k = 2, \ldots n, \ , \ j = 1, \ldots, m$. The above nonlinear system consists of $2m + 3m$ unknowns and $m(n-1)$ equations. If we in addition require that the w_j all lie on a circle then we have the additional equations $a_j^2 + b_j^2 = 1, \ j = 1, \ldots, m$, making for a total of mn equations. In general the system is overconstrained, which may be helpful because in an experiment the angular measurements will of course contain errors. One may try to find an solution in the "least squares sense" by minimizing the sum of the squares of these equations. This method has numerous problems. Most prominent is that one may not find the global minimum of the function being minimized. Additionally, this approach is computationally expensive.

In this chapter we are primarily concerned with a special case of the above problem. We will assume that the first two points, z_1 and z_2, are known and that the points w_i all lie on

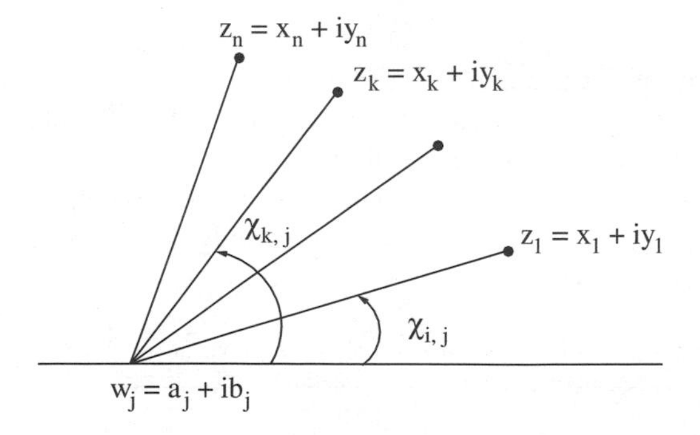

Figure 10.2 Here we have the sensor placed at the point $w_j = a_j + ib_j$ in the presence of landmarks at $z_1 = x_1 + iy_1, \ldots, z_n = x_n + iy_n$. The angle formed between the segment connecting the landmark at z_k and the sensor at w_j, and the horizontal line through w_j is denoted at $\chi_{k,j}$. **This is not an angle that our sensor can measure.** Rather, the sensor provides the angles between the landmarks, since it does not have a means of detecting an absolute direction, i.e., it does not have a compass. Thus it measures the angles $\chi_{k,j} - \chi_{1,j}$.

a circle, i.e., $|w_i| = 1, i = 1, \ldots, m$. Additionally, we will assume that the w_i are densely distributed in the circle (but not necessarily in a uniform fashion). Notice that if the problem can be solved for $n = 3$ with z_1, z_2 known, then it can be solved for any $n > 3$, as long as z_1, z_2 are known. To do this, use the method used to find z_3 given z_1 and z_2, but instead using any z_j in place of z_3. From an experimental point of view, this means we may reconstruct an entire scene if we lay down two known reference points.

10.1.2 Signatures

In the general problem stated above, we are given measurements $m_{2,1}, \ldots, m_{n,m}$ and we want to solve a system of equations

$$F_1(m_{2,1}, \ldots, m_{n,m}, z_1, \ldots, z_n, w_1, \ldots, w_n, \chi_{1,1}, \ldots, \chi_{1,m}) = 0$$

$$\vdots$$

$$F_m(m_{2,1}, \ldots, m_{n,m}, z_1, \ldots, z_n, w_1, \ldots, w_n, \chi_{1,1}, \ldots, \chi_{1,m}) = 0$$

where the unknowns are $z_1, z_2, \ldots, z_n$, and $w_1, w_2, \ldots, w_n$.

The method we describe below allows for the equations to be decoupled, (unlike the above system), resulting in a single equation of the form

$$G(\mathbf{f}(\mathbf{m}), z_1, z_2, z_3) = 0$$

where $\mathbf{m} = (m_{2,1}, \ldots, m_{2,m}, m_{3,1}, \ldots, m_{3,m})$ and $\mathbf{f}$ is a smooth complex-valued function defined on the space that $\mathbf{c}$ lives in. We will take $\mathbf{f}$ to be the composition of an integral operator with an analytic function that acts pointwise on m. We say that $\mathbf{f}(\mathbf{m})$ is a *signature* of (z_1, z_2, z_3). The fact that $\mathbf{f}$ is smooth makes the system robust to errors in measurement, while taking advantage of global information that may be embedded in the measured data.

10.1.3 Panoramic Sensors

Recently, many researchers in the robotics and vision community have begun to investigate the use of curved mirrors to obtain panoramic and omnidirectional views. Typically, such systems consist of a standard CCD camera pointing upward at a convex mirror, as in Fig. 10.3. How to interpret and make use of the information obtained by such sensors, e.g., how it can be used to control robots, is not immediately clear. It does seem likely that panoramic systems may be able to handle some problems that are difficult or impossible to address with a standard camera.

A panoramic sensor provides a natural means of addressing the above problem experimentally, since it acts very much like an overhead camera. A natural feature to consider for extraction from an image taken on such a system is any edge that is vertical in the real world, because these edges appear as radial lines in the image (this is clear in Fig. 10.3). In the center of this image the lens of the camera is clearly visible. To measure the angles between the vertical edges we extend them to the center of the lens and compute the angle at their intersection.

10.1.4 Overview of the Solution

For each point on the circle, z, we have two real numbers $\theta(z)$ and $\phi(z)$. (These are the values that are measured experimentally.) Thus we have a function from the unit circle to the plane. The area bounded by this curve depends of course on z_1, z_2, and z_3, i.e., we can consider it as a function $A(z_1, z_2, z_3)$, and in this case this number is the signature discussed above.

Figure 10.3 On the left we see one way to create a panoramic sensor: point a camera upwards at a curved mirror. On the right we see a view from the panoramic sensor demonstrating how it distorts the environment. A checkerboard pattern was placed on the floor made from 8-in. squares. Several vertical cylinders were placed outside this pattern, illustrating how vertical lines become radial in the image.

Using a panoramic sensor as described above, the curve can be sampled experimentally. Hence the signature, m, can be computed numerically, which is equal to $A(z_1, z_2, z_3)$ can be calculated. On the other hand a general formula for $A(z_1, z_2, z_3)$ can be determined using the methods of complex analysis, and so the equation $A(z_1, z_2, z_3) = m$ gives a relationship between z_1, z_2 and z_3. If z_1 and z_2 are known then the equation $A(z_1, z_2, z_3) = m$ can be solved for z_3. Because the angles are integrated to calculate m, the value of z_3 obtained from solving the equation is very insensitive to errors in the original measurements.

We can make a slight modification to the above procedure that simplifies the equations greatly. The integral that represents the average of the angles is not a simple one. A much better quantity (for reasons discussed in Subsection 10.3.2) to consider is the average of certain complex exponentials of the angles. This poses no experimental difficulty—one simply takes the measured angles and applies the appropriate transformation to them before averaging.

10.1.5 Related Work

There are several works from the eighties to early nineties which perform reconstruction using vertical edges and known motion (Kriegman [1] and Kak [2]). There is work in structure from motion from circular trajectories (Shariat and Price [3], Sawhney [4]). This work, however, uses a set of equations with the constraint that the projection centers are on a circle.

Recently, a number of approaches to navigation and reconstruction using omnidirectional systems have been proposed (Nayar [5], Svoboda [6], Onoe [7], Srinivasan [8], Yagi [9], Medioni [10]). The work by Yagi and Medioni is very similar but uses an already known environmental map. The work most relevant to this chapter is that done on omnidirectional multibaseline stereo done by Kang and Szeliski [11], but this work uses conventional cameras and measurement equations.

10.1.6 Contributions

Our reconstruction method has a number of unusual features that the standard methods do not.

1. Rather than constructing one or more equations for each measurement, we first process the measured data to produce a single number, which is then used to create exactly

two equations in the two real variables representing the coordinates of the unknown landmark. **These equations are quadratic** [see (10.8)].

2. Due to the simplicity of the equations, optimization methods are not needed, and when the solutions are found they are known to be true solutions, as opposed to possible spurious solutions produced by numerical optimization methods.
3. The positions of the camera never appear as variables, although they can be reconstructed after the position of the unknown landmark is determined.
4. Due to the stable means of processing the data (integration), our method handles noise well. In addition, if one desired to carry out a detailed error analysis, it is possible since the equations are exactly solvable (this is the topic of ongoing work). This is not the case with a method that produces a very large number of nonlinear equations.

10.2 APPLICATIONS TO ROBOTICS

The motivation for the above is the problem of having a robot enter an unknown environment and use unknown landmarks to estimate its pose.

Consider the a 2D situation where a robot can detect and track three fixed points in the plane and measure the angles θ and ϕ, between them, as in Fig. 10.4.

In the formal sense, θ and ϕ define a local coordinate system on the plane, and so even if the robot does not have a means of measuring its Cartesian coordinates with respect to some frame, it can measure its (θ, ϕ) coordinates.

To estimate the position of the robot one must convert from angular coordinates to Cartesian coordinates. (Of course, to do this one must know the positions of the landmarks.) This is an old problem, familiar to sailors and surveyors. A method used by surveyors can be found in [12]. One approach is simply to write down the equation of each of the three circles that contain the robot and two of the landmarks and the position of the robot is where they all intersect. Despite the fact that this method does not provide an explicit formula for converting from angular to Cartesian coordinates, it is effective from a numerical viewpoint. This problem and the effects of noise on it are considered in [13]. A fast linear algorithm is given in [14]. In [15] this method of robot pose estimation is studied experimentally using a panoramic sensor constructed from a conical mirror.

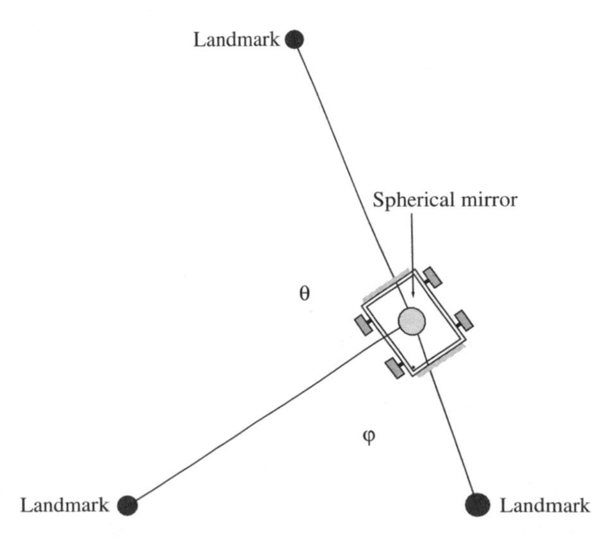

Figure 10.4 The angles between three fixed landmarks provide a natural coordinate system.

Thus, if a robot is able to calculate the position of three landmarks, it can use this information for future pose estimation. In the next section we demonstrate our method for finding the positions of unknown landmarks given that two known reference points are available.

10.3 CALCULATING SIGNATURES

10.3.1 Angles and Complex Logarithms

If z is a complex number, then there are real numbers Θ and r such that $z = re^{i\Theta}$. Even if $r > 0$, the number Θ is not uniquely determined—adding any multiple of 2π to Θ will give the same value of z. For a given z the set of all such angles is called the **argument** of z and denoted $arg(z)$. Note that arg is not a real valued function in the usual sense, but up to an additive factor of 2π, "the" argument of a complex number is well defined.[2] If $z > 0$ we define the logarithm of z by

$$\log(z) = \ln(r) + i\arg(z)$$

where ln is the real natural logarithm. Therefore, $\Theta = \Im(\log(z))$ and since $\Im(w) = \frac{w-\overline{w}}{2i}$ for any w, we have that $\Theta = \frac{\log(z)-\log(\overline{z})}{2i} = \frac{1}{2i}\log(\frac{z}{\overline{z}})$. This allows us to define the angle between the complex numbers z_1 and z_2 by

$$\angle(z_1, z_2) = \frac{1}{2i}\left(\log\left(\frac{z_1}{\overline{z_1}}\right) - \log\left(\frac{z_2}{\overline{z_2}}\right)\right) = \frac{1}{2i}\log\left(\frac{z_1\overline{z_2}}{\overline{z_1}z_2}\right)$$

10.3.2 Applying the Residue Theorem

We denote the unit circle in the complex plane as $S^1 = \{e^{it} | t \in [0, 2\pi]\}$. Let z_1, z_2, and z_3 be complex numbers. Then we define the angle between z_1 and z_2 with respect to $z = z(t) = e^{it}$ as

$$\theta(z) = \frac{1}{2i}\log\left[\frac{(z_1 - z(t))(\overline{z_2} - \overline{z}(t))}{(\overline{z_1} - \overline{z}(t))(z_2 - z(t))}\right] \tag{10.2}$$

and likewise we define the angle between z_2 and z_3 with respect to $z = z(t) = e^{it}$ as

$$\phi(z) = \frac{1}{2i}\log\left[\frac{(z_2 - z(t))(\overline{z_3} - \overline{z}(t))}{(\overline{z_2} - \overline{z}(t))(z_3 - z(t))}\right] \tag{10.3}$$

Notice that $\overline{z} = \frac{1}{z}$ because $z = e^{it}$, so that in fact

$$\theta(z) = \frac{1}{2i}\log\left[\frac{(z_1 - z)(\overline{z_2} - \frac{1}{z})}{(\overline{z_1} - \frac{1}{z})(z_2 - z)}\right] \tag{10.4}$$

and

$$\phi(z) = \frac{1}{2i}\log\left[\frac{(z_2 - z)(\overline{z_3} - \frac{1}{z})}{(\overline{z_2} - \frac{1}{z})(z_3 - z)}\right] \tag{10.5}$$

Considered as functions of z, (10.4) and (10.5) define functions which are analytic at all points of their domain. This is a necessary hypothesis in our next step, in which we apply the residue theorem.

---

[2]This makes for certain technical problems when computing with the logarithm, but we will for simplicity ignore them since they will have cause no inconsistencies in our calculations.

A natural quantity to compute is

$$\oint_{S^1} \theta d\phi \tag{10.6}$$

which represents the area of the curve discussed in Section 10.1.4. Unfortunately, calculating this integral is problematic due to the "branch cut" of the complex logarithm. The origin of this difficulty is due to the ambiguity in the choice of angle in the polar representation. But the ambiguity disappears of course if one applies to the logarithm an exponential function. Therefore a reasonable form to integrate would appear to be $e^{2i\theta} d\phi$. From an experimental point of view, what this means is that the angular data are gathered in the form of two lists of angles, and then one of these lists is transformed by the above function and then the two lists are used to calculate the integral. Since $d\phi = \frac{d\phi}{dz} dz$ the integral can be written as the complex contour integral

$$\oint_{S^1} e^{2i\theta} \frac{d\phi}{dz} dz \tag{10.7}$$

Clearly,

$$e^{2i\theta} = \frac{(z_1 - z)(\overline{z_2} - \frac{1}{z})}{(\overline{z_1} - \frac{1}{z})(z_2 - z)}$$

and

$$\frac{d\phi}{dz} = \frac{1}{2i} \left[\frac{(\overline{z_2} - \frac{1}{z})(z_3 - z)}{(z_2 - z)(\overline{z_3} - \frac{1}{z})} \right] \cdot \frac{d}{dz} \left[\frac{(z_2 - z)(\overline{z_3} - \frac{1}{z})}{(\overline{z_2} - \frac{1}{z})(z_3 - z)} \right]$$

and so the integrand in the above integral is a rational function in z. Although in principle it is possible to compute the residues of this by hand, for this calculation and another below we found it useful to employ the symbolic computing package Maple to find the location of the singularities of this function and compute their residues. Singularities were found to occur at $\frac{1}{\overline{z}_1}$, $\frac{1}{\overline{z}_3}$, z_2, and z_3. For completeness we now state the residue theorem.

The Residue Theorem. Let C be a positively oriented simple closed contour within and on which a function f is analytic except at a finite number of singular points $w_1, w_2, \ldots, w_n$ interior to C. If $R_1, \ldots, R_n$ denote the residues of f at those respective points then

$$\int_C f(z)dz = 2\pi i(R_1 + \cdots + R_n)$$

Thus the value of (10.7) depends on whether or not z_1, z_2 and z_3 lie inside or outside of S^1. Assuming that the three landmarks lie outside of the circle implies that the contour encloses the singularities at $\frac{1}{\overline{z}_1}$ and $\frac{1}{\overline{z}_3}$.(If z_1, z_2, or z_3 lie on S^1 then this method does not apply.) By using Maple we have that the value of (10.7) is

$$\begin{aligned}
&\pi(\overline{z}_1 z_1 z_2 \overline{z}_3 \overline{z}_2 z_3 + 2\overline{z}_3 z_3 \overline{z}_1 z_2 - \overline{z}_1^2 z_2^2 \overline{z}_3 z_3 - z_3 \overline{z}_1 \overline{z}_3 z_1 - \overline{z}_3 z_3 \overline{z}_2 z_2 + z_3 \overline{z}_2 \\
&+ \overline{z}_1^2 z_1 z_3 - z_3 \overline{z}_1 + \overline{z}_1^2 z_2^2 z_3 \overline{z}_2 - \overline{z}_1 z_2 z_3 \overline{z}_2 - \overline{z}_1^2 z_1 z_2 \overline{z}_2 z_3 - z_2 \overline{z}_3 - \overline{z}_3 z_1 \overline{z}_1 z_2 \\
&- \overline{z}_1 z_1 z_2^2 \overline{z}_3 \overline{z}_2 + z_1 \overline{z}_3 + z_2^2 \overline{z}_3 \overline{z}_2 + \overline{z}_1^2 z_2^2 \overline{z}_3 z_1 - \overline{z}_2 z_1 + \overline{z}_1 z_2 - z_2^2 \overline{z}_2 \overline{z}_1 + 2 \\
&\overline{z}_1 z_2 \overline{z}_2 z_1 - z_1 \overline{z}_1^2 z_2)/(-1 + z_3 \overline{z}_1)(-1 + \overline{z}_1 z_2)^2 (z_2 \overline{z}_3 - 1)
\end{aligned} \tag{10.8}$$

Although the above expression is a rational function of $z_1, z_2, \overline{z}_1, \overline{z}_2$, and $\overline{z}_3$, when we equate it to the experimentally measured value and take it's real and imaginary parts, equations for two conic sections result. This is clear from examining the denominator and seeing that the unknown, z_3, occurs only with constant coefficients, or possibly a coefficient of $\overline{z}_3$.

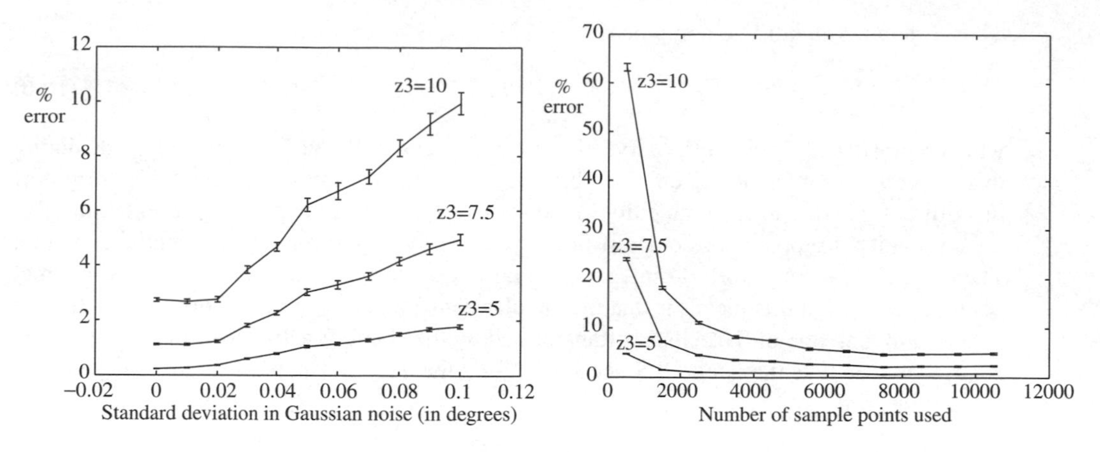

Figure 10.5 Simulation results giving the percent error or reconstruction versus the standard deviation of the noise (left) and versus the number of sample points (right).

10.4 SIMULATION RESULTS

The main question to be addressed is the effect of noise on the above method. Thus we need to know what the nature of the noise that we expect from a panoramic sensor when measuring angles is. From experiment we have found the noise in the device when measuring angles to have a standard deviation of about 0.05° at a range of 1 meter.

There are many parameters that can be varied in simulation, such as the magnitude of the point being reconstructed, the number of data points, the numerical integration scheme, and so on. Therefore, we first fixed the two known points to be $z_1 = 2i$ and $z_2 = 2 + 2i$, fixed the number of measurements taken on the circle to be 5000, and considered the effect of zero mean Gaussian noise with σ ranging from 0 to 0.1°, and chose $z_3 = 5, 7.5$, and 10. (Note that our circle determines the scale—it has radius 1.) Given all of the above parameters 100 trials were performed and the error in each case was averaged and a percent error computed. We measured error by computing the distance in the complex plane between the estimate and the true value, and then dividing by the magnitude of the true value. For each choice of z_3 this yields a plot of percent error versus the standard deviation in the noise, which can be seen in the left plot of Fig. 10.5.

The plot in the right of Fig. 10.5 illustrates the percent error in reconstruction versus the number of sample points used, i.e., the number of angular pairs measured. In this case we fixed $\sigma = 0.05$ considered three separate choices for z_3 to illustrate the effect of range. We see that in this plot there is rapid convergence to a constant percent error in each case.

10.5 CONCLUSION

In this chapter we have introduced a new method for recovering the environmental lay-out from a monocular image sequence. To do this we have introduced certain spatiotemporal signatures that give rise to quadratic equations for the solution of the problem. These signatures are based on the principle of reconstruction from controlled motion where the precise motion is not known, but the shape of the motion is known. The method we propose is robust to noise and outliers and while taking into account global information.

REFERENCES

[1] D. Kriegman, E. Triendl, and T. Binford, "Stereo vision and navigation in buildings for mobile robots," *Transactions on Robotics and Automation*, vol. 5, pp. 792–804, 1989.

[2] A. Kosaka and A. Kak, "Fast vision-guided mobile robot navigation using model-based reasoning and prediciton of uncertainties," *Computer Vision Image Understanding*, vol. 56, pp. 271–329, 1992.

[3] H. Shariat and K. Price, "Motion estimation with more then two frames," *IEEE Transactions on Pattern Analysis and Machine Intelligence*, vol. 12, pp. 417–434, 1990.

[4] H. Sawhney, J. Oliensis, and A. Hanson, "Description and reconstruction from image trajectories of rotational motion," in *International Conference Computer Vision*, pp. 494–498, 1990.

[5] S. Nayar, "Catadioptric omnidirectional camera," in *Proceedings of Computer Vision Pattern Recognition*, pp. 482–488, 1997.

[6] T. Svoboda, T. Padjla, and V. Hlavac, "Epipolar geometry for panoramic cameras," in *Proceedings of European Conference on Computer Vision*, 1998.

[7] Y. Onoe, H. Yokoya, and K. Yamazawa, "Visual surveillance and monitoring system using an omnidrectional system," in *Proceedings of International Conference on Pattern Recognition*, 1998.

[8] J. Chahl and M. Srinivasan, "Range estimation with a panoramic sensor," *Journal Optical Society America A*, vol. 14, pp. 2144–2152, 1997.

[9] Y. Yagi, S. Nishizawa, and S. Tsuji, "Map-based navigation for a mobile robot with omnidirectional image senso," *Transactions on Robotics and Automation*, vol. 11, pp. 634–648, 1995.

[10] F. Stein and G. Medioni, "Map-based localization using the panoramic horizon," *Transactions on Robotics and Automation*, vol. 11, pp. 892–896, 1995.

[11] S. Kang and R. Szeliski, "3-d scene data recovery using omnidirectional multibaseline stereo," *International Journal of Computer Vision*, vol. 25, pp. 167–183., 1997.

[12] R. Davis and F. Foote, *Surveying: Theory and Practice*. New York, McGraw-Hill, 1953.

[13] M. Betke and L. Gurvits, "Mobile robot localization using landmarks," *IEEE Transactions on Robotics and Automation*, vol. 13, pp. 251–263., 1997.

[14] U. Hanebeck and G. Schmidt, "Absolute localization of a fast mobile robot based on an angle measurement technique," in *Proceedings of Service Robots*, pp. 164–166, 1995.

[15] C. Pegard and E. Mouaddib, "A mobile robot using a panoramic view," in *IEEE Conference on Robotics and Automation*, pp. 89–94, 1996.

Chapter 11

USING FOVEATED VISION FOR ROBUST OBJECT TRACKING: THREE-DIMENSIONAL HOROPTER ANALYSIS

Naoki Oshiro
University of the Ryukyus

Atsushi Nishikawa
Osaka University

Fumio Miyazaki
Osaka University

Abstract

Foveated vision system has a multiresolution structure in which the resolution is high in the central area, and this resolution continuously drops toward the periphery. In this chapter it is shown that using this structural characteristics makes the stereo matching performance better than the conventional methods based on the use of uniform spatial scales. A method for "robust" object tracking using foveated stereo vision will be also described. We analyze the horopter shape for foveated stereo vision in detail.

11.1 INTRODUCTION

A human possesses foveated vision and the retina has an invariant structure in which the resolution is high in the central area and it drops toward the periphery. It is well-known that Log Polar Mapping (LPM) is a good approximation of the mapping between the retina and the visual cortex. In our previous work [1], we have developed a binocular tracking system which uses Log Polar Mapped stereo images for the gaze holding of a moving object in the cluttered scene, without a priori knowledge about the target object. Our tracking scheme is based on the following observation as was first made by Coombs et al. [2]: an object lying in the horopter can be easily picked up from the background only by extracting image features of zero disparity from the stereo image. Through various experiments and simple mathematical analysis, it was verified that the foveated vision-based method can track a target object in complex environments more robustly than the conventional methods using uniform sampling images (about details, see [1]). In this chapter, the "robustness" of object tracking using foveated stereo vision is also discussed through more detailed numerical analysis of the 3D horopter shape.

11.2 PRELIMINARIES

First of all, we explain the horopter geometry and make some definition.

11.2.1 Horopter

When a binocular camera fixates on a 3D point, we can consider a circle passing through the two nodal points of the camera and the fixation point. This circle corresponds to the set of points whose binocular disparity is zero and is in general referred to as the *(geometric) horopter* (see Fig. 11.1).

11.2.2 Virtual Horopter

The horopter generated by shifting the left (or right) image horizontally by some pixels is referred to as the *virtual horopter*. In Rougeaux et al. [4], it was pointed out that, near the image center, small shifts of the left image are almost equivalent to small virtual rotations of the left camera. For convenience, we refer to a pair of the right image and the shifted left image as the *virtual stereo image*. Furthermore, to avoid the confusion, we sometimes call the geometric horopter (described in Subsection 11.2.1) the *true horopter*.

11.2.3 Pixel Horopter

By the above geometric definition, the horopter corresponds to the set of intersection points of the paired lines of sight passing through the left and right image point with identical screen coordinates. In practice, however, the horopter should be plotted and analyzed by taking the width of a pixel into account as shown in Fig. 11.2. In this chapter, the horopter generated

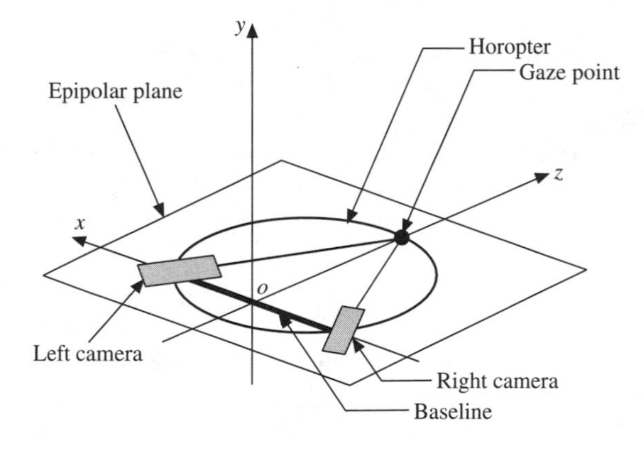

Figure 11.1 Geometric horopter.

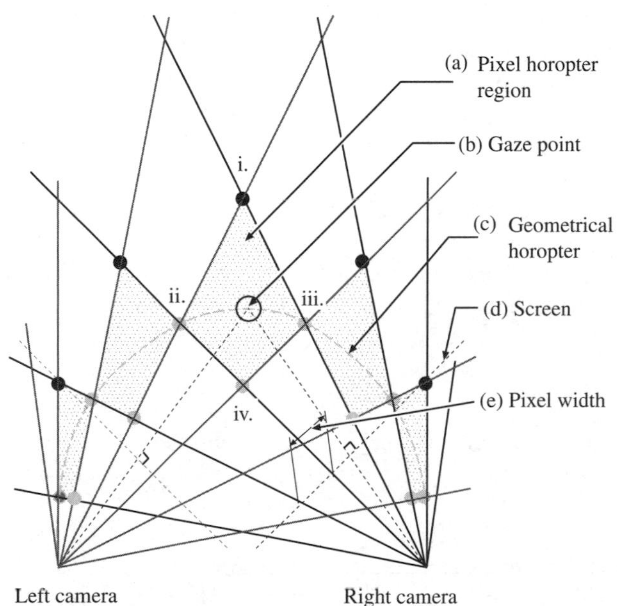

Figure 11.2 Concept of pixel horopter.

by giving consideration to that point is referred to as the *(true) pixel horopter*. Also, the pixel horopter generated from the virtual stereo image is referred to as the *virtual pixel horopter*.

A target object lying in the true or virtual pixel horopters can be easily picked up from the background only by extracting image features of zero disparity from stereo images and virtual stereo images, respectively. Using this horopter characteristics makes it possible to "robustly" track a moving object in cluttered environments, without a priori knowledge about the target object. In this case, the tracking performance (especially, target extraction performance) depends on the shape of pixel horopters, that is, the width of the image pixels extracted from stereo images.

In the following section, the 3D shape of the pixel horopter is numerically analyzed to examine the influence of the width of pixels on the target extraction performance.

11.3 HOROPTER ANALYSIS

11.3.1 Method

The method for drawing a true/virtual pixel horopter is summarized as follows. Let 3D position be described through the positive orthogonal system of coordinates $\Sigma_U : o - xyz$ with the origin at the middle point of the line connecting two image center points of the left and right cameras, as shown in Fig. 11.3.

For each grid point in space $(x, y, z)_U$,

1. The grid point is expressed in terms of the left camera coordinate system Σ_l and the right camera coordinate system Σ_r respectively (see Fig. 11.3),
$$(x, y, z)_U \mapsto (x, y, z)_l,$$
$$(x, y, z)_U \mapsto (x, y, z)_r. \tag{11.1}$$
2. The projection onto the left/right screen is calculated (see Fig. 11.3),
$$(x, y, z)_r \mapsto (i, j)_r$$
$$(x, y, z)_l \mapsto (i, j)_l \tag{11.2}$$
Below we denote the resolution of the original image screen by $i_{\max} \times j_{\max}$(pixel).
3. In order to generate *virtual stereo image*, coordinates of the left screen are shifted horizontally. Denoting the shift value by Δ_i (pixel), we have
$$(i, j)_{l'} = (i - \Delta_i, j)_l$$
Notice that we should set $\Delta_i = 0$ in case of drawing the "true" horopter.

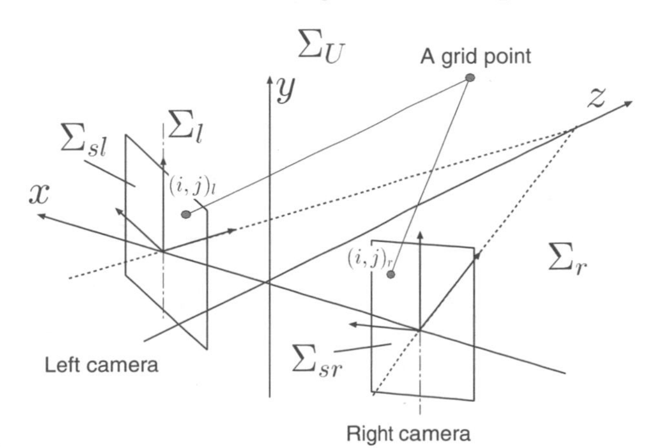

Figure 11.3 Definition of coordinate system.

4. A transformation $\mathcal{F}$ is applied to the stereo pair $\{(i, j)_r, (i, j)_{l'}\}$. That is,

$$\begin{aligned}(i, j)_r^{\mathcal{F}} &= \mathcal{F}(i, j)_r \\ (i, j)_l^{\mathcal{F}} &= \mathcal{F}(i, j)_{l'}\end{aligned} \tag{11.3}$$

   As shown in Fig. 11.4, we consider the following three cases for the transformation $\mathcal{F}$: (a) no-transformation, (b) uniform shrink, (c) LPM (Log Polar Mapping) (see Fig. 11.4). Details of the LPM transformation can be found in [3].
5. A point is plotted at 3D position $(x, y, z)_U$ if and only if $(i, j)_r^{\mathcal{F}} = (i, j)_l^{\mathcal{F}}$.
   We consider only horizontal disparity (Δ_i), that is, do not deal with vertical disparity. It is equivalent to vertical edge-based stereo matching.

11.3.2 Results and Discussion

As described above, we tested three kinds of image transformation (original image, uniformly shrunken image, and LPM image). The size of the original image is $(i_{\max}, j_{\max}) = (200, 200)$. On the other hand, the size of the LPM image is $(u_{\max}, v_{\max}) = (30, 32)$, that of

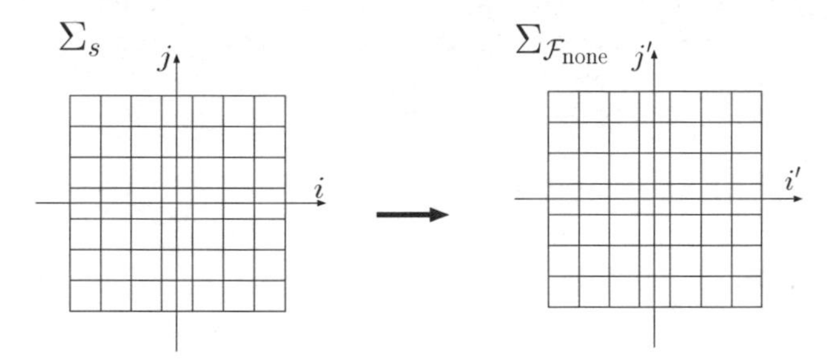

(a) Original resolution (no-transformation) $i_{\max} \times j_{\max}$

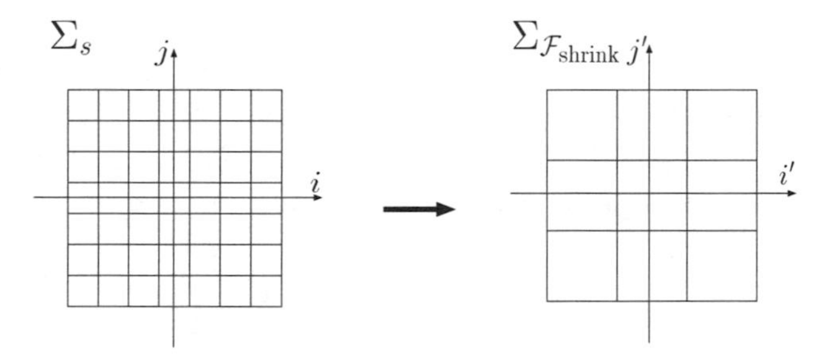

(b) Shrink $s_{\max} \times t_{\max}$

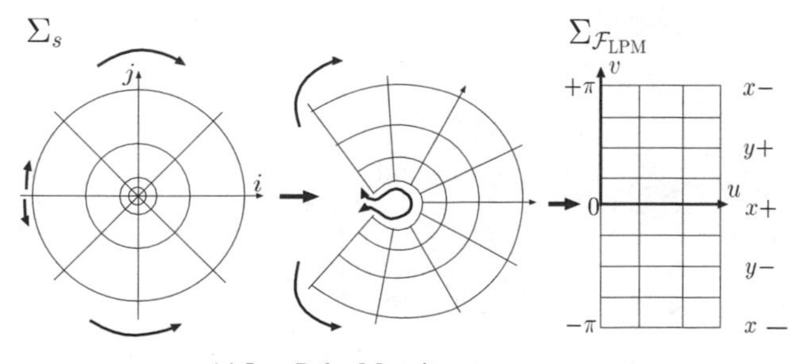

(c) Log Polar Mapping $u_{\max} \times v_{\max}$

Figure 11.4 Three kinds of image transformation.

the uniformly shrunken image is $(s_{\max}, t_{\max}) = (29, 29)$ (shrink ratio is $1/7$)[1]. Camera and image transformation parameters are shown in Table 11.1.

Figure 11.5 illustrates the 3D shape of LPM pixel horopter. We can see that it has a hyperbolic shape like a hand drum with the origin on the gaze point and an axis along z direction. Figures 11.6(a), (b), and (c) show the orthogonal projection of the pixel horopter and two virtual pixel horopters (shift value $\Delta_i = -10, +10$) onto the $x - z$ plane for each image type, respectively. On the other hand, Figs. 11.7(a), (b), and (c) show the intersection of these three horopters with the plane $x = 0$ for each image type respectively. In the diagrams, '+' means the pixel horopter, '◇' indicates a virtual pixel horopter ($\Delta_i = -10$), and '□' also indicates another virtual one ($\Delta_i = +10$).

From these figures, we can see that there are many gaps along the "vertical" direction (y-axis) between the true and virtual horopters generated by a stereo pair of uniform sampling images [see Fig. 11.6(b)]. In contrast, we can also see that there are several gaps along the "horizontal" direction (x-axis) between the horopters generated by the LPM stereo pair and they concentrate in the vicinity of the gaze point [see Figs. 11.5 and 11.7(c)]. This is mainly caused by the difference between orthogonal and (log-)polar transformation. In

TABLE 11.1 Camera and image transformation parameters.

Focal length	4.8mm
Original pixel width	0.0123mm
Baseline length	120mm
Original image $(i_{\max}, j_{\max})$	(200, 200) pixels
LPM image $(u_{\max}, v_{\max})$	(30, 32) pixels
Uniformly shrunken image $(s_{\max}, t_{\max})$	(29,29) pixels
Position of gaze point	(0, 0, 340) mm

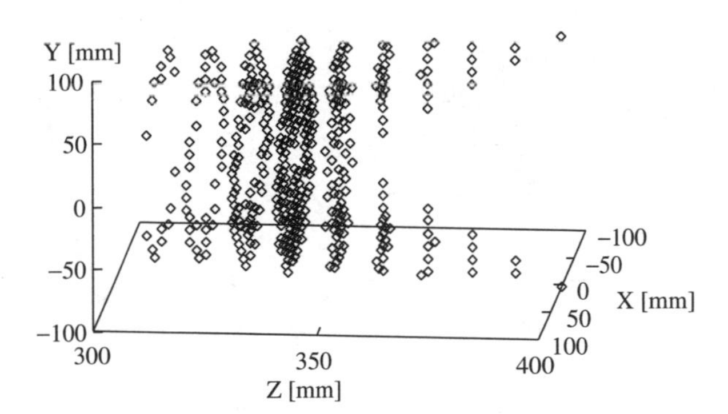

Figure 11.5 LPM horopter.

[1] We selected the image size parameters such that the width and height of uniformly shrunken image: (s_{max}, t_{max}) are almost equal to those of LPM image: (u_{max}, v_{max}). The reason for our selection is the following: we are investigating the influence of image resolution (i.e., width of image pixels) on the performance of horopter-based object tracking, by comparing the uniform sampling image with the LPM image; in so doing, we must apply the same measure to the former as to the latter. By definition, the shape of pixel horopter is closely related with the pixel width. It is, however, very difficult to select LPM transformation parameters that correspond to the pixel width of uniformly shrunken image because the pixel of LPM image has various width that depends on position in the image. We are so far coping with this problem by making the pixel number in the image to be transformed equal. In this case, however, another problem arises: since LPM is one of polar coordinate transformation, a circular region is clipped from the original image as the transformed region. Therefore, for comparison, we should also apply the same clipping to uniform sampling image. That is, strictly speaking, the pixel number of uniformly shrunken image to be transformed is not coincident with that of LPM image. We do not tackle this problem, so far.

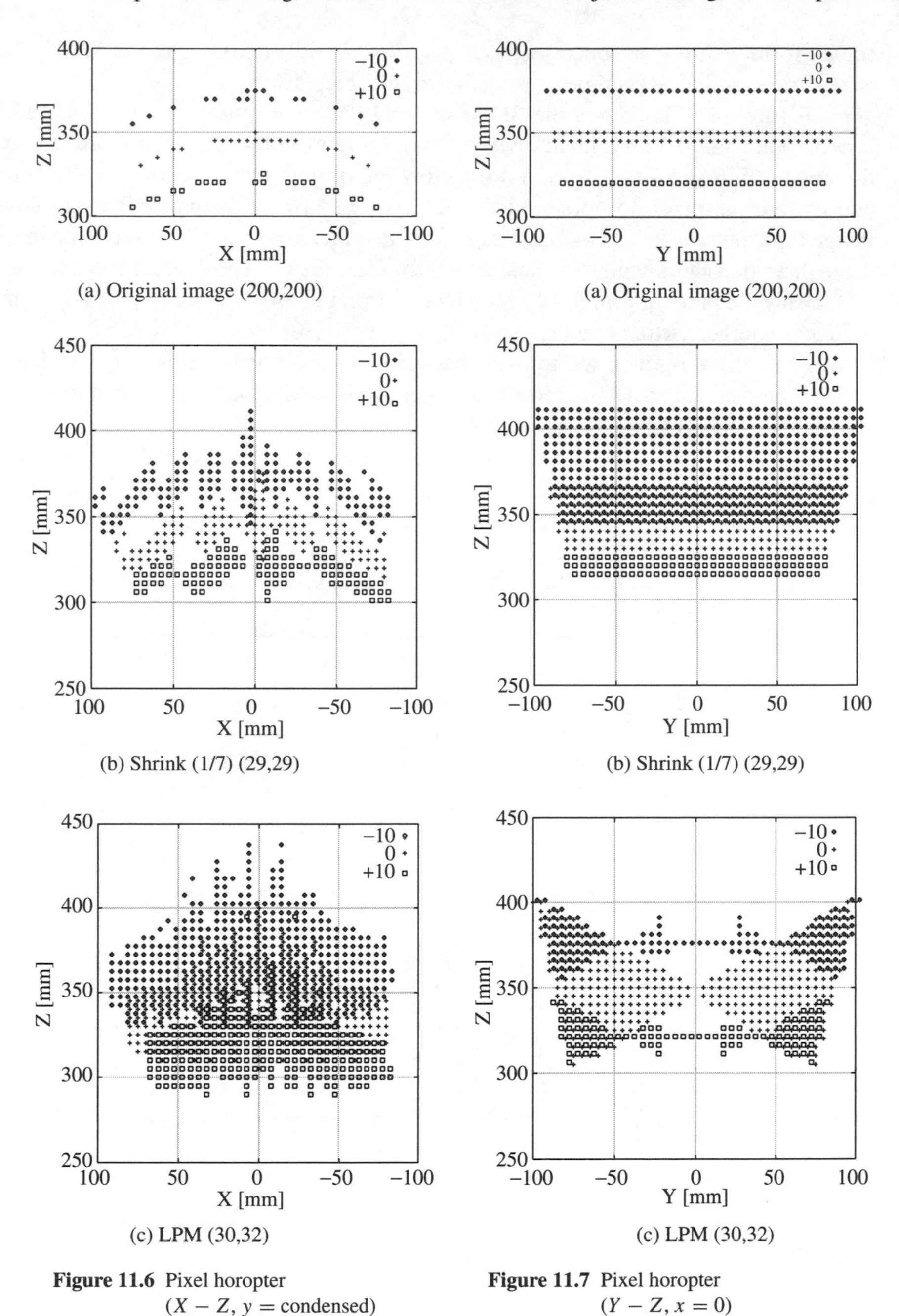

(a) Original image (200,200)

(b) Shrink (1/7) (29,29)

(c) LPM (30,32)

Figure 11.6 Pixel horopter ($X - Z$, $y =$ condensed) disparity $= \{-10,0,10\}$pixels.

Figure 11.7 Pixel horopter ($Y - Z$, $x = 0$) disparity $= \{-10,0,10\}$pixels.

general, the LPM horopter cannot extract the region in the vicinity of the gaze point because LPM image has a low resolution at the central region. This problem, however, can be easily resolved by adding uniform resolution fovea with an appropriate radius to the LPM image. Therefore, these results suggest that LPM image is more suitable for detection of horizontal disparities than uniform sampling image.

When using uniform images, the horopter forms a cylindrical shape. This is because vertical disparities were ignored in this analysis as described in Subsection 11.3.1. If vertical disparities are also taken into account, the target extraction region will become narrower (that is, gaps will become larger). By contrast, when using LPM images, we cannot ignore vertical disparities because of the property of polar transformation. Nonetheless, the target extraction region covers a wide range of the scene. It demonstrates the setting-simplicity and robustness of LPM image-based object extraction.

In uniform resolution image, characteristics of the target detection region heavily depend on the shrinking ratio. In contrast, as for LPM image, it is insensitive to the selection of LPM transformation parameters because the horopter region spreads from the gaze point toward the periphery.

11.4 CONCLUDING REMARKS

In this chapter, we analyzed the shape of horopter(zero disparity area) in detail by giving consideration to the width of an image pixel. The analysis were conducted numerically through the transformation of 3D space coordinates into the screen coordinates of stereo cameras. We applied this analysis to the three kinds of images whose resolution differs from one another: (1) original resolution image, (2) uniformly shrunken image, and (3) LPM image. In conclusion, we got two major results:

1. To extract a target "robustly" from the scene using Zero Disparity Filtering techniques, original stereo images with high resolution must be shrunken or their pixel width must be spread appropriately.
2. LPM horopter covers a wide range of the scene and consequently the target extraction area generated by both true and virtual LPM horopters is large and has few gaps, as compared to the case of using "uniformly" shrunken images whose pixel width is almost *equivalent* to that of the LPM images. Therefore, LPM-ZDF combination approach has capabilities as a "robust" object tracking method.

REFERENCES

[1] N. Oshiro, N. Maru, A. Nishikawa, and F. Miyazaki. "Binocular Tracking using Log Polar Mapping." *Proceedings of the 1996 IEEE/RSJ International Conference on Intelligent Robots and Systems* (IROS'96), vol. 2, pp. 791–798, Osaka, Japan, November 1996.

[2] T. J. Olson, and D. J. Coombs. "Real-Time Vergence Control for Binocular Robots." *International Journal of Computer Vision*, vol. 7, no. 1, pp. 67–89, Kluwer Academic Publishers, 1991.

[3] N. Oshiro, A. Nishikawa, N. Maru, and F. Miyazaki. "Foveated Vision for Scene Exploration." *Third Asian Conference on Computer Vision* (ACCV'98), vol. I, pp. 256–263, Hong Kong, China, January 1998.

[4] S. Rougeaux, N. Kita, Y. Kuniyoshi, S. Sakane, and F. Chavand. "Binocular Tracking Based on Virtual Horopters." *Proceedings of the 1994 IEEE/RSJ International Conference on Intelligent Robots and Systems* (IROS'94), vol. 3, pp. 2052–2057, Munich, Germany, August 1994.

Chapter 12 EVALUATION OF THE ROBUSTNESS OF VISUAL BEHAVIORS THROUGH PERFORMANCE CHARACTERIZATION

João P. Barreto, Paulo Peixoto, Jorge Batista, and Helder Araujo
University of Coimbra

Abstract

The robustness of visual behaviors implemented in an active vision system depends both on the vision algorithms and control structure. To improve robustness an evaluation of the performance of the several system algorithms must be done. Performance analysis can be done within the framework of control theory because notions such as stability and controllability can contribute to a better understanding of the algorithms and architectures weaknesses. In this chapter we discuss the generation of reference test target trajectories and we characterize the performance of smooth pursuit and vergence. The responses to motion are used to accurately identify implementation problems and possible optimizations. The system evaluation leads to solutions to enhance the global performance and robustness.

12.1 INTRODUCTION

Robust visual control of motion depends on issues related both to vision processing and control. Robustness of a specific visual behavior is a function of the performance of vision and control algorithms as well as the overall architecture [1]–[3]. Performance characterization of both vision and control aspects should be performed within a common framework. This would enable a global view of the performance of a specific approach. For example, when dealing with the problem of uncertainties and coping with varying environments (which are difficult or impossible to model) one can, in principle, choose to use more complex vision algorithms and/or more robust control algorithms. Good decisions and choices can only be made if all the aspects can be characterized in a common framework. Control theory has a number of tools that enable a common global characterization of the performance in visual servoing and active vision systems [4]. Several different measures and concepts can be used to perform such common characterization.

Many aspects related to visual servoing have been studied, and several systems have been demonstrated [5, 6]. One of these aspects is the issue of system dynamics. System dynamics is essential to enable the performance optimization of the system [7, 8]. Other aspects are related to stability and the system latencies [9]–[11]. In [11] Corke shows that dynamic modeling and control design are very important for the improved performance of visual closed-loop systems. One of his main conclusions is that a feedforward type of control strategy is necessary to achieve high-performance visual servoing. Nonlinear aspects of system dynamics have also been addressed [12, 13]. In [12] Kelly discusses the nonlinear aspects of system dynamics and proves that the overall closed loop system composed by the full nonlinear robot dynamics and the controller is Lyapunov stable. In [13] Hong models the dynamics of a two-axis camera gimbal and also proves that a model reference adaptive controller is Lyapunov stable. In [14] Rizzi and Koditschek describe a system that takes into account the dynamic model of the target motion. They propose a novel triangulating state estimator and prove the convergence of the estimator. In [15, 16] the control performance of the Yorick head platform is presented, considering especially the problem of dealing with the inherent delays and in particular with variable delays. Problems associated with overcoming system latencies are also discussed in [17]–[19].

12.2 CONTROL OF THE MDOF BINOCULAR TRACKING SYSTEM

The binocular MDOF robot head (see Fig. 12.1) is a high-performance active vision system with a high number of degrees of freedom [20]. Real-time complex visual behaviors were implemented after careful kinematics modeling and adequate selection of basic visual routines [21, 22]. Binocular tracking of 3D motion was achieved by controlling neck pan/tilt and vergence.

In most cases visual servoing systems are analyzed as servo systems that use vision as a sensor [23, 24]. Therefore, the binocular tracking system should be considered as a servomechanism whose reference inputs are the target coordinates in space and whose outputs are the motor velocities and/or positions. In this case binocular vision is used to directly estimate target 3D motion parameters.

However, in the case of this system, and as a result of both its mechanical complexity and its goal (tracking of targets with unknown dynamics), we decided to relate the system outputs with the data measured from the images. Thus this system can be considered as a regulator whose goal is to keep the target in a certain position in the image (usually its center). As a result of this framework target motion is dealt with as a perturbation. If the perturbation affects the target position and/or velocity in the image it has to be compensated for. The changes in the head geometry during the tracking process can be used to estimate the target 3D motion parameters.

12.2.1 Monocular Smooth Pursuit: Pan/Tilt Block Diagram

Each camera joint has two independent rotational degrees of freedom: pan and tilt. Even though pure rotation cannot be guaranteed, we model these degrees of freedom as purely rotational. A schematic for one of the these degrees of freedom is depicted in Fig. 12.2 (both degrees of freedom are similar and decoupled). Notice that two inputs and two outputs are considered. Both position and velocity of the target in the image are to be controlled or

Figure 12.1 The MDOF binocular system.

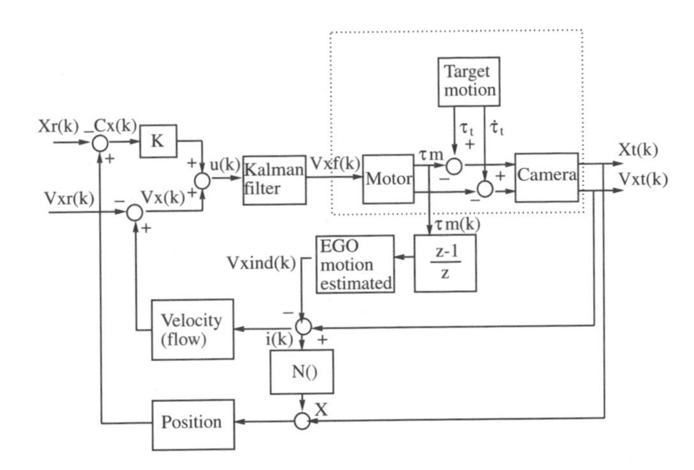

Figure 12.2 Monocular smooth pursuit block diagram. The dotted box encloses the analog components of the structure. Block $N(i(k))$ represents a nonlinear function. $V_{xf}(k)$ is the command sent to the motor, obtained by filtering $u(k)$, the sum of the estimated velocity with the position error multiplied by a gain $K.V_{xind}(k)$ is the velocity induced in image by camera motion.

regulated. Even though the two quantities are closely related, this formal distinction allows for a better evaluation of some aspects such as nonlinearities and limitations in performance.

$$\begin{cases} i(k) = V_{xt}(k) - V_{xind}(k) \\ N(i(k)) = 1 \Longleftarrow i(k) \neq 0 \\ N(i(k)) = 0 \Longleftarrow i(k) = 0 \end{cases} \tag{12.1}$$

Considering that the motion computed in the image is caused by target motion and by camera motion, the computation of the target velocity requires that the effects of egomotion are compensated for. The egomotion is estimated based on the encoder readings and on the inverse kinematics. Once egomotion velocity ($V_{xind}(k)$) is compensated for, target velocity in the image plane is computed based on an affine model of optical flow. Target position is estimated as the average location of the set of points with nonzero optical flow in two consecutive frames (after egomotion has been compensated for). This way what is actually computed is the center of motion instead of target position. The estimated value will be zero whenever the object stops, for it is computed by using function $N(i(k))$ [see (12.1)].

12.2.2 Vergence Block Diagram

In this binocular system, pan and tilt control align the cyclopean Z (forward-looking) axis with the target. Vergence control adjusts both camera positions so that both target images are projected in the corresponding image centers. Retinal flow disparity is used to achieve vergence control. Vergence angles for both cameras are equal, and angular vergence velocity is computed in (12.2) where Δv_{xf} is the horizontal retinal motion disparity and f the focal length [25].

$$\frac{\partial \beta}{\partial t} = \frac{\Delta v_{xf}}{2f} \tag{12.2}$$

A schematic for vergence control is depicted in Fig. 12.3. Horizontal target motion disparity is regulated by controlling the vergence angle.

Both in smooth pursuit and vergence control, target motion acts as a perturbation that has to be compensated for. To study and characterize system regulation/control performance

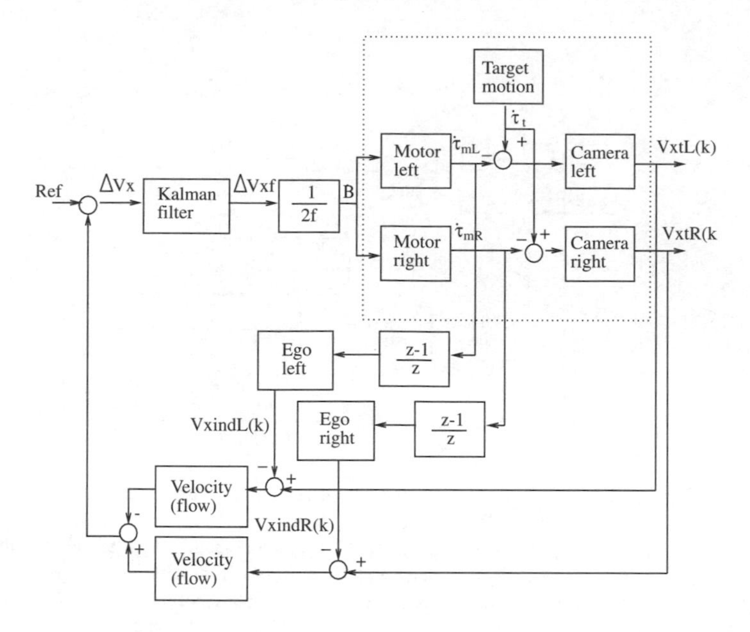

Figure 12.3 Vergence block diagram. Egomotion is estimated for each camera. After that target velocities in both left and right images are computed using differential flow. Estimated horizontal disparity (Δv_{xf}) is obtained by filtering the difference of measured velocities in both images.

usual control test signals must be applied. Two problems have to be considered:

- The accurate generation of perturbation signals
- The generation of perturbation signals functionally defined, such as steps, ramps, parabolas, and sinusoids

12.3 REFERENCE TRAJECTORIES GENERATION USING SYNTHETIC IMAGES

To characterize the system ability to compensate for the perturbations due to target motion, specific signals have to be generated. Instead of using real targets, we decided to use synthetic images so that the mathematical functions corresponding to reference trajectories could be accurately generated. These images are then used as inputs in the binocular active vision system. Given a predefined motion, captured frames will depend, not only on the target position, but also on the camera orientation. Due to the change on the system geometry as a result of its operation, images have to be generated online to take into account the specific geometry at each time instant. Therefore at each time instant both target position and camera orientation have to be known in the same inertial coordinate system. The former is calculated using a specific motion model that enables the computation of any kind of motion in space. Camera orientation is computed by taking into account the motor encoders readings and the inverse kinematics. The inertial coordinate system origin is placed at optical center (monocular case) or at the origin of the cyclopean referential (binocular case).

To accurately describe the desired target motion in space the corresponding equations are used. Motion coordinates are converted into inertial Cartesian coordinates by applying the suitable transformation equations. Target coordinates in the inertial system are converted in camera coordinates. This transformation depends on motor positions that are known by reading the encoders. Perspective projection is assumed for image formation. These computations are performed at each frame time instant.

12.4 REFERENCE TRAJECTORIES EQUATIONS

To characterize control performance, target motion correspondent to a step, a ramp, a parabola, and a sinusoid should be used to perturb the system.

12.4.1 Smooth Pursuit: Pan/Tilt Control System

12.4.1.1 Reference Trajectories Defined for the Actuators. Consider the perturbation at actuator/motor output. The reference trajectories are studied for both a rotary and a linear actuator.

In the former the actuator is a rotary motor and the camera undergoes a pure rotation around the Y (pan) and X (tilt) axis. Consider target motion equations defined in spherical coordinates (ρ, ϕ, θ), where ρ is the radius or depth, ϕ the elevation angle and θ the horizontal angular displacement. The target angular position $\theta(t)$ at time t is given by one of

$$\theta(t) = \begin{cases} Const \Longleftarrow t > 0 \\ 0 \Longleftarrow t = 0 \end{cases} \tag{12.3}$$

$$\theta(t) = \omega t \tag{12.4}$$

$$\theta(t) = \frac{\gamma}{2} t^2 \tag{12.5}$$

$$\theta(t) = A \sin(\omega t) \tag{12.6}$$

Equations (12.3), (12.4), (12.5) and (12.6) describe a step, a ramp, a parabola, and a sinusoid for the pan motor. For instance, if the target moves according to (12.4), the motor has to rotate with constant angular velocity ω to track the target. These definitions can be extended to the tilt motor by making $0 = 0$ and varying ϕ according to (12.3) to (12.6).

Assume now a linear actuator and camera moving along the X axis. Cartesian equations (12.7) to (12.10) are the equivalent to spherical equations (12.3) to (12.16). In all cases the depth z_i is made constant.

$$x_i(t) = \begin{cases} Const \Longleftarrow t > 0 \\ 0 \Longleftarrow t = 0 \end{cases} \tag{12.7}$$

$$x_i(t) = vt \tag{12.8}$$

$$x_i(t) = \frac{a}{2} t^2 \tag{12.9}$$

$$x_i(t) = A \sin(vt) \tag{12.10}$$

12.4.1.2 Reference Test Signals Defined in Image: Static Camera Situation. To relate the system outputs with the data measured from the images, control test signals must be generated in the image plane. Thus a step (in position) is an abrupt change of target position in image. A ramp/parabola (in position) occurs when the 3D target motion generates motion with constant velocity/acceleration in the image plane. And a sinusoid is generated whenever the image target position and velocity are described by sinusoidal functions of time (with a phase difference of 90°).

$$x_{img} = f \tan(\theta) \tag{12.11}$$

$$\frac{dx_{img}}{dt} = f \frac{d\theta}{dt} \frac{1}{\cos^2(\theta)} \tag{12.12}$$

$$\frac{d^2 x_{img}}{dt^2} = f \frac{d^2\theta}{dt^2} \frac{1}{\cos^2(\theta)} + 2f \left(\frac{d\theta}{dt}\right)^2 \frac{\tan(\theta)}{\cos^2(\theta)} \tag{12.13}$$

Consider that the camera is static. The observations depicted in Fig. 12.4 agree with equations derived in (12.11), (12.12), (12.13) which relate angular position (θ) in space with target image coordinates ($x_{img}, y_{img}, z_{img}$) ($f$ is the focal length and perspective projection is assumed). Notice in Fig. 12.4 that, despite the inexistence of an angular acceleration, a residual acceleration can be observed in target image motion due to the second term of (12.13). Target motion described by spherical equations (12.4) to (12.6), does not generate the desired perturbations in the image plane when the camera is static. Fig. 12.4 shows that image motion distortion is significant when θ is above 50°.

$$x_{img} = f \frac{x_i}{z_i} \tag{12.14}$$

$$\frac{dx_{img}}{dt} = \frac{f}{z_i} \frac{dx_i}{dt} - \frac{f x_i}{z_i^2} \frac{dz_i}{dt} \tag{12.15}$$

$$\frac{d^2 x_{img}}{dt^2} = \frac{f}{z_i} \frac{d^2 x_i}{dt^2} - \frac{f x_i}{z_i^2} \frac{d^2 z_i}{dt^2} - \frac{2f}{z_i^2} \frac{dx_i}{dt} \frac{dz_i}{dt} - \frac{2 f x_i}{z_i^3} \left(\frac{dz_i}{dt}\right)^2 \tag{12.16}$$

However, in the case of a target moving in a rectilinear trajectory parallel to the image plane (constant depth), the standard perturbations are obtained. Whenever images are obtained with a static camera, linear motion described by (12.7) to (12.10) is adequate to generate the standard control test signals. This conclusion is confirmed by (12.14), (12.15), and (12.16) (z_i remains constant) that relate image coordinates with Cartesian motion coordinates. This result is still true if camera moves along a linear path.

12.4.1.3 Reference Test Signals Defined in Image: Camera Undergoing Pure Rotation. The MDOF binocular system uses rotary eye joints. Thus, considering the monocular situation, the camera moves along a circular trajectory. We assume camera rotation around X_c axis (pan), the target moving along a circular/spherical path (see Fig. 12.5) and perspective

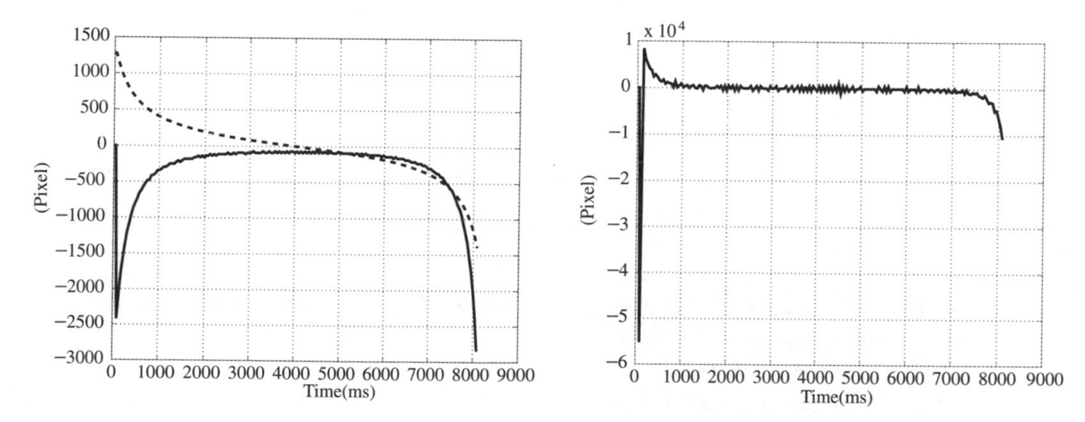

Figure 12.4 Motion projection in a static camera. The target moves along a circular path with constant angular velocity $\omega = 5$(degrees) [(12.4)]. Left: Target position (- -) and velocity (-) in image. Right: Target acceleration (-) in image.

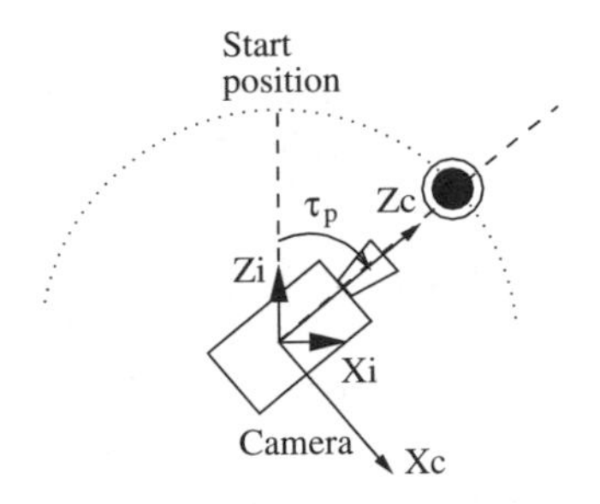

Figure 12.5 Tracking using the pan motor. α_p is the motor angular position and $\theta(t)$ the target angular position.

projection modeling image formation.

$$x_{img} = f \tan(\theta - \alpha_p) \tag{12.17}$$

$$\frac{dx_{img}}{dt} = f\frac{d\theta}{dt}\frac{1}{\cos^2(\theta - \alpha_p)} - f\frac{d\alpha_p}{dt}\frac{1}{\cos^2(\theta - \alpha_p)} \tag{12.18}$$

$$\frac{d^2x_{img}}{dt^2} = f\left(\frac{d^2\theta}{dt^2} - \frac{d^2\alpha_p}{dt^2}\right)\frac{1}{\cos^2(\theta - \alpha_p)} + 2f\left(\frac{d\theta}{dt} - \frac{d\alpha_p}{dt}\right)^2\frac{\tan(\theta - \alpha_p)}{\cos^2(\theta - \alpha_p)} \tag{12.19}$$

Target position ($x_{img}(t)$) in the image is dependent both on the camera angular position ($\alpha_p(t)$) and target angular position ($\theta(t)$) [(12.17)]. To compute the target velocity in the image, (12.18) is derived by differentiating equation (12.17). Notice that the target and camera angular positions are time dependent. By differentiating (12.18) the expression for target acceleration in image is obtained [(12.19)].

As can be noticed in these equations, motion in the image is caused both by target motion and camera motion. For a perfect tracking situation the former is compensated by the latter and no motion is detected in the image. Whenever perfect tracking does not happen there will be image motion as a result of tracking error. Therefore, the objective of tracking is to move the camera in such a way that egomotion compensates for the motion induced in the image by the target. From this point of view the system perturbation will be the motion induced by the target.

$$\omega_i = f\frac{d\theta}{dt}\frac{1}{\cos^2(\theta - \alpha_p)} \tag{12.20}$$

$$\gamma_i t = f\frac{d\theta}{dt}\frac{1}{\cos^2(\theta - \alpha_p)} \tag{12.21}$$

$$A\omega_i \cos(\omega_i t) = f\frac{d\theta}{dt}\frac{1}{\cos^2(\theta - \alpha_p)} \tag{12.22}$$

The reference trajectories that generate a perturbation in ramp, parabola, and sinusoid are derived by solving the differential equations (12.20), (12.21), and (12.22) in order to $\theta(t)$ (ω_i, γ_i and A are the desired induced velocity, acceleration and amplitude). The difficulty is that the reference trajectories ($\theta(t)$) will depend on the system reaction to the perturbation ($\alpha_p(t)$). That is because the image is not only function of target position in space, but also of camera orientation. Thus to induce a constant velocity in image during operation, target angular velocity must be computed at each frame time instant as a function of the the tracking error.

Consider that perfect tracking is going to occur. The tracking error will be null and $\alpha_p(t) = \theta(t)$. With this assumption the solutions of differential (12.20) to (12.21) are given by (12.4) to (12.6) (making $\omega = \frac{\omega_i}{f}$ and $\gamma = \frac{\gamma_i}{f}$). These are the reference trajectories that we are going to use to characterize the system. It is true that for instance, trajectory of (12.4) (the ramp) only induces a constant velocity in image if tracking error is null (small velocity variation will occur otherwise). However, it is independent of the system reaction and the generated perturbation allows the evaluation of system ability to recover from tracking errors.

12.4.2 Vergence Control System

Taking into account the considerations of last section, the reference trajectories for vergence control characterization of the binocular system depicted in Fig. 12.6 are now discussed. The distance between the cameras is $2b$ and symmetric vergence is assumed. The Z coordinate of the target position (in the cyclopean coordinate frame) is ρ.

$$\Delta x_{img} = 2f\frac{-\rho\sin(\beta) + b\cos(\beta)}{\rho\cos(\beta) + b\sin(\beta)} \tag{12.23}$$

$$\beta = \arctan\left(\frac{b}{\rho}\right) \tag{12.24}$$

Vergence control is achieved using retinal disparity. The differences of target position and velocity in the images of both cameras are the system stimuli. The position retinal disparity is calculated in (12.23). Perfect tracking is achieved when β is computed by (12.24). In this case $\Delta x_{img} = 0$.

$$\Delta V x_{img} = -\frac{2fb}{\sqrt{\rho^2 + b^2}}\frac{d\rho}{dt} \tag{12.25}$$

Deriving (12.23), the expression for velocity retinal disparity is obtained. Suppressing the egomotion effect (considering $\frac{d\beta}{dt} = 0$), the stimulus generated by target motion is computed in (12.25) assuming a perfect tracking situation.

$$2fb\frac{d\rho}{dt} + v\rho^2 = -vb^2 \tag{12.26}$$

$$a = -\frac{2fb}{\rho^2 + b^2}\frac{d^2\rho}{dt^2} + \rho\frac{4fb}{(\rho^2 + b^2)^2}\left(\frac{d\rho}{dt}\right)^2 \tag{12.27}$$

$$2fb\frac{d\rho}{dt} + Aw\cos(wt)\rho^2 = -Aw\cos(wt)b^2 \tag{12.28}$$

The target motion equation $\rho(t)$ that generates a motion corresponding to a ramp in image target position (constant velocity disparity v) is determined solving (12.26) derived from (12.25). For a parabola (constant acceleration disparity a) (12.27) must be solved. In the case of a sinusoidal stimulus, the relevant target motion equation $\rho(t)$ can be computed by solving (12.28). Test signals obtained by solving differential equations (12.26) and (12.28) are depicted in Fig. 12.7. Notice that to induce a constant velocity disparity in the images the 3D target velocity increases with depth. This is due to the perspective projection.

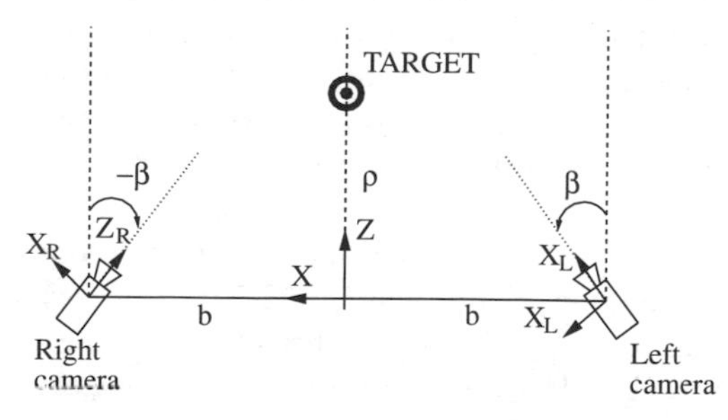

Figure 12.6 Top view of the binocular system. The distance between the cameras is $2b$ and symmetric vergence is assumed. The $\rho(t)$ is the target Z coordinate.

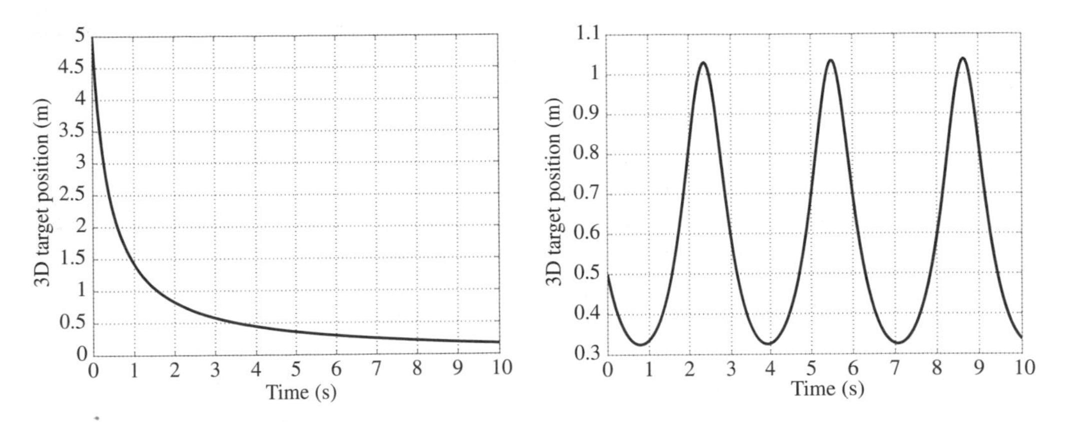

Figure 12.7 Left: Ramp perturbation. Target motion to generate a constant disparity of 1 pixel/frame ($\rho(0) = 5$(m)). Right: Sinusoidal perturbation. Target motion that generates a sinusoidal velocity disparity in images ($A = 2$(pixel), $\omega = 2$(rad/s), and $\rho(0) = 1$(m)).

12.5 SYSTEM RESPONSE TO MOTION

In this section we analyze the system ability to compensate for perturbations due to target motion.

12.5.1 Smooth Pursuit: Pan/Tilt Control Algorithm

As shown previously, spherical/circular target motion must be used to generate the standard control test signals. Pan and tilt control algorithms are identical except for some of the parameter values. Therefore we will consider only the pan axis.

12.5.1.1 Step Response. A step in position is applied to the system. Figure 12.8 shows the evolution of the target position (X_t) in the image. An overshoot of about 10% occurs. The regulation is done with a steady-state error of about 1.5 pixels. These observations are in agreement with the observed positional servo-mechanical performance. This is a typical second order step response of a type 0 system. In experiments done with smaller amplitude

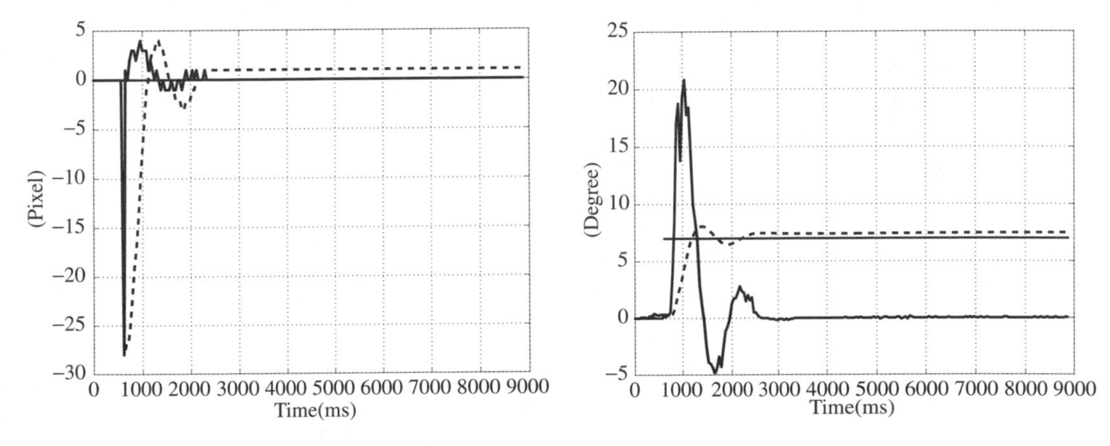

Figure 12.8 Left: Regulation performance. Target position (- -) and velocity (-) in the image. Right: Servo-mechanical performance. Target angular position (.), motor position (- -) and velocity (-).

steps the system fully compensates for target motion. In these situations the regulation error is 0 and we have a type 1 system. The type of response depends on the step amplitude which clearly indicates a nonlinear behavior. One of the main reasons for the nonlinear behavior is the way position feedback is performed. After compensating for egomotion, target position is estimated as the average location of the set of points with nonzero optical flow in two consecutive frames. Thus the center of motion is calculated instead of the target position. If the target stops, any displacement detected in the image is due to camera motion. In that case target velocity ($V_{xt}(k)$) is equal to the induced velocity ($V_{xind}(k)$) and the position estimate C_x will be 0. Therefore target position would only be estimated at the step transition time instant. Only with egomotion as a pure rotation would this occur. In practice, sampling and misalignment errors between the rotation axis and the center of projection introduce small errors.

A step in position corresponds to an impulse perturbation in velocity. Figure 12.8 shows the system ability to cancel the perturbation. Note that only the first peak velocity is due to real target motion.

12.5.1.2 Ramp Response. Figure 12.9 exhibits the ramp response for a velocity of 10°/s (1.5 pixels per frame). The target moves about 6 pixels off the center of image before the system starts to compensate for it. It clearly presents an initial inertia where the action of the Kalman filter plays a major role. The Kalman filtering limits the effect of measurement errors and allows smooth motion without oscillations.

Considering the motor performance we have a type 1 position response to a ramp and a second order type 1 velocity response to a step. The position measurement error

$$e(k) = X_t(k) - C_x(k) \tag{12.29}$$

will be directly proportional to the speed of motion.

The algorithm for velocity estimation using optical flow only performs well for small velocities (up to 2 pixels per frame). For higher motion speeds the flow is clearly underestimated. This represents a severe limitation that is partially compensated for by the proportional position error component on the motor commands. Experiments were performed that enabled us to conclude that the system only follows motions with constant velocities of up to 20°/s.

12.5.1.3 Parabola Response. The perturbation is generated by a target moving around the camera with a constant angular acceleration of 5°/s^2 and an initial velocity of 1°/s. When the

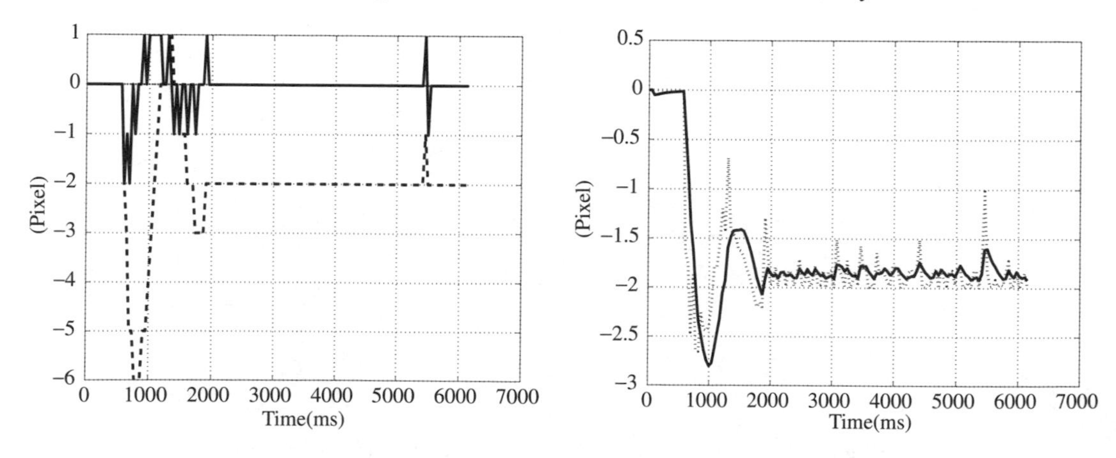

Figure 12.9 Left: Regulation performance.Target position (- -) and velocity (-) in the image. Right: Kalman filtering. Kalman input $u(k)$ (.) and output $V_{xf}(k)$(-).

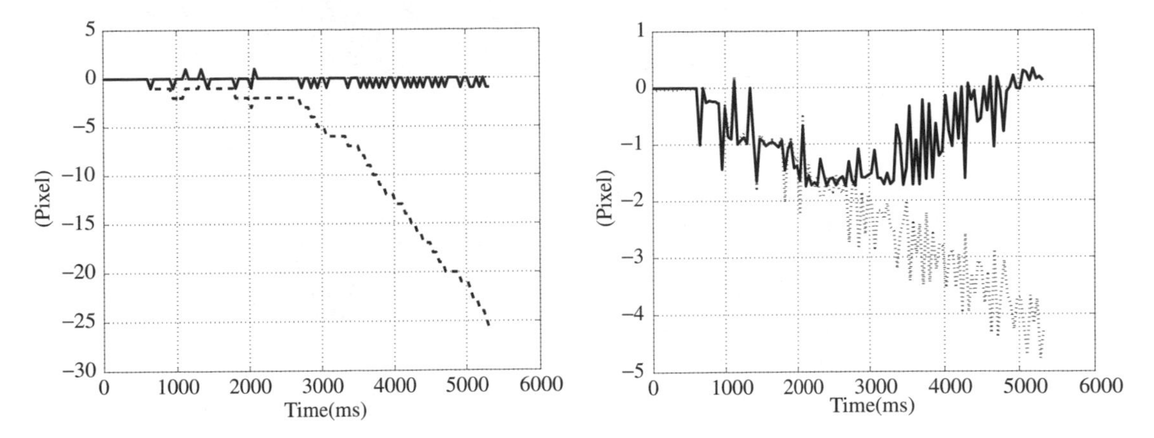

Figure 12.10 Left: Regulation performance. Target position (- -) and velocity (-) in the image). Right: Velocity estimation. Target velocity (.) and flow (-).

velocity increases beyond certain values flow underestimation bounds the global performance of the system. The system becomes unable to follow the object and compensate for its velocity. As a consequence the object image is increasingly off the image center and the error in position increases.

12.5.1.4 Sinusoidal Response. System reaction to a sinusoidal perturbation of angular velocity 2 rad/s is studied. Figure 12.11 shows target position X_t and velocity V_x in the image. Nonlinear distortions, mainly caused by velocity underestimation, can be observed. Notice the phase lag and the gain in position motor response in Fig. 12.11.

12.5.2 Vergence Control System

Figure 12.12 depicts the vergence performance in compensating for a perturbation in step and ramp. The test signals are obtained as explained in Subsection 12.4.2. These new observations confirm that visual processing is limiting the global performance. In the next section we describe some solutions that were explored to improve position and velocity estimation in image.

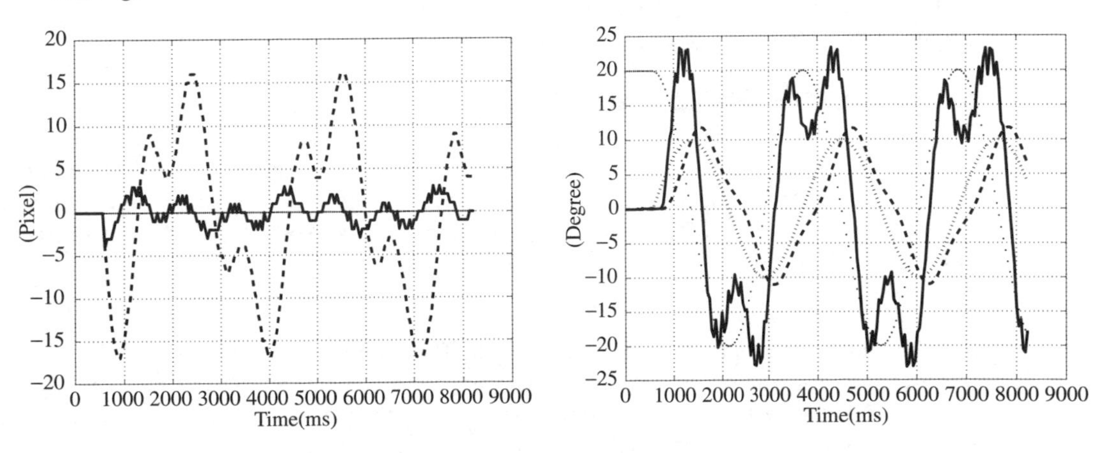

Figure 12.11 Left: Regulation Performance–Target position (- -) and velocity (-) in the image. Right: Servo-mechanical performance in position. Motor position (- -) and velocity (-). Target position (:) and velocity (.).

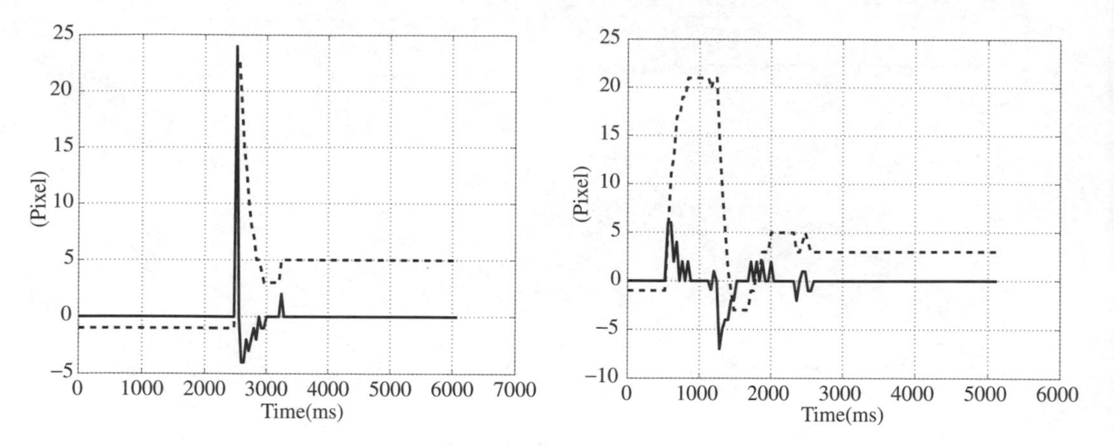

Figure 12.12 Left: Vergence regulation performance for a step perturbation. Right: Vergence regulation performance for a ramp perturbation. Target position (- -) and velocity (-) disparity in the image.

12.6 IMPROVEMENTS IN THE VISUAL PROCESSING

12.6.1 Target Position Estimation in Image

Target position estimation in the image is fundamental to keep the position regulation error small and to reduce the effects of occasional erroneous velocity prediction.

$$C_x[k] = C_x[k-1] + V_{xind}[k] \tag{12.30}$$

Some problems in position estimation, that interfere with global system performance, were detected. The center of motion is estimated only when the target induces motion in the image. When no target motion is detected (after egomotion compensation) it can be considered that the target did not move. Thus the new position estimate should be equal to the previous estimate compensated for the induced displacement due to camera motion [(12.30)]. Another problem is that the center of motion is computed instead of the target position. The position estimate is computed as the average location of the set of points with nonzero optical flow in two consecutive frames. If this set is restricted to the points of the most recently acquired frame that have nonzero spatial derivatives, the average location will be near the target position. The improvements in position estimation can be observed in Fig. 12.13.

12.6.2 Target Velocity Estimation in the Image

To estimate the target velocity in the image, the brightness gradient is calculated in all pixels of the acquired frame. Considering the flow constraint equation and assuming that all points in the image move with the same velocity, the velocity vector (u, v) is estimated using a least squares method.

$$I_x u + I_y v + I_t = 0 \tag{12.31}$$

The flow constraint equation (12.31) is valid for a continuous brightness function (under the assumption of brightness constancy). However, the actual function $I(x, y, t)$ is discrete in time and space. Aliasing problems in partial derivatives computation can compromise a correct velocity estimation. When the target image moves very slowly high spatial resolution is needed in order to correctly compute the derivatives Ix and Iy and to estimate the velocity. On the other hand, if the the target image moves fast, there are high frequencies in time and I_t must be computed with shorter sampling periods. However, the sampling frequency is limited to 25

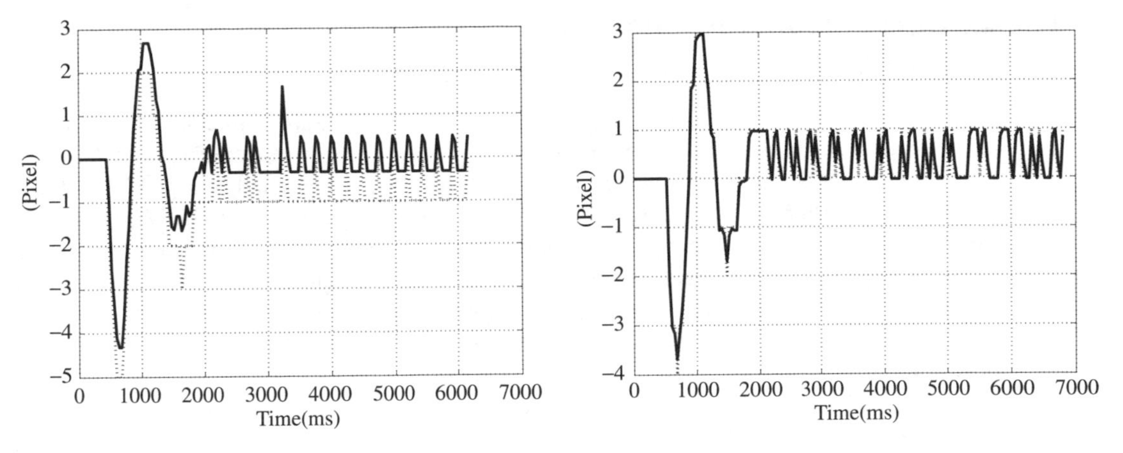

Figure 12.13 Response for a ramp perturbation of 1.5 pixels per frame (10°/s). Left: Position estimation using the original method. Target position (:) and target position estimation (-). Right: Position estimation using the improved method. Target position (:) and target position estimation (-).

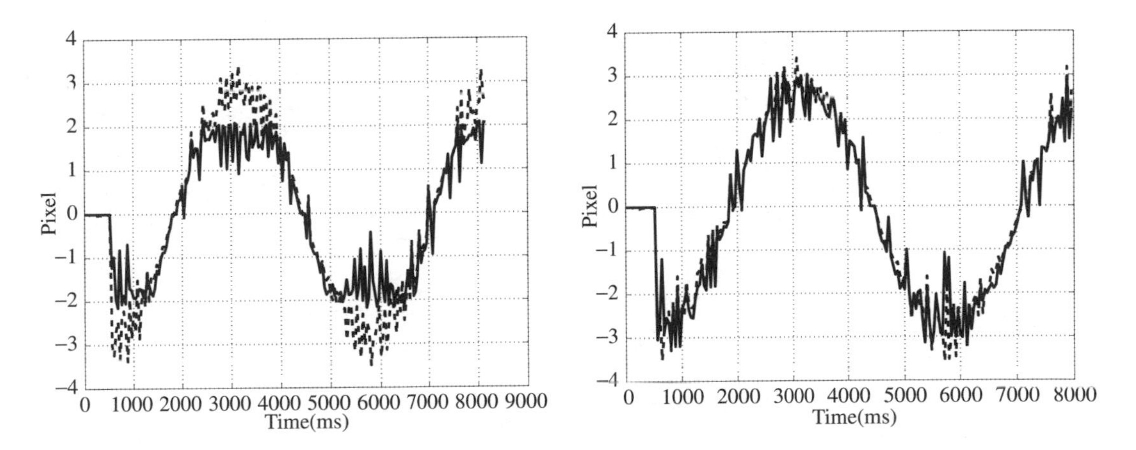

Figure 12.14 Response to a sinusoidal perturbation. Left: Velocity estimation using the original method. Right: Velocity estimation using the new method with a two-level pyramid . The target velocity in the image (–) and the estimated value (-). Both methods perform a correct estimation of velocity.

Hz. One solution to estimate high target velocities is to decrease the spatial resolution. The drawback of this approach is that high frequencies are lost, and small target movements will no longer be detected. We tested several methods to increase the range of target velocities in the image that the system is able to estimate. The method that gave the best results is next presented.

The method starts by building a pyramid of images with different resolutions. Two levels are considered: the lower with a 64×64 image, and the higher with a 32×32 resolution. The flow is computed at the high level using a 2×2 mask. The value of the computed flow (V_{high}) is used to define the size of the mask that is employed to estimate target velocity at the 64×64 level (V_{low}). The mask can have a size of 2, 3, or 4 pixels depending on the value of V_{high} at each time instant. Notice that the final velocity is always given by V_{low}. The rule that controls the mask size is based on intervals between predefined threshold values. To each interval corresponds a certain mask size that it is chosen if the value of V_{high} belongs to that interval. The threshold values are determined experimentally. The range of velocity estimation can be increased by using more levels in the pyramid.

12.7 MOTOR PERFORMANCE AND GLOBAL SYSTEM BEHAVIOR

The latency of the actuators is an important issue to achieve high-speed tracking. In the MDOF robot head actuation is done using DC motors with harmonic drives controlled by Precision Microcontrol DCX boards. The implemented control loop is depicted in Fig. 12.15. Motor position is controlled using a classic closed-loop configuration with a digital PID controller running at 1 KHz. For velocity control the reference inputs (in position) are computed by a profile generator. This device integrates the velocity commands sent by the user process. Acceleration and deceleration values can be configured to assure more or less smoothness in velocity changes. Due to the fact that each board controls up to six axis, the user process can only read the encoders and send commands for 6 ms time intervals.

The PID of the inner position loop must be "tight" in order to minimize the position error and guarantee small velocity rise times. Figure 12.16 exhibits the motor response to successive velocity commands. The rise time is about 1 frame time instant. The overshoot is not constant (nonlinear behavior), and the global performance decreases for abrupt changes in input. Therefore, during operation, abrupt changes in velocity commands must be avoided to maximize motor performance.

A decrease in processing time from 38 ms to 8 ms was achieved by upgrading the processor. The effects in global performance can be observed in Fig. 12.16. In the first implementation, the frame was captured and the actuating command was sent just before the following frame grabbing. Considering a rise time of 1 frame time interval, the motor

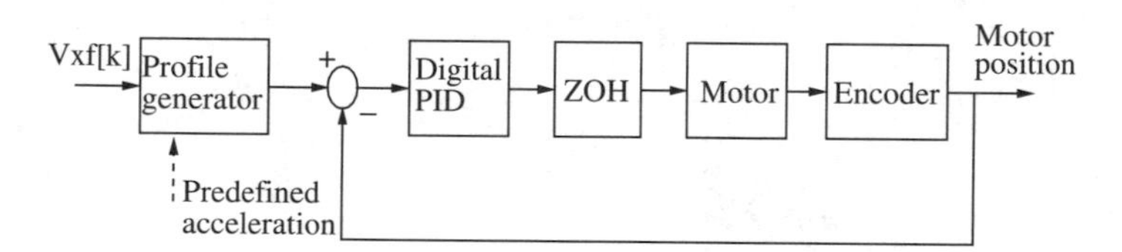

Figure 12.15 Motor control loop. A PID is used to control motor position. The sampling frequency in the closed-loop is 1 KHz. A profile generator allows to control the motor in velocity.

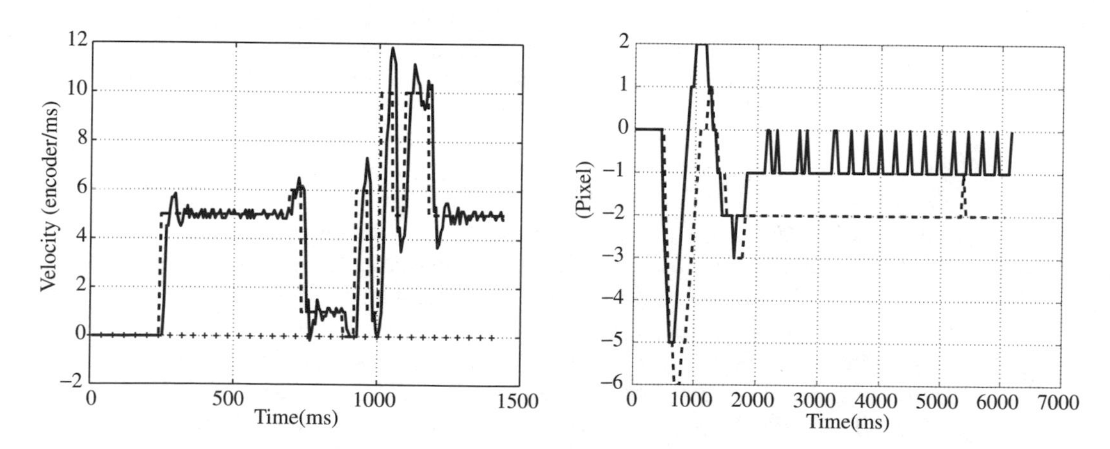

Figure 12.16 Left: Motor response to sudden changes in velocity. The velocity command (–) and the motor velocity response measured for a sampling interval of 6 ms (-). The dashes along zero axis mark the frame time instants (40 ms). Right: Regulation performance for a ramp perturbation of 1.5 pixel/frame (10°/s). Processing time of 38 ms (–) and 8 ms(-).

only reached the velocity reference 80 ms after the capture of the corresponding frame. By decreasing the image processing time the reaction delay is reduced to almost half the value and the system becomes more responsive. When the second frame is grabbed, the camera is approximately moving with the target velocity estimated in the previous iteration.

12.8 IMPROVEMENTS IN GLOBAL PERFORMANCE—EXPERIMENTAL RESULTS

The performance and robustness of the pan/tilt smooth pursuit improved by decreasing the computation time and enhacing the visual processing (position and velocity estimation) (see Fig. 12.17). Figure 12.18 shows that vergence control performance improved as well.

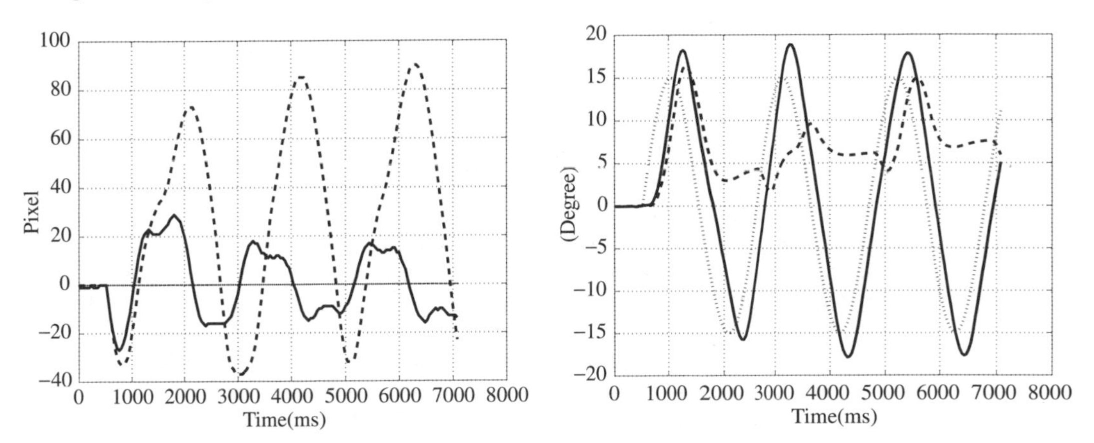

Figure 12.17 Pan/Tilt Control System. Response to a sinusoidal perturbation. Left: Regulation performance. Target position in the image for the original (- -) and improved implementation (-). Right: Servo-mechanical performance. Target angular position (.), motor position in the original ( ) and improved (-) algorithm.

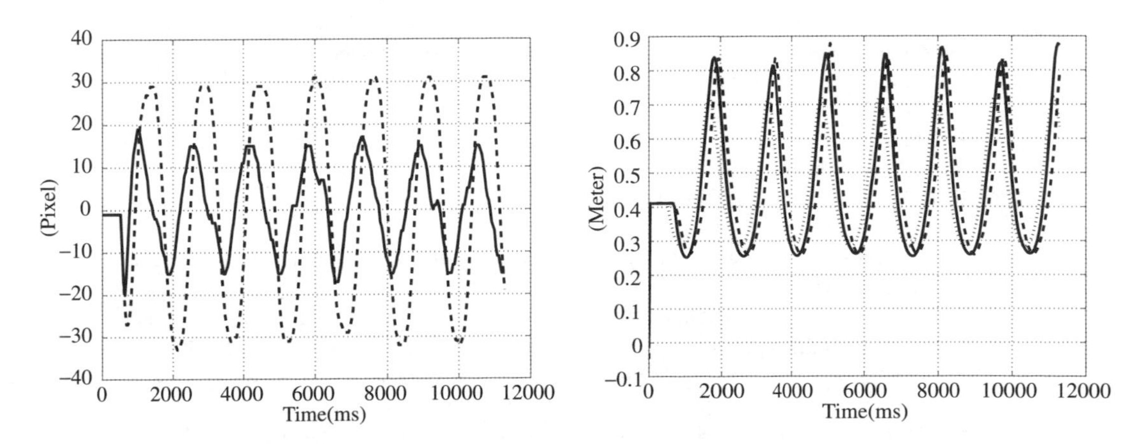

Figure 12.18 Vergence control system. Response to a sinusoidal perturbation. Left: Regulation performance. Target position disparity in image for the original (- -) and improved implementation (-). Right: Servomechanical performance. Target depth position (.), vergence depth position in the original (- -), and improved (-) algorithm.

12.9 SUMMARY AND CONCLUSIONS

In this chapter we address the problem of improving the performance and robustness of tracking performed by a binocular active vision system. In order to enable the evaluation of the robustness of both vision and control algorithms in a common framework, we decided to use a methodology inspired by control techniques. The different subsystems were characterized by their responses to test inputs. Due to the specific features of an active vision system several questions related to the definition of system reference inputs had to be addressed. As a result we propose and justify a methodology for the definition and generation of such reference inputs.

System identification of some modules of the system, including the visual processing routines (which required their linearization) was also done. The results enabled us to identify elements that should be improved. Specifically, we described the improvements in the visual processing algorithms. These improvements enable the system to track targets in a much larger range of depths.

REFERENCES

[1] J. L. Crowley and editors H. I. Christensen, *Vision as a process*, Springer-Verlag, 1995.

[2] J. L. Crowley, J. M. Bedrune, M. Bekker, and M. Schneider, "Integration and control of reactive visual processes," *Third European Conference in Computer Vision*, vol. 2, pp. 47–58, May 1994.

[3] H. I. Christensen, J. Horstmann, and T. Rasmussen, "A control theoretic approach to active vision," *Asian Conference on Computer Vision*, pp. 201–210, December 1995.

[4] Y. Aloimonos, I. Weiss, and A. Bandyopadhay, "Active vision," *International Journal of Computer Vision*, vol. 1, no. 4, pp. 333–356, January 1988.

[5] G. Hager and S. Hutchinson, "Special section on vision-based control of robot manipulators," *IEEE Transactions on Robotics and Automation*, vol. 12, no. 5, October 1996.

[6] R. Horaud and F. Chaumette (Eds.), *Workshop on New Trends in Image-Based Robot Servoing*, September 1997.

[7] E. Dickmanns, "Vehicles capable of dynamic vision," in *Proceedings of the 15th International Conference on Artificial Intelligence*, August 1997.

[8] E. Dickmanns, "An approach to robust dynamic vision," in *Proceedints of the IEEE Workshop on Robust Vision for Vision-Based Control of Motion*, May 1998.

[9] P. I. Corke and M. C. Good, "Dynamic effects in visual closed-loop systems," *IEEE Transactions on Robotics and Automation*, vol. 12, no. 5, pp. 671–683, October 1996.

[10] P. I. Corke, "Visual control of robot manipulators-a review," In K. Hashimoto (ed.), *Visual Servoing*, pp. 1–31. World Scientific, New York, 1993.

[11] P. I. Corke, "Visual Control of Robots: High-Peformance Visual Servoing," *Mechatronics*, New York, John Wiley & Sons, 1996.

[12] R. Kelly, "Robust asymptotically stable visual servoing of planar robots," *IEEE Transactions on Robotics and Automation*, vol. 12, no. 5, pp. 697–713, October 1996.

[13] W. Hong, *Robotic catching and manipulation using active vision*, Master's thesis, MIT, Cambridge, MA, September 1995.

[14] A. Rizzi and D. E. Koditschek, "An active visual estimator for dexterous manipulation," *IEEE Transactions on Robotics and Automation*, vol. 12, no. 5, pp. 697–713, October 1996.

[15] P. Sharkey, D. Murray, S. Vandevelde, I. Reid, and P. Mclauchlan, "A modular head/eye platform for real-time reactive vision," *Mechatronics*, vol. 3, no. 4, pp. 517–535, 1993.

[16] P. Sharkey and D. Murray, "Delays versus performance of visually guided systems," *IEE Proceedings–Control Theory Applications*, vol. 143, no. 5, pp. 436–447, September 1996.

[17] C. Brown, "Gaze controls with interactions and delays," *IEEE Transactions on Systems, Man and Cybernetics*, vol. 20, no. 2, pp. 518–527, 1990.

[18] D. Coombs and C. Brown, "Real-time binocular smooth pursuit," *International Journal of Computer Vision*, vol. 11, no. 2, pp. 147–164, October 1993.

[19] J. Dias, C. Paredes, I. Fonseca, H. Araujo, J. Batista, and A. Almeida, "Simulating pursuit with machines: Experiments with robots and artificial vision," *IEEE Transactions on Robotics and Automation*, vol. 14, no. 1, pp. 1–18, 1998.

[20] H. Araujo J. Batista, J. Dias and A. Almeida, "The isr multi-degrees-of-freedom active vision robot head: Design and calibration," *M2VIP'95–Second International Conference on Mechatronics and Machine Vision in Practice*, Hong–Kong, September 1995.

[21] J. Batista, P. Peixoto, and H. Araújo, "Real-time visual behaviors with a binocular active vision system," *ICRA97–IEEE International Conference on Robotics and Automation*, New Mexico, April 1997.

[22] J. Batista, P. Peixoto, and H. Araújo, "Real-time vergence and binocular gaze control," *IROS97–IEEE/RSJ International Conference on Intelligent Robots and Systems*, Grenoble, France, September 1997.

[23] B. Espiau, F. Chaumette, and P. Rives, "A new approach to visual servoing in robotics," *IEEE Transactions on Robotics and Automation*, vol. 8, no. 3, pp. 313–326, June 1992.

[24] P. Allen, A. Timcenko, B. Yoshimi, and P. Michelman, "Automated tracking and grasping of a moving object with a robotic hand-eye system," *IEEE Transactions on Robotics and Automation*, vol. 9, no. 2, pp. 152–165, 1993.

[25] J. Batista, P. Peixoto, and H. Araujo, " Real-time active visual surveillance by integrating peripheral motion detection with foveated tracking," *Proceedings of the IEEE Workshop on Visual Surveillance*, pp. 18–25, 1998.

Chapter 13

ROBUST IMAGE PROCESSING AND POSITION-BASED VISUAL SERVOING

W. J. Wilson and C. C. Williams Hulls
University of Waterloo
F. Janabi-Sharifi
Ryerson Polytechnic University

Abstract

This chapter presents image processing techniques and issues that are related to *position-based* robot visual servo control. It examines the trade-offs between the requirements of speed, accuracy and robustness. To achieve the appropriate balance of these requirements, directed image processing is implemented which uses windowing techniques, prediction techniques, and a priori information regarding the features of the target object. Details of the binary image processing used to measure hole and corner feature locations on the image plane are presented. Kalman filtering is used to estimate the relative pose of the target object with respect to the camera, and this information in conjunction with the object description, is used to predict the window size and location for each feature on the next sample. In addition, feature planning techniques are used to provide a database which forms the basis for real-time feature switching along the relative motion trajectory to maintain a suitable set of visible features for visual servoing. Finally, sensor fusion techniques, based on the Kalman filter, are presented to show how measurements from multiple cameras, or a camera and range sensor can be properly integrated to improve the robustness and accuracy of the pose estimation, based on the image processing results.

13.1 INTRODUCTION

Visual control of robots has received a significant amount of attention over the last ten years and has matured as a research subject in robotics and automation. The tremendous increase in computation speeds, and therefore image processing speed, has made it feasible to perform dynamic control of robot motion based on vision as the prime feedback sensor. Although several successful visual servoing systems have been demonstrated [1], this technology has not yet been transferred to real industrial problems. Some of the main reasons for this are the complexity of the visual servoing systems and concerns about their robustness or reliability in real life visual environments. This chapter will focus on the image processing issues of robustness, accuracy, and speed in the context of the visual servoing environment.

In visual servoing applications, image processing is usually directed toward the measurement of specific geometric feature characteristics rather than general scene analysis and interpretation. For example, a distinguishable image area, which is generically referred to as a hole, will have a center, an area, a perimeter, and a major and minor axis as potential geometric measurements. For a distinguishable edge, the geometric measurements are the slope and intercept, and for a corner, the geometric measurement is the intersection point of the two edges. Image feature attributes such as texture, reflectivity, shape and color may be used to distinguish features from the surrounding image, but are not directly used as measurements for visual servoing.

It follows that the two main image processing functions required in visual servoing are the segmentation and identification of selected features in the image plane, and the subsequent image processing required to obtain the geometric measurements associated with the segmented features. In this sense, the camera is being used a *directed sensor* rather than a general purpose vision system. The three main required characteristics of these vision sensors are *accuracy*, *speed*, and *reliability* of the feature measurements. Accuracy is of course critical, since the ultimate accuracy of the visual servoing operation is directly dependent on the accuracy of the raw measurements. The measurement accuracy depends upon the type of features being used, the size of the feature image, the point of view of the feature, the lighting conditions, the spatial quantization, the gray level quantization, and the image processing methods used to obtain the geometric measure.

Image processing speed is also critical since the camera acts as a sensor in the feedback loop of a sample-data, dynamic control system. To obtain good dynamic responses from the visual servoing system, the sample rates should be at least 50–100 Hz, which means that all of the image processing and all the controller calculations must be carried out in less than 10–20ms [2]. Also, the processing time affects the robustness of the image plane feature tracking. The sample rate must be high enough relative to the rate of motion of the target object, such that the image features do not move large distances between samples. Even with the availability of fast processors, the time constraints on the image processing present significant challenges, and trade-offs between sample time and the complexity of image processing must be made.

The reliability of obtaining the geometric measurements for the selected set of image features is very important in maintaining the integrity of the visual servo loop. There is a minimum measurement set required that will depend on the DOF of the visual servo loop. For example, for full six DOF motion, the visual servo loop requires the x-y coordinates of at least three non-coplanar feature points for a pose estimate, and at least four feature points for a *unique* pose estimate [3]. Without this number of measurements, some DOFs will become uncontrollable. For these reasons, it is very important to use well defined features and robust image processing so that features are not lost, and to also use redundant sets of image features to minimize the effect of lost features and enhance the reliability of the system.

Although the image processing problem could be defined as a stand-alone function, there can be significant interaction between the image processing requirements and the visual servoing technique that uses the resulting measurements. For example, the image processing requirements for a manipulator mounted camera can be significantly different from those of a stationary mounted camera. The relative motion, perspective, and field of view will vary much more for a manipulator mounted camera than for a stationary camera. The number and type of feature measurements required may also depend on the method being used for visual servoing. Visual flow methods for detecting image motion are quite different from methods that track a number of feature points to determine image motion. Also, the type and amount of feedback information that is available to make the image processing function more efficient and robust depends on the visual servoing methods being employed.

This chapter focuses on the image processing issues associated with the *position-based* class of visual servoing. In these methods the visual servoing problem is separated into two major functions. The first is to implement high-quality real-time estimation of the position and orientation of the target object relative to the camera coordinate frame. The second function is the implementation of a Cartesian coordinate robot controller using the estimated position and orientation defined in the camera frame. In this visual servoing method, there is significant interaction between the position and orientation (*pose*) estimation process and the image processing that provides the feature measurements.

The next section gives a brief overview of position-based visual servoing and discusses the interaction between the image processing and the pose estimation functions. Then Section 13.3 discusses the details of the adaptive image processing used to determine hole and corner image plane feature measurements for manipulator mounted cameras. Section 13.4 discusses feature planning methods for defining the optimum set of features to use for visual servoing as the manipulator moves relative to the target object. Then in Section 13.5, methods for dealing with redundant sensors, including multiple cameras, are presented, and the effect on the robustness of the image processing and the visual servoing is discussed. Section 13.6 presents some conclusions and discussion of some problems that need future investigation.

13.2 POSITION-BASED VISUAL SERVOING AND IMAGE PROCESSING

13.2.1 Position-Based Visual Servoing Overview

A brief description of the position-based visual servoing method [4] is presented here to provide some background for understanding the interaction between this method and the image processing functions. Figure 13.1 shows a block diagram of the overall system. It is assumed that a camera is mounted on the end-effector of the robot and that its world coordinate position and orientation at time k, is represented by the pose vector, $\mathcal{X}_{\mathbf{robot,k}}$. If the object motion is described in the world coordinate frame by $\mathcal{X}_{\mathbf{object,k}}$ the camera sees the relative pose of the target object in the camera frame. Then the measurements of the image plane coordinates of the selected object features, $\tilde{\mathcal{F}}_{\mathbf{k}}$, are used as the input to the separate pose estimation block.

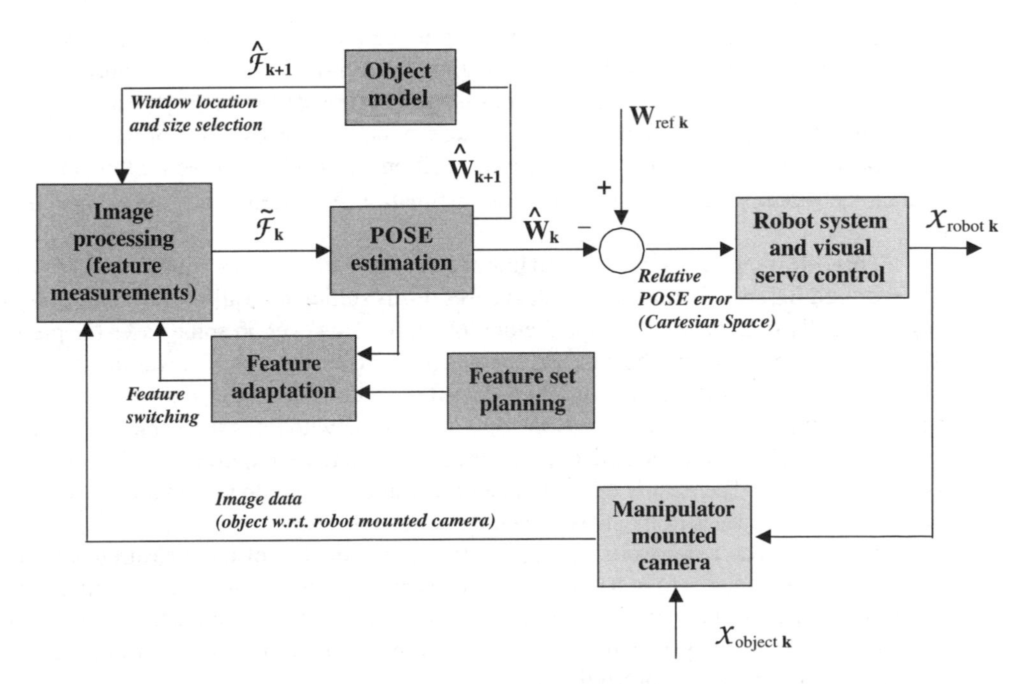

Figure 13.1 Image processing with position-based visual servoing and an endpoint mounted camera.

13.2.2 Adaptive Image Processing and Position-Based Visual Servoing

In the position-based visual servoing, it is assumed that the feature characteristics of the object are known, which allows the solution of the photogrammetric equations given four image plane feature coordinate measurements. This results in the relative pose estimate, $\hat{\mathbf{W}}_{\mathbf{k}}$ defined in the camera Cartesian coordinate frame. A Kalman filter is used for the implicit solution of these equations in a recursive manner. This solution choice provides a sound basis for statistically combining redundant feature measurements from the camera, or in fact, from several cameras in the environment. It also has the flexibility of dropping and adding feature measurements without losing the appropriate weighting of previous measurements in the pose estimate. This is achieved by dropping or adding feature measurements in the state output vector, and by making the appropriate adjustments to the measurement noise covariance matrix. In addition, the Kalman filter method generates predictions of the relative pose at the next sample time, $\hat{\mathbf{W}}_{\mathbf{k+1}}$, which is combined with the object model information to generate good estimates of the image feature locations at the next sample time, $\hat{\mathcal{F}}_{\mathbf{k+1}}$.

The visual servo loop is completed by comparing the estimated relative pose, $\hat{\mathbf{W}}_{\mathbf{k}}$, with the desired relative pose, $\hat{\mathbf{W}}_{\mathbf{ref,k}}$ to generate the relative pose error vector in the camera Cartesian coordinate frame. This pose error is then used to provide fairly standard Cartesian control of the robot. This method has been tested experimentally [4] and has demonstrated very good dynamic estimation of the relative pose, good tracking of a free moving object based on these estimates, and a good ability to execute desired relative trajectories even while the target object is moving. The sample rate in these experiments was 61 Hz and pose estimation accuracies were in the range of 0.5 mm and 0.5 degrees for position and orientation, respectively.

To achieve the required trade-off between accuracy, speed, and reliability, *directed* image processing is required. In most cases windowing techniques are used to focus the image processing around the features of interest and to minimize the number of pixels that need to be processed. This means that at each sample, a certain amount of a priori information must be assumed, which typically includes a specification of the image features to be used, their type, their estimated location on the image plane, and an estimate of the required window size around the feature. In the case of position-based visual servoing, this information is obtained directly from the predicted pose estimation, $\hat{\mathbf{W}}_{\mathbf{k+1}}$, and the target object description.

This information provides a good initial estimate of the window coordinates and window size required for each feature, and in most cases this is sufficient to allow the image processing to successfully determine the image feature locations. However, in some cases the prediction is not accurate enough and the window is not centered over the feature, resulting in a partially captured feature or lost feature. In other cases the window size may be too large resulting in multiple features in the window, or too small so that the whole feature cannot be seen in the window. Therefore, some adaptive image processing may be required to adjust the window location and size. The details of the image processing and window adaptation used for hole and corner feature measurements is presented in Section 13.3.

There are also a number of issues relating to the quality of the measurements and the accuracy of the results. These include focusing, spatial quantization, the use of binary images instead of grey scale, field of view limitations, and the ability to discriminate between two near by features. Some analysis and discussion of these issues in the context of hole and corner features are also presented in Section 13.3.

13.2.3 Feature Set Planning and Feature Switching

It is clear that with a manipulator mounted camera, the field of view of the target object and the point of view with respect to the features will change as the robot moves, and this will usually result in occluded features or poor quality images of the selected features. It makes sense then to consider feature set planning which will provide information on visual constraints and visual quality of the feature images at various points along the desired relative trajectory. In some cases this could be done heuristically; however, it is much better to make a systematic evaluation of the best feature sets to use to enhance the accuracy and reliability of the feature measurements. The first part of this feature set planning is to determine the best feature sets to use over a range of relative poses, using criteria based on geometric constraints, visual constraints and visual quality measures. This planning requires the model of the target object and the lighting conditions which are consistent with the assumptions made in the position-based visual servoing.

Once the planning database has been calculated, it is used in conjunction with the real-time predicted pose estimates, to select the best feature sets at any sample time along the path. The implementation of this depends on the ability of the image processing methods to switch feature sets from one sample to the next, and the ability of the pose estimation algorithm to use different feature set measurements without losing the past estimation information so that a smooth transition can be made. Again, the Kalman filter structure allows measurements to be dropped and added from one time step to the next without degrading the pose estimation process. This is a very important capability when the relative trajectories start at a distance and then move in close to the object to perform an operation. A more detailed description of this feature set planning and real-time feature switching is presented in Section 13.4.

13.2.4 Integration of Multiple Sensors and Redundant Image Measurements

The pose estimator is based on Kalman filtering which will properly integrate redundant image plane feature measurements to give improved pose estimation accuracy. In addition, the pose estimation includes a check for the consistency of the set of image plane feature measurements so that outlier measurements can be identified, and if necessary rejected from the estimate. If the same feature measurements are consistent outliers, then they are rejected from further feature sets. The combination of redundant information and the feedback of estimator predictions leads to a more accurate and robust visual servoing system.

Further enhancement can be obtained through the integration of several sensors, providing further redundancy and flexibility in the image measurements and resulting pose estimation. During some parts of the visual servoing trajectory, there may be insufficient information from one camera sensor to allow full six DOF control. If the measurements from another sensor, such as a second camera or a range sensor, are properly integrated with the measurements from the first camera, failure of the operation can be avoided. These topics are discussed in more detail in Section 13.5.

13.3 DIRECTED IMAGE PROCESSING AND ADAPTIVE WINDOWING

Section 13.2 described a pose-based visual servoing system. The system can be divided into a low-level execution layer and a high-level planning layer. At each level, different techniques can be used to improve the robustness of the image processing. This section describes techniques that can be used as part of the low-level image processing that is performed by the execution layer. First, the image processing requirements associated with hole and corner features are discussed. Then a discussion of methods to improve the robustness of the feature extraction process is described. Next, the problem of defocused image features is explored, particularly in the context of binary image processing. Finally, two low-level approaches to perform adaptive windowing and feature selection are presented. High-level planning-based approaches are presented in Section 13.4.

13.3.1 Image Processing Associated with Different Feature Types

An image feature is any structural feature that can be extracted from an image. Image features can generally be divided into region-based features such as holes, planes, and surfaces, and line segment-based features such as edges and corners. Hutchinson et al. [1] defines an image feature parameter to be "any real-valued quantity that can be calculated from one or more image features." The image plane coordinates of a feature have been used in many visual servoing systems [1]. In addition, parameters such as area, perimeter, thickness, curvature, straightness, complexity, and elongatedness can be used to describe a region [5]. A region can also be described by moments which are invariant to translation, rotation, and scaling [6]. Segment-based features can also be defined in terms of length and orientation [7]. The discussion in this section focuses on hole and corner features as these are the features that are currently used by the pose-based visual servoing system described in Section 13.2.

13.3.1.1 Hole Feature Extraction. A hole is defined to be any bounded region of an object which has an intensity level different from is surrounding region. Convex hole features can be classified into three types: circular and elliptic holes, symmetrical and asymmetrical polygonal holes, and irregular holes. Estimation of parameters from hole features requires that a single window be placed so as to include the entire feature point. Parameters that can be extracted from a hole include the area, centroid, orientation, and perimeter.

The area of a hole, A_h, in the image plane is defined as the zeroth moment,

$$A_h \sum_i \sum_j I(i, j) \tag{13.1}$$

where

$$I(i, j) = \begin{cases} 1 & \text{if intensity of pixel at } (i, j) < \text{ threshold} \\ 0 & \text{otherwise} \end{cases} \tag{13.2}$$

The area parameter can be used in several ways. Area information when combined with the object model information can be used to help determine the pose of an object. Also, it can be used for feature selection. If a feature is becoming too large or small, it can be dropped in favour of more suitable features. In addition, it can be used to perform adaptive window sizing. When the system detects that the feature area is changing, the window size can be adjusted accordingly. Finally, it can be used for outlier rejection. When the measured area is significantly different from the expected area, then it is likely that the feature has been misidentified or poorly extracted.

The centroid of a hole feature, (X_c, Y_c), is determined as the first moment of the image which is, the average of the x and y coordinates of the pixels in the image that represent the hole.

$$X_c = \frac{1}{A_h} \sum_i \sum_j i I(i, j) \tag{13.3}$$

$$Y_c = \frac{1}{A_h} \sum_i \sum_j j I(i, j)$$

where $I(i, j)$ is given by (13.2). The image plane coordinates of hole features are the primary feedback measurements used by the pose-based visual servoing system. When combined with information about the object model, they allow the system to accurately determine the relative pose of the object with respect to the robot end-effector. For pose estimation and control in 3D space, the centroid of the projected image of the hole feature must correspond to the centroid of the object hole feature. Thus, irregularly shaped holes cannot be used. The projection of an elliptical feature is always a circle or an ellipse. Thus the projected centroid will always correspond to the centroid of the object feature [8]. The centroid location can also be used for feature selection and planning. As features move into difficult to extract positions they can be replaced by better-positioned features. Finally, outlier rejection can be performed by comparing the expected centroid location with the measured location.

Orientation is not uniquely defined for symmetrical polygonal holes and is undefined for circular holes. Hence this parameter is evaluated with respect to unsymmetrical polygonal holes. The orientation of a 2D connected convex region can be determined from just the boundary pixels (x_b, y_b). When the pixel coordinates are defined with respect to the centroid, the orientation, θ, is defined as [9],

$$\theta = \frac{1}{2} \arctan \left[\frac{2m_{1,1}}{m_{2,0} - m_{0,2}} \right] \tag{13.4}$$

where

$$m_{j,k} = \frac{1}{N_b} \sum \sum_{R_b} (x_{b_i} - X_c)^j (y_{b_i} - Y_c)^k) I(x_{b_i}, y_{b_i}) \tag{13.5}$$

where $I(x_{b_i}, y_{b_j})$ is given by (13.2), N_b is the number of boundary pixels, and R_b is the perimeter of the area. The orientation can be used as a measurement parameter to determine the object pose, and it can be used for outlier rejection.

The perimeter of a hole is defined by the boundary pixels R_b. It can be used for feature selection, outlier rejection, and adaptive window sizing in a manner similar to the area parameter. The location of the perimeter can also be used to perform adaptive window sizing. This is the approach used for the system described in this chapter. The details of this approach are given in Subsection 13.3.5. The perimeter can be obtained using edge detection methods. Alternatively, it can be extracted during a hole pixel classification process such as the one described in Subsection 13.3.2.

13.3.1.2 Corner Feature Extraction. A corner is defined as the point of intersection of two straight edges. Estimation parameters obtained from a corner feature include the corner angle, the corner orientation, and the corner location. There are two main approaches used for corner extraction, two-window approaches and single-window approaches [10].

Two window corner extraction methods consist of three stages [11]:

1. Window placement
2. Edge detection
3. Corner estimation

The first step is to place a window along each of the two *observable* edges that form the corner. Then edge detection must be performed within each window. When using binary images, edges are detected as discontinuities in the intensity. Factors such as nonuniform illumination and reflectivity will introduce spurious discontinuities. Thus as part of the edge detection process a linking algorithm is used to assemble the edge pixels into a meaningful set of boundaries. Linking methods can be classified as local and global methods [9]. Local methods use characteristics of the gradient to determine if the pixel satisfies a particular set of conditions. These methods require relatively extensive processing. Using global methods, edge pixels are linked by evaluating their distance from a curve of specified shape. The method of least squares is suitable for this process as the edges involved are straight lines [9]. Once the edges have been extracted, the corner location is calculated as the intersection of the two edges.

The advantage of a two-window approach is that the algorithm is the simplest to implement. There are several inherent problems with this approach [10]. First, it is not always possible to place two windows such that each window encloses only one edge. This is often the case for small corner angles. Second, it can be difficult to determine the appropriate window locations. Of the three edges that determine a simple corner in 3D space, only two edges are likely to provide images with a noticeable discontinuity. Also, if the windows are placed some distance from the corner, small errors in the slope and intercept of the two least-square line fits can cause large errors in the estimated corner position. Finally, the processing of two windows makes it difficult to achieve the fast sampling rates which are necessary for dynamic control. Fortunately, these problems can be overcome by using the single-window approach.

A single-window approach uses one window which covers the corner feature itself. Single-window approaches consist of four stages [10]:

1. Window placement
2. Window division based on a rough estimate of the corner position
3. Edge detection within each window part
4. Corner estimation

Two single-window approaches were introduced by Madhusudan [9]. The only difference between the approaches is the procedure to obtain the rough estimate of the corner position. However, both of the methods presented involve numerous special cases which depend on the orientation and position of the corner feature. In addition, the method of moments only works well for corners with angles near 90° that are located near the center of the window, while the pivot point approach does not work for angles greater than 90° and it is very sensitive to noise.

The discussion presented here focuses on the maximum distance approach described by Wong [10]. Figure 13.2 shows the corner extraction process using this method. Once the window has been obtained, the first step of the algorithm involves locating the two intersection points of the corner edges and the window borders, (x_1, y_1) and (x_2, y_2). The window is then scanned to locate the pixel (x_c, y_c) which has the greatest vertical or horizontal distance from the line defined by (x_1, y_1) and (x_2, y_2). Since the scanning process is time consuming, a high-pass filter is implemented such that only the edge pixels are left to process. To obtain a finer estimate of the corner position, a line is drawn from the pixel (x_c, y_c) to the midpoint (x_m, y_m) of the line defined by (x_1, y_1) and (x_2, y_2). The line defined by (x_c, y_c) and (x_m, y_m) is used to divide the window into two parts. Edge detection and least squares line fits are performed to extract the two edges which form the corner. As shown in Fig. 13.2(b), the estimated corner position is the intersection of these two edges resulting from the least squares line fits.

The distance between the corner and the line joining the intersections is always the maximum regardless of the corner angle and the corner location in the window. There is only

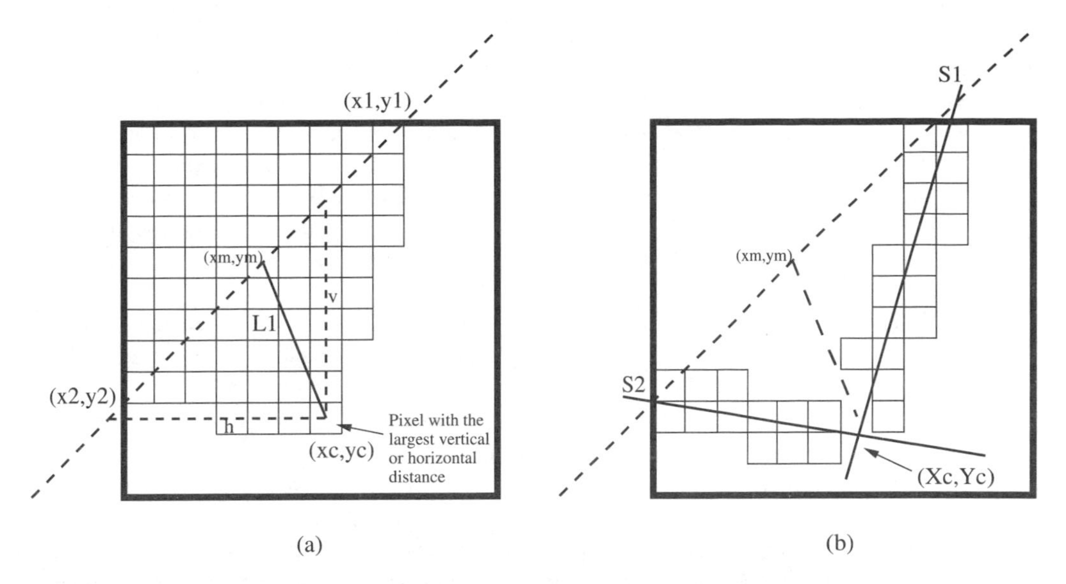

Figure 13.2 Corner estimation using the maximum distance approach.

one special case associated with this approach and it occurs when the two corner edges cut the window border at the same x or y coordinate. Figure 13.3 shows this case. In this case, only the horizontal or vertical distance can be used for searching for the corner pixel (x_c, y_c) as the other distance is undefined. Once (x_c, y_c) has been located, the algorithm proceeds as described above.

13.3.2 Accuracy of Feature Measurements

This section considers the suitability of hole and corner parameters for object pose estimation. For hole features, the area, orientation, and centroid parameters can be used as measurements for pose estimation. Experiments have shown that the area of a hole is sensitive to the relative position and the size [9]. The percentage error was shown to increase from

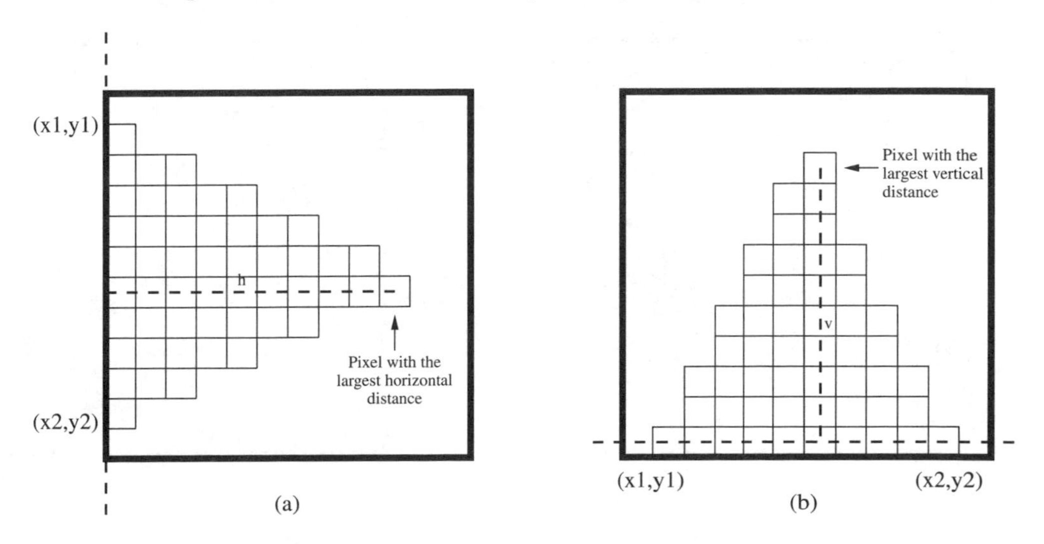

Figure 13.3 Special case of the maximum distance approach.

18% to 38% with an increase in size from 3 to 8 pixels. The orientation is only defined for asymmetrical holes and thus cannot be used for the circular holes that have (to date) been used for the experimental system described in this chapter. The centroid measurements are the most suitable for pose estimation. Experimentally, errors in the centroid of circular holes have been characterized as near normal with a zero mean and a variance of 0.06 pixel^2 [9, 11]. For hole feature extraction however, the window must be placed such that the entire hole is included. Subsection 13.3.5 discusses the use of adaptive window relocation and sizing. In addition, multiple holes within a single window can cause errors in the centroid calculations if all window pixels are considered.

The occurrence of capturing multiple features can be reduced by using a window size that is just large enough to contain the single feature. However, this increases the likelihood of only capturing a partial hole. Subsection 13.3.5 discusses an algorithm for calculating an appropriate window size that attempts to balance these two issues. To deal with the possibility of multiple features, the hole of interest can be identified and separated before the centroid calculations are executed. The feature identification method can be based upon factors such as the shape or size of the feature, or the location of the features within the window. Assuming reasonable accuracy for the window prediction process, the desired hole should be located near the center of the image window. Thus, the system described in this chapter selects the hole that is closest to the center of the image. While this method has the advantage of handling multiple holes, it has the disadvantage that a single noise pixel close to the center could be selected as the hole feature over the actual hole which is located further away. Figure 13.4(a) shows a correctly identified hole, while Fig. 13.4(b) shows an incorrectly identified hole. Thus, the reliability is dependent upon the accuracy of the window prediction method.

Once the hole pixel that is closest to the center of the window has been located, the pixels associated with it need to be classified. A search of adjacent pixels is used to completely identify and separate the hole feature. Figure 13.5 shows an original image and a processed image where the hole has been identified and separated.

In some cases, however, searching adjacent pixels may not be sufficient. As shown in Fig. 13.6, for a particularly noisy image this could result in a portion of the hole being excluded. Thus in a noisy environment, the search could be widened to include pixels within a small neighborhood. This would increase the processing time. A more suitable threshold level or the use of grey scale images could also alleviate some of this problem.

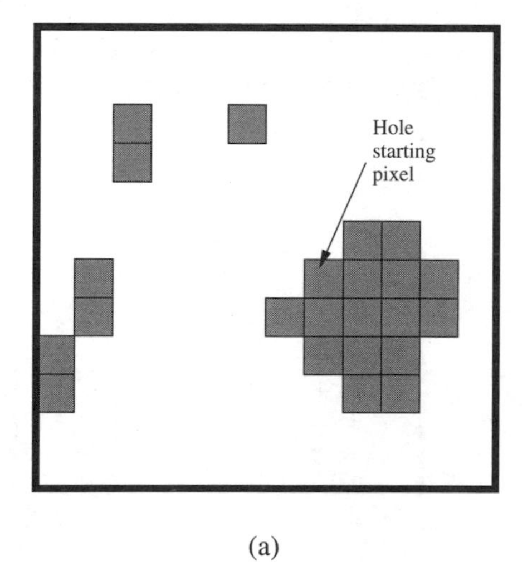

(a)

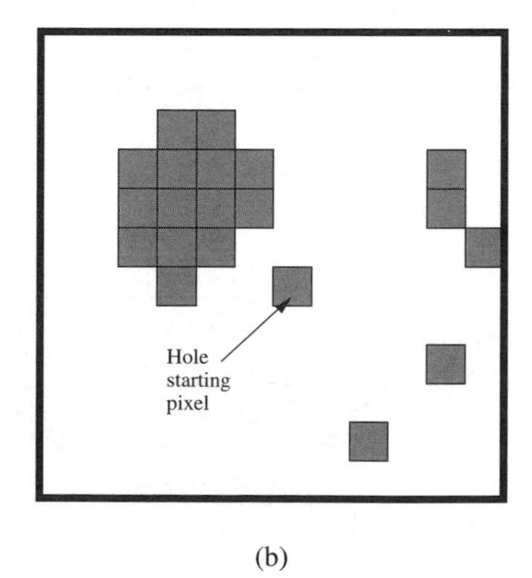

(b)

Figure 13.4 Hole identification.

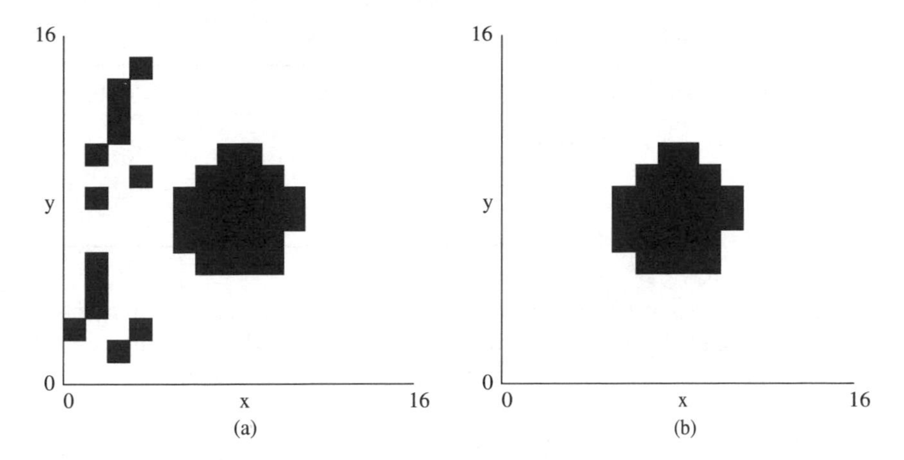

Figure 13.5 Hole identification and separation.

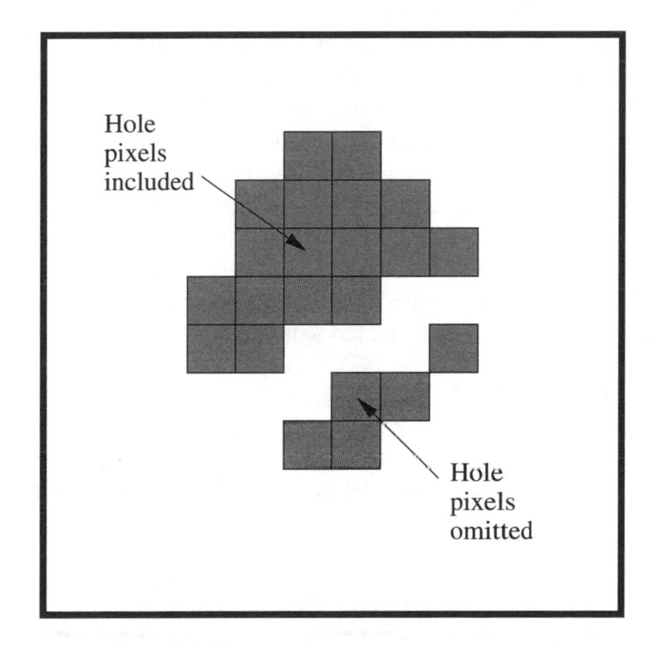

Figure 13.6 Hole identification and separation error.

The advantage of identifying and separating the desired hole feature is that the measurement accuracy is improved as noisy pixels and other features are not included as part of the measurement calculation. However, in addition to the potential problems described above, a major disadvantage of the method is that the window must be analyzed in detail in order to find the desired hole and classify the pixels associated with it. Table 13.1 compares the time required to extract five hole features from 16×16 pixel windows for the simple case when all black pixels are assumed to be hole pixels, and for the case when a hole identification and separation algorithm is utilized. When adaptive windowing is performed, the algorithm may be performed several times until a suitable image window is obtained.

Performing hole identification and separation significantly increases the execution time involved in calculating the centroid of a hole. Thus, for well-separated holes that are well-illuminated it is better to obtain a higher sample rate by simply declaring all dark pixels to be

TABLE 13.1 Comparison of hole extraction times.

Algorithm step	Simple Case (ms)	Using Hole Id (ms)
Pre- and postprocessing	1.7	1.7
Find and classify hole pixels	0.0	2.1
Feature extraction	1.8	1.7
Total time	**3.5**	**5.5**

hole pixels and then using outlier rejection to eliminate outlier measurements resulting from noise or from partial captures of a hole. When these conditions cannot be met, or if some other part of the system is the computational bottleneck, then identifying and separating a hole feature can be worthwhile.

The location of an object corner is also a suitable parameter to use for 3D pose estimation. Experiments have shown that the error associated with corner feature extraction is orientation dependent, with maxima occurring near orientations that are multiples of 45°. In addition, the use of larger windows results in more accurate measurements. For a 16×16 window the variances were found to be 0.0533 pixel^2 in the x direction and 0.0884 pixel^2 in the y direction [10]. For a 32×32 window the variances were found to be 0.0161 pixel^2 in the x direction and 0.0262 pixel^2 in the y direction [10].

Like the hole separation algorithm discussed above, the maximum distance corner estimation approach can suffer from noise sensitivity. As shown in Fig. 13.7, a noisy pixel in the background region can cause an error in corner estimation. This problem can be reduced by executing a corner classification algorithm before the feature extraction occurs. For example, the pixels that are classified as part of the largest area of object pixels can be defined as object pixels while all other pixels are declared to be part of the background. This will, however, increase the computation time associated with the feature extraction process.

13.3.3 The Effects of Lens Focus

When a visual servoing system operates over a large range of object-camera distances, the issue of image defocusing must be considered. For a monofocal lens, this is the only

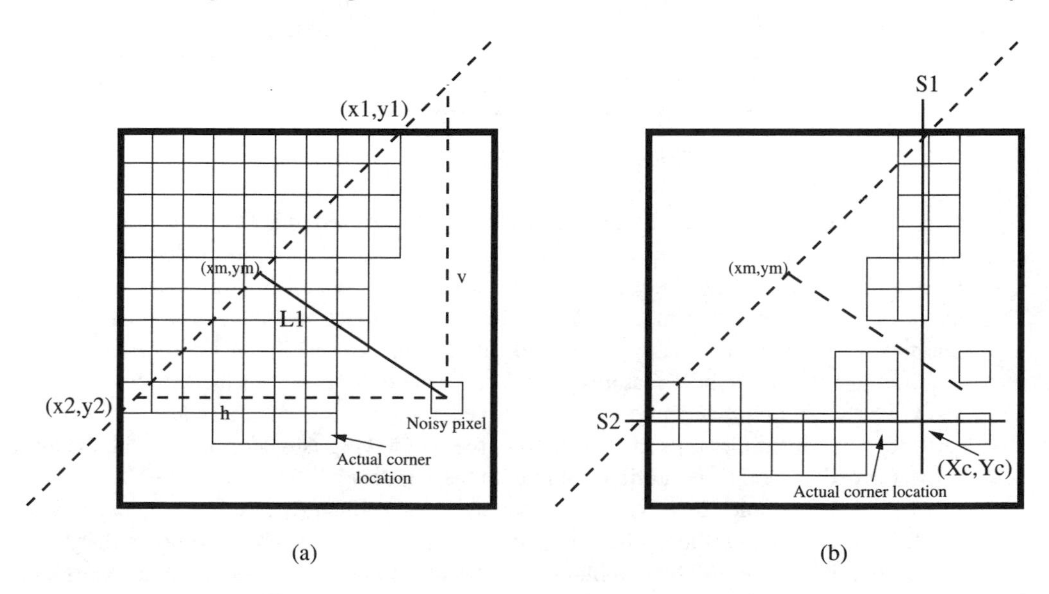

Figure 13.7 Maximum distance approach error due to noise.

disturbance that cannot be reduced, even for a perfect lens. A defocused image occurs when light rays originating from the same source do not converge at the same position on the image plane. The light rays will instead spread around and form a disc on the image plane. For a perfect lens, this disc is circular in shape and it is commonly known as the disc of confusion. Figure 13.8 shows the case where the camera is too close to properly focus upon an object.

Image defocusing can be represented by a two-dimensional convolution of the image intensity with a radially symmetric function [12]. Bishop et al. [12] presented a method of simulating defocus by applying a discrete convolution to the image multiple times. For the simulation experiments described in this chapter, a common linear smoother known as the 2D Gaussian filter is used to simulate the effects of defocusing [10].

Using simulation experiments, the effects of defocusing on a small hole feature (6 pixel diameter), a large hole feature (10 pixel diameter), and a corner feature were studied. Figures 13.9, 13.10, and 13.11 show the results of defocusing for each case.

To determine the magnitude of the error introduced by defocusing, simulation experiments were involving a camera placed at a distance of 0.32 m from an object that is approximately 12 cm×12 cm with hole diameters in the neighborhood of 1cm, were carried out. Table 13.2 shows the results of these experiments.

The first set of rcsults shows the estimate error that occurs when the image is in focus. Note that there is a significant difference in the estimation error of the x and y coordinates. The object features used to estimate the relative position are further apart in the x direction than in the y direction. The smaller separation in the y direction increases the resulting estimate error. For these experiments, the location of features within the object feature set has a greater impact on the error than the effects of image defocusing.

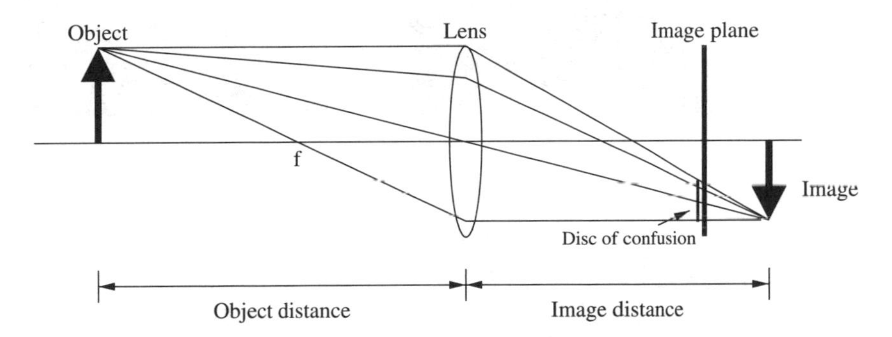

Figure 13.8 Ray diagram of an out of focus image.

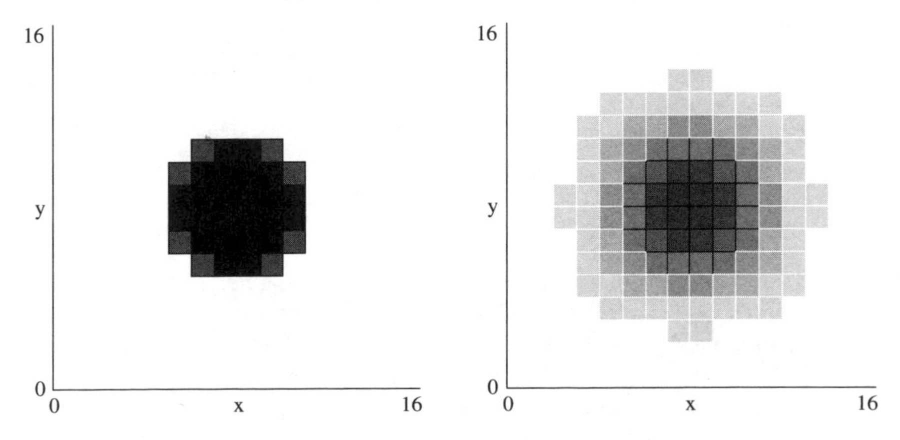

Figure 13.9 Defocusing of a small hole feature.

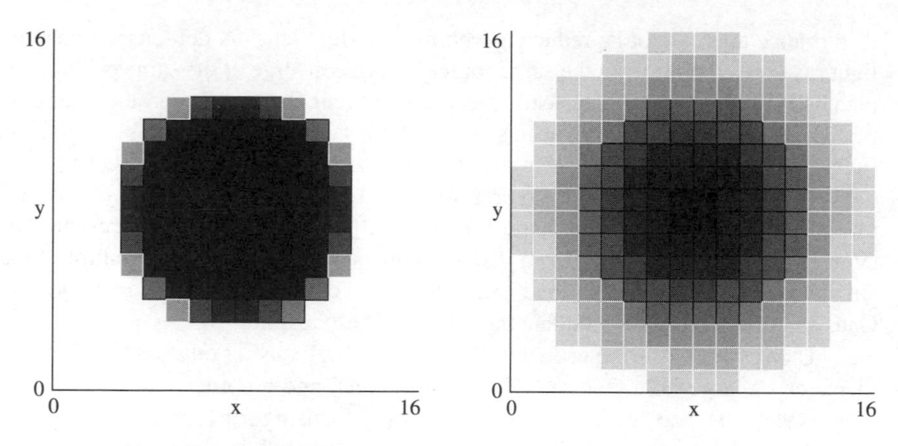

Figure 13.10 Defocusing of a large hole feature.

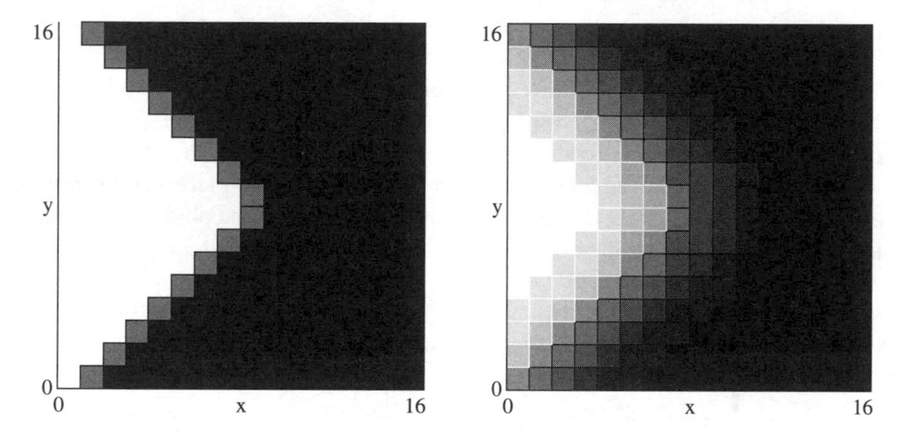

Figure 13.11 Defocusing of a corner feature.

TABLE 13.2 Estimation error changes after defocusing.

	FP Change (pixel)		**Mean Est Err (mm)**		**Percentage Increase**	
Conditions	x	y	x	y	x	y
4 focused FP	0.0	0.0	0.136	0.672	0%	0 %
4 defocused holes	−0.0658	0.0132	0.146	0.673	7%	0.1%
4 defocused corners	−0.6576	0.6888	0.679	0.859	400%	28 %
4 defocused corners	−0.6576	0.6888				
Different	−0.6935	−0.6985	0.183	0.702	34%	5 %
orientations	0.4237	0.4547				
	−0.0666	0.7623				

To simulate defocusing, a Gaussian 2D filter with a neighborhood of 4 pixels in each direction and $\sigma = 2$ pixels is applied to the image. This defocusing introduces a small change in the x and y measurements of each feature. When this change is added as bias to the hole measurements there is a small increase in the estimation error.

A more noticeable effect can occur in the calculation of a corner feature location. The magnitude of the defocusing error is dependent upon the orientation of the feature and the threshold setting of the image. Assuming that each of the four corner features is oriented in the same direction, the estimation error increases significantly. If the corners are oriented at 0°, 90°, 180°, and 270° the effect is diminished.

In addition to reducing the accuracy of feature measurements, another problem which results from a defocused image is the resolvability of features that are close together. If two or more features are very close to each other, defocusing of the image may cause these features to merge with each other. Figure 13.12 shows the effect of defocusing on two nearby hole features.

13.3.4 Binary Image Processing

For black and white cameras, the image processing system can either use grey scale images or binary images. The main advantage of a binary image over a grey scale image is the great reduction in the amount of data associated with the image. This reduces the amount of computation involved when processing the image, and hence increases the system sample rate. However, the loss of information inherent in a binary image as compared to a grey scale image can result in reduced measurement accuracy. This loss of information can lead to the misidentification of object features or the incorrect classification of feature pixels. This issue is discussed in greater detail in Subsection 13.3.2.

A nonideal threshold level can increase the errors resulting from defocusing. Table 13.3 shows the effects of different threshold levels on the position measurements obtained from defocused images of a small hole, a large hole, and a corner at various angles with respect to the image window [10]. The results show that the threshold level has a negligible effect on the measurement of defocused hole features, except when the small hole disappeared when using a high threshold level. The effect of the threshold level is more significant for defocused corner features. The measured position of a corner feature tends to shift to the direction of the corner orientation for low threshold settings, while the opposite is true for high threshold settings.

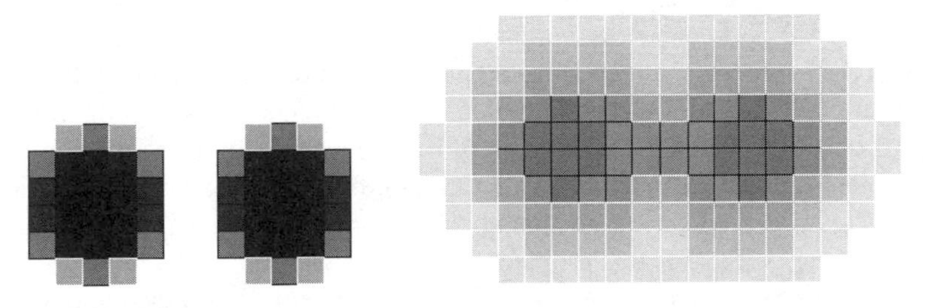

Figure 13.12 Feature merging due to defocusing.

TABLE 13.3 Position changes (in pixels) after defocusing.

Feature Type and Orientation	Threshold = 20%		Threshold = 50%		Threshold = 80%	
	x	y	x	y	x	y
Hole (10 pixels)	0.0655	−0.1276	−0.0658	0.0132	−0.0217	−0.0652
Hole (6 pixels)	0.0333	0.1333	0.0500	0.0000	∞	∞
Corner (at 0°)	1.2992	0.6957	−0.6576	0.6888	−1.7046	0.7060
Corner (at 90°)	0.1836	1.4582	−0.6935	−0.6985	−0.2022	−2.2956
Corner (at 180°)	−1.1676	0.0467	0.4237	0.4547	1.6857	0.3111
Corner (at 270°)	1.0414	−1.0977	−0.0666	0.7623	−0.5412	1.6795

These results were verified using an experimental system. The experiments used a hole of 10 pixels in diameter, and a 90° corner oriented at angles of 0°, 90°, 180°, and 270°. The camera was placed at a position 30 cm above the object. The focus was initially set to 30 cm, and then was changed to infinity, causing the image to become defocused. The change in position was recorded for three different threshold levels. Table 13.4 shows the experimental results.

As expected, defocusing has a small effect on hole extraction due to the nature of the centroid calculations. The maximum change in the position of the hole centroid is 0.561 pixels. Measurements of the corner feature tend to shift it toward the corner orientation for low threshold settings, and away from the corner orientation for high threshold settings. The maximum change in position is 4.6715 pixels.

The above results emphasize the need for the system to use an appropriate threshold level in order to obtain accurate measurements of object feature parameters. However, it is not always possible to determine a suitable threshold due to nonuniform illumination within the environment, and due to changing conditions as the visual servoing operations occur. Thus, some method of performing adaptive thresholding is desirable. The threshold could be adjusted based on whether a particular parameter is producing outliers. For example, if the area of a hole is either too small or large when compared to the predicted value of the area, the threshold would be adjusted accordingly. In addition to adjusting the global threshold levels, if the image is originally captured using grey scale, individual threshold levels could then be used for each image window. This increases the robustness of the system in the case of nonuniform illumination.

13.3.5 Adaptive Windowing

For hole feature extraction, ideally an image window should completely contain a single hole feature. To reduce the likelihood of capturing multiple features, and to reduce the overall processing time which is dependent on window size, adaptive window sizing can be used to adjust the window sizes at each time step. Ideally, the window size would be set such that there is a single pixel between each window border and the hole perimeter. However, this greatly increases the probability of capturing an incomplete hole due to inevitable inaccuracies in the prediction of the appropriate window locations. The choice of appropriate border size is a trade-off based on the expected amount of hole movement prediction error and the closeness of the object features. If partial captures cannot be avoided, there are techniques, which are described later in this section, that can be used to relocate the window during the given time step in order to correct for this problem.

There are two approaches that can be used to adaptively set the window size, preplanned and reactive. A preplanned approach to window sizing is described in Subsection 13.18.

TABLE 13.4 Actual position changes (in pixels) after defocusing.

Feature Type and Orientation	Threshold = 17%		Threshold = 24%		Threshold = 31%	
	x	y	x	y	x	y
Hole (10 pixels)	0.4550	−0.2599	0.5610	−0.3528	0.5541	−0.3584
Corner (at 0°)	1.3610	0.0968	−3.2425	−0.1418	−6.6029	0.0511
Corner (at 90°)	0.5165	2.8588	0.7012	−1.7425	−1.9112	−5.3228
Corner (at 180°)	−3.1873	−0.8355	1.1159	−0.6182	4.6715	0.1393
Corner (at 270°)	0.9309	−1.9809	−1.0313	2.5374	−1.7127	5.7321

Alternatively, or as a complement to a preplanned approach, the window size for the *next* time step can be set based upon the size of the border that was extracted from the *current* window. Figure 13.13 illustrates this approach. At each sample time, the distance from each edge to the hole perimeter is obtained. If the sum of the top distance D_t and the bottom distance D_b (or left D_l and right D_r distances) is greater than twice the desired border width then the window is reduced in size in the appropriate direction. Similarly, the window is increased in size if the sum of the border distances is less than twice the desired border width.

For the experimental system described in this chapter, a hole border of 5 pixels appears to work well. Experiments show that while tracking and then grasping an object that followed a moderate speed random trajectory, the system did not have any occurrences of partially captured holes. The main drawback of this method is the execution time associated with identifying the hole pixels nearest each of the four window borders in order to calculate the hole border size.

As mentioned above, another problem which affects the accuracy of hole parameter estimation occurs when a hole is not completely bounded by the window. This problem occurs when the hole feature is larger than the window, and when there has been an unpredicted change in the hole feature location. A fast sample rate combined with a suitable border size for the predicted windows can reduce the probability of this problem occurring. When it does occur, the window can be relocated during the current time step in order to fully capture the hole. It may also be necessary to resize the window.

The partial capture of a hole feature can be detected by checking the distance from the hole perimeter to the window edge. When this value is zero, the hole feature is incomplete. To attempt to capture the complete feature, the window can be moved in the direction of the incomplete perimeter. Figure 13.14 illustrates this process. The new window location (X_n, Y_n)

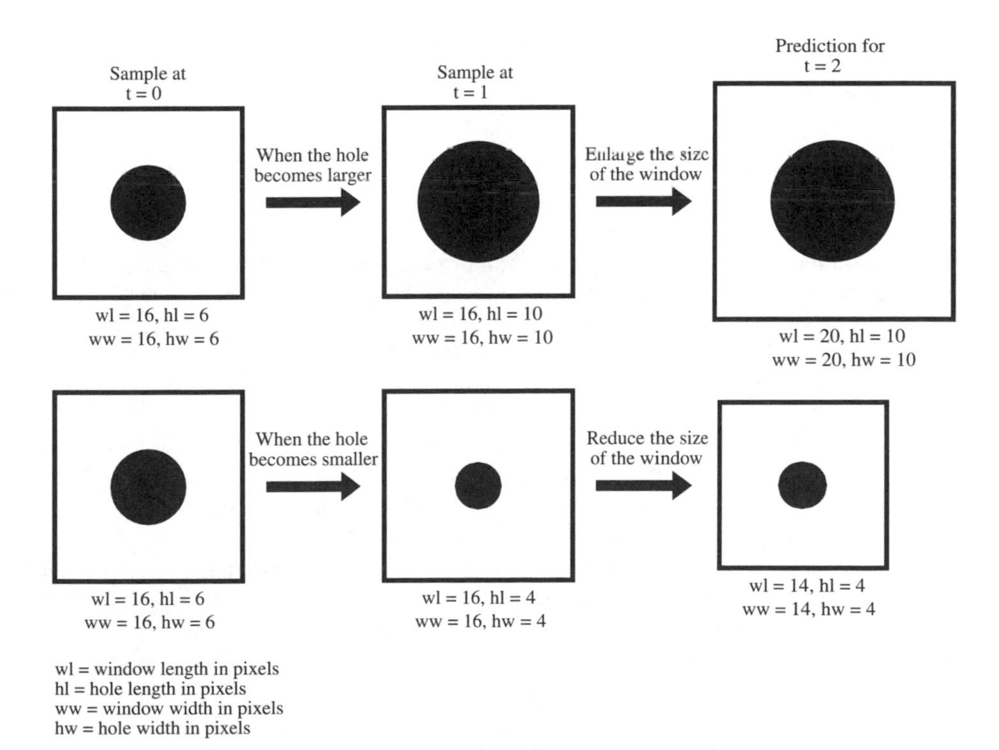

Figure 13.13 Adaptive window sizing using the previous window borders.

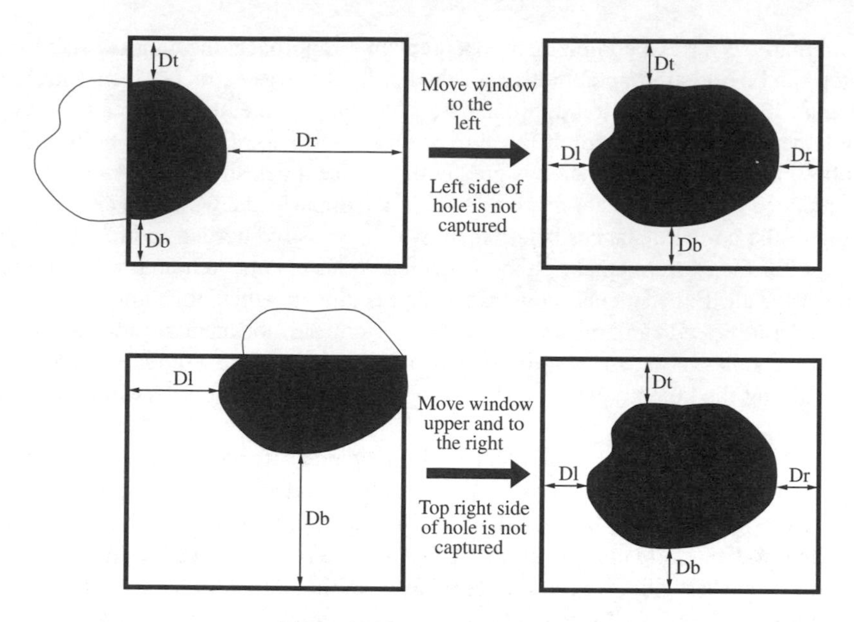

Figure 13.14 Window relocation.

is adjusted to be

$$X_n = X - \frac{1}{2}D_r, \text{ if } D_l = 0 \tag{13.6}$$
$$X_n = X - \frac{1}{2}D_l, \text{ if } D_r = 0$$
$$Y_n = Y - \frac{1}{2}D_b, \text{ if } D_t = 0$$
$$Y_n = Y - \frac{1}{2}D_t, \text{ if } D_b = 0$$

where (X, Y) is the current window location.

A special case occurs when $D_r + D_l < 1$ or $D_t + D_b < 1$. In this case, relocating the window is not sufficient to completely capture the hole, and the window must be resized as well as relocated. An example of this case is shown in Fig. 13.15. The new window location (X_n, Y_n) is adjusted to be

$$X_n = X - \frac{1}{2}K \text{ if } D_l + D_r < 1 \tag{13.7}$$
$$Y_n = Y - \frac{1}{2}K \text{ if } D_t + D_b < 1$$

while the window size in the appropriate direction is increased by K pixels. The system described in this chapter uses a window size adjustment factor of 4 pixels.

After the window is relocated, and possibly enlarged, the borders need to be recalculated and checked to see whether the hole is now completely captured. This process repeats until the hole is completely captured, until the window has been enlarged to its maximum size, or until a specified number of iterations has occurred.

While this approach can handle partial captures, it can significantly increase the overall sample time, particularly if it is combined with the hole feature identification and separation technique described in Subsection 13.3.2. For the experimental system, each iteration, which involves reobtaining the window and then identifying and separating the hole, takes 0.4 ms for a 16×16 pixel hole.

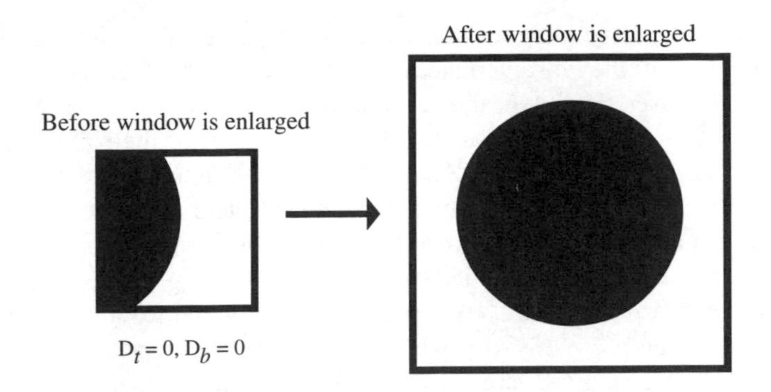

Figure 13.15 Window enlargement and relocation.

13.3.6 Feature Switching

There are two situations where the set of features to be used for pose estimation are adjusted. The first is when the measurement has been declared to be an outlier. In the second case, the feature set is adjusted based on preplanned information. Both of these situations can be handled easily within the framework of the Kalman filter estimation process.

Consider the linear state space system

$$\mathbf{x}_{k+1} = A_k \mathbf{x}_k + w_k \tag{13.8}$$

$$\mathbf{z}_k = C_k \mathbf{x}_k + \eta_k \tag{13.9}$$

where
- $\mathbf{x}_k$ is the state vector at time k
- A_k is an $n \times n$ system matrix
- w_k is a zero mean Gaussian disturbance noise vector with covariance Q_k
- $\mathbf{z}_k$ is the observation vector at time k
- C_k is a $q \times n$ observation matrix
- η_k is a zero mean Gaussian measurement noise vector with covariance R_k

The Kalman filter is then given by
Prediction equations:

$$\hat{\mathbf{x}}_{k,k-1} = A_{k-1}\hat{\mathbf{x}}_{k-1,k-1} \tag{13.10}$$

$$P_{k,k-1} = A_{k-1}P_{k-1,k-1}A_{k-1}^{\mathrm{T}} + Q_{k-1} \tag{13.11}$$

Estimate update equations:

$$K = P_{k,k-1}C_k^{\mathrm{T}}(C_k P_{k,k-1} C_k^{\mathrm{T}} + R_k)^{-1} \tag{13.12}$$

$$\hat{\mathbf{x}}_{k,k} = \hat{\mathbf{x}}_{k,k-1} + K(\mathbf{z}_k - C_k\hat{\mathbf{x}}_{k,k-1}) \tag{13.13}$$

$$P_{k,k} = P_{k,k-1} - KC_kP_{k,k-1} \tag{13.14}$$

where
- $\hat{\mathbf{x}}_{k,k-1}$ is the state prediction at time k based on $k-1$ measurements
- $\hat{\mathbf{x}}_{k,k}$ is the state estimate at time k based on k measurements
- $P_{k,k-1}$ is the state prediction error covariance matrix
- $P_{k,k}$ is the state estimate error covariance matrix

For the system described in this chapter, the observation equation (13.9) is nonlinear. Thus, the system uses the extended Kalman filter where a linearization is performed at each time step.

At each time step, outlier rejection can be performed on each of the feature measurements. Subsection 13.5.4 describes outlier rejection methods. Outliers, by definition, are measurements that have a higher than expected variance. Thus, for the given time step, it is reasonable to adjust the noise variances $r_i \in R$ to reflect this fact. By setting the variances to infinity for the given time step, the rows in the Kalman gain matrix, K, associated with the feature are set to zero for the time step due to the nature of the Kalman gain calculations as given in (13.12). The end result is that the feature measurement is dropped for the given time step. Experiments where image features were unexpectedly occluded for a short time, showed that the system is able to continue to perform visual servoing with only minor perturbations in the relative pose estimation accuracy [4].

For the second case, the system must drop and/or add features to the feature set used at each time step. This process is somewhat more involved as the output equation (13.9), changes depending on which features are selected. When the feature set changes, the first step is to adjust the set of predicted feature locations so that new window locations can be calculated. Then, the linearized observation matrix, C, is restructured to reflect the new feature set. Finally, the covariance matrix R is restructured to reflect changes in the output equations.

The low-level algorithms associated with feature switching have been implemented as part of the experimental system that is described in this chapter. However, the higher level planning techniques described in Section 13.4 have yet to be implemented as part of the system. Instead, an off-line heuristic approach has been taken to determining an appropriate feature set. A set of features has been associated with certain time periods that are defined for the each relative pose trajectory. For example, during a grasping operation, feature sets have been defined for the beginning of the operation when the object is far away, for the middle as the object nears the end-effector, and for the end of the trajectory when the object is very close to the end-effector or is within its grasp.

13.4 FEATURE PLANNING AND SELECTION

Prior to or during visually servo-controlled motions, a set of visual features must be selected. The accuracy and robustness of visual servoing systems are highly affected by the selected features. For example, during visual servoing, some of the features may become occluded or move out of the field of view or provide weak information which will lead to the task failure. A remedy of using a large feature set will not be acceptable due to the computational considerations. Therefore, minimum but discriminatory, information intensive and reliable feature sets should be automatically selected and used dynamically depending on the status of the vision system with respect to the task object. Some of the issues related to visual feature selection have been reported in [13, 14, 15, 16]. With the exception of the first paper, the rest of these approaches have been written in the context of image-based visual servoing. In the following section we will introduce the main criteria used for feature selection. In this section the reduced set of holes, circles, and corners, whose parameters are their image plane coordinates, is basically used as the features. This is due to the availability of these features in many industrial parts and, ease and robustness of their extraction. For the list of other image feature parameters, one can refer to [1, 17].

In the following section we will discuss the methods for automating the feature selection and finally we will give an overview of off-line and on-line feature planning techniques. Our preference in presenting the material is to prescribe those techniques that have some generic applicability. Although the methods introduced in this article predominantly relate to pose-based visual servoing using extended Kalman filter (EKF) algorithm [2], many of them, with minor modifications, could be applied to other control schemes such as image-based control [16]. The proposed framework offers a systematic method of synthesizing the task object features and, consequently, removes the ad-hoc analysis and selection of features for the robust performance of a visual servoing task. This also eliminates possible human errors and inconsistencies, resulting in a more autonomous robotic system.

13.4.1 Criteria for Feature Selection

The quality of features for a specific operation must be judged by considering not only the geometric constraints but also the constraints imposed by the feature extraction and pose estimation processes (in pose-based visual control). In the following two sections, we will analyze these constraints, respectively.

13.4.2 Criteria for Task Constraints

The task constraints can be categorized into *radiometric* and *geometric* constraints. The radiometric constraints are imposed by the illuminability conditions and the contrast of the features with their surroundings. Although the use of highly textured background and artificial lighting could be restrictive for some practical applications, it has proved to work appropriately for our experimental purposes [2]. Therefore, radiometric constraints will not be discussed any further; however, they could be formulated using the conventional techniques, e.g., the methods given in [18]. The geometric constraints are imposed by the geometry of the workspace, including the object and vision sensor. The geometric constraints are usually *hard* in the sense that the features which do not satisfy any of these constraints, i.e., *nonfeasible features*, must be rejected. The projection model of Fig. 13.16 will be used throughout this analysis. Here $\mathbf{P}_j^c = (X_j^c, Y_j^c, Z_j^c)^T$ and $\mathbf{P}_j^o = (X_j^o, Y_j^o, Z_j^o)^T$ are the coordinate vectors of the jth object feature center in the camera and object frames, respectively; $\mathbf{T} = (X, Y, Z)^T$ denotes the relative position vector of the object frame with respect to the camera frame. We will also denote $\mathbf{\Theta} = (\phi, \alpha, \psi)^T$ as the relative orientation vector with Roll, Pitch and Yaw parameters, respectively. The point $\mathcal{T} : (\mathbf{T}, \Theta)$ will also be used as the viewpoint or a relative trajectory node. The coordinates of the projection of this feature center on the image plane will be x_j^i, y_j^i. Also, since each corner can be extracted by two possible combination of its edges, we define the notion of required edge list.

Definition 1 A required edge list, denoted by RE, is the set of all possible combinations of feature edges taken two at a time. That is, $RE = \{RE_1, RE_2, \ldots, RE_n\}$ with $RE_i = \{e_1^i, e_2^i\}$.

Here e_j^i denotes the jth edge associated with RE_i. If k is the total number of edges associated with corner feature, the binomial coefficient $n = \frac{k!}{(k-2)!2!}$. For clarification and demonstration of the constraints, we will use the following example.

EXAMPLE

The object (Fig. 13.17) has 36 image features and the parameters of our experimental camera [8] are: $F = 1.723$ (cm) as the focal length, $a = 1.077$ (cm) as the diameter of the lens aperture, and $P_x = P_y = 0.006$ (cm) as interpixel spacing, and $N_x = N_y = 128$ (pixels) as the number of pixels along X^i and Y^i directions. The relative pose vectors corresponding to nodes 1–6 of our trajectory are: (4,30,1.5,90,0,0); (4,20,1.5,90,0,0); (6,25,1.5,90,0,0); (8,25,1.5,90,0,0); (8,12,1.5,90,0,0); and (4,10,1.5,90,0,0), respectively, expressed in cm and degrees. However, for most constraints we will take node 2 of our relative trajectory as the example viewpoint.

In the remaining part of this section, we will analyze geometric constraints in the order of their importance and priority in our automatic feature selection algorithm. The measures associated with these constraints have been summarized in Table 13.5.

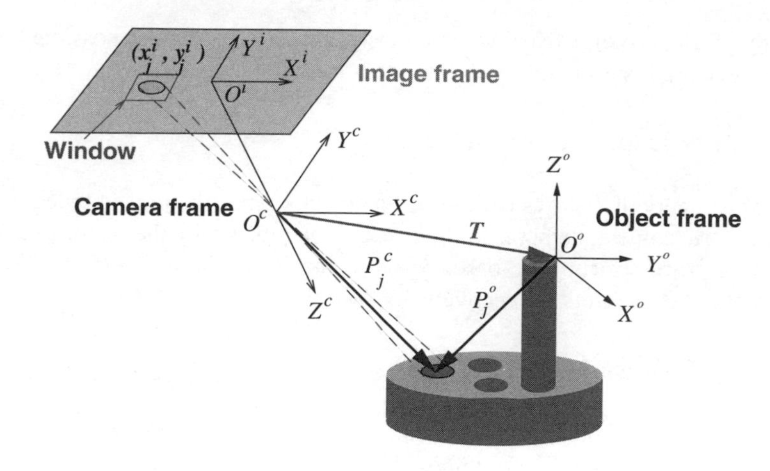

Figure 13.16 The geometry of the object feature projection on image plane.

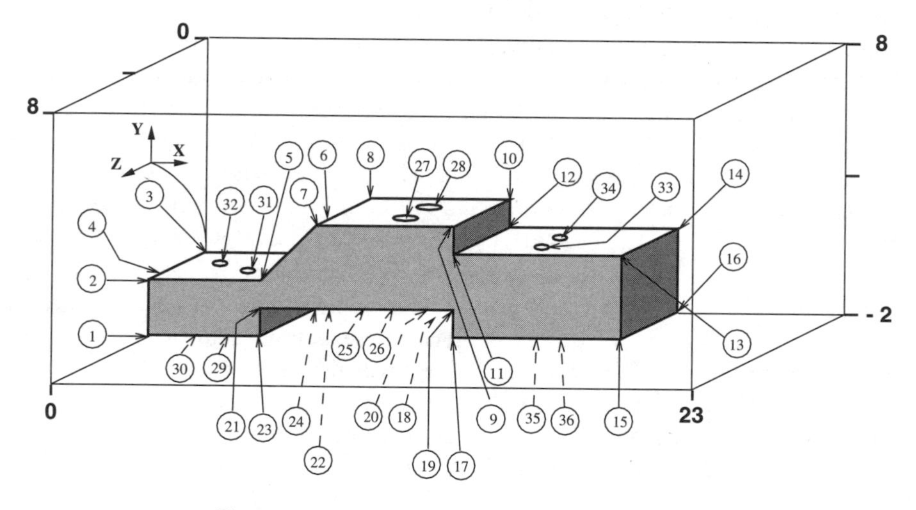

Figure 13.17 The object and Id numbers of its features.

13.4.2.1 Visibility. The visibility constraint is the requirement that the target feature(s) to be observed from a selected sensor location without being obstructed by occluding object(s). There are many approaches to visibility constraint analysis, e.g., [19, 13]. However, the simple approach in [13] gives unoccluded portions of the feature. The visibility measure for both corner and hole features has been given in Table 13.5. For the definition of the visibility measure for a corner feature we have considered *accuracy* and *robustness* criteria. The accuracy requires that unoccluded portions of the corner edges to be as close as possible to the corner point. The robustness needs larger unoccluded portions with higher probability of remaining unoccluded for a longer servoing time. The formulation [(3)] given in [13] takes the above criteria in measuring the quality of corner features. The measure for the visibility constraint is given in Table 13.5. The visibility locus can be calculated using the method given in [19].

13.4.2.2 Resolution. The resolution criterion requires the minimum image resolution of the target feature to be observed. A simple method in [13] can be used to measure the quality of a feature for resolution constraint. In order to avoid the need for prior knowledge of relative orientation, we assume the focal axis direction to coincide with the bisector of view angle

TABLE 13.5 Geometric criteria applied to node 2 for hole and corner features, with $\nu_t = 1$, $RC = 1.2$, $C_f = 1.1$, $N_W = 1$, $\delta x = \delta y = 0.03$ (cm), $(S_{\min})_x = (S_{\min})_y = 0.091$ (cm), $(S_{\max})_x = (S_{\max})_y = 0.288$ (cm).

Criteria	Measure	Value
Visibility	$J_{f_1}(\varepsilon, \mathbf{T}) = \begin{cases} 1 & \text{if } \varepsilon \text{ fully visible} \\ 0 & \text{otherwise} \end{cases}$ or $J_{f_1}(\varepsilon, \mathbf{T}) = \max_{RE_i \in RE} \prod_{e^i_j \in RE_i} J_{f_1}(e^i_j, \mathbf{T})$; $J_{f_1}(e^i_j, \mathbf{T})$ from [13]	1 for (2, 3, 5–10, 13, 14, 27, 28, 31–34)
Resolution	$J_{f_2}(\varepsilon, \mathbf{T}) = \max_{RE_i \in RE} \quad J_{f_2}(RE_i, \mathbf{T})$ and $J_{f_2}(RE_i, \mathbf{T}) = \begin{cases} 1 & \text{if } \delta > \delta_f \quad \text{for } \forall e \in RE_i \\ 0 & \text{otherwise} \end{cases}$	1 for (2, 3, 5–10, 13, 14, 27, 28, 31–34)
Field of view	$J_{f_3}(\varepsilon, \mathbf{T}, \Theta) = \max_{RE_i \in RE} J_{f_3}(RE_i, \mathbf{T}, \Theta)$ and $J_{f_3}(RE_i, \mathbf{T}, \Theta) = \begin{cases} 1 & \text{if (13.16) holds } \forall e \in RE_i \\ 0 & \text{otherwise} \end{cases}$	1 for (2,3,5–8,31,32)
Depth of field	$J_{f_4}(\varepsilon, \mathbf{T}, \Theta) = \max_{RE_i \in RE} J_{f_4}(RE_i, \mathbf{T}, \Theta)$; $J_{f_4}(RE_i, \mathbf{T}, \Theta) = \begin{cases} 1 & \text{if (13.17) holds for } \forall \text{ vertices of } \forall e \in RE_i \\ 0 & \text{otherwise} \end{cases}$	1 for (2,3,5–8,31,32)
Window	$J_{f_5}(\varepsilon, \mathbf{T}, \Theta) = \begin{cases} 1 & \text{if (13.19) is satisfied} \\ 0 & \text{otherwise} \end{cases}$	1 for (5–8,31,32)

(Fig. 13.18). The Proposition in [13] shows that this assumption will not lead to inclusion of nonfeasible features, rather it is a conservative approach. Therefore, if the resolution constraint is satisfied for this specific orientation, it will be satisfied for the other orientations. Now, since the minimum resolution occurs along the diagonal of each pixel, resolution can be defined as

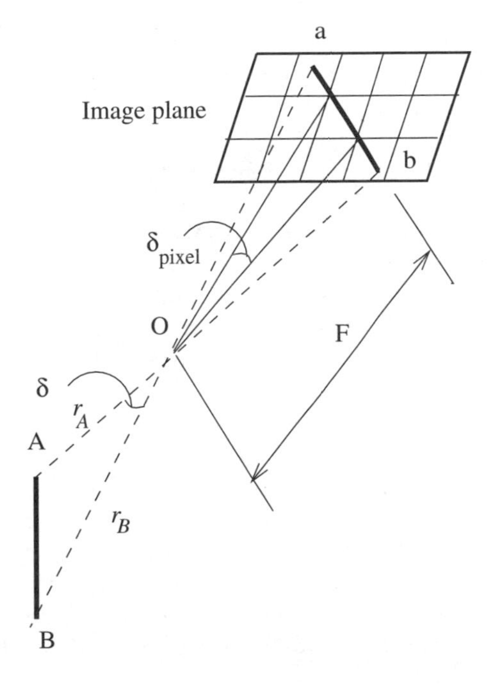

Figure 13.18 The image of a line segment.

a length l that must have at least angle δ_f corresponding to the diagonal of N_δ pixels. The number of pixels N_δ is $\frac{L_f}{RC}$, where L_f is the feature length and RC is the minimum required resolution in (length unit/pixels). Therefore, if

$$\delta = \arccos \frac{r_A^2 + r_B^2 - L_f^2}{2 r_A r_B} > \delta_f = 2 \frac{L_f}{RC} \arctan \left(\frac{\sqrt{P_x^2 + P_y^2}}{2F} \right) \tag{13.15}$$

the resolution constraint will be satisfied. The measure for the resolution constraint is given in Table 13.5. Also the resolution locus for the corner and hole features has been formulated in [20].

13.4.2.3 Field of view. The field of view constraint requires that the image of the feature lie completely inside the image plane, for a given camera and a viewpoint. In contrast to the resolution constraint that puts an upper bound on the distance between the feature center and viewpoint, the field of view constraints provides the lower bound. Given a point feature ε at ($\mathbf{P}\varepsilon^c = (X_\varepsilon^c, Y_\varepsilon^c, Z_\varepsilon^c)^T$), and using Fig. 13.16, it can be shown that the field of view constraint will be satisfied if

$$\left| \frac{X_\varepsilon^c}{Z_\varepsilon^c} \right| \leq \frac{N_x P_x}{2F}, \qquad \left| \frac{Y_\varepsilon^c}{Z_\varepsilon^c} \right| \leq \frac{N_y P_y}{2F} \tag{13.16}$$

For high accuracy, at in-close distances, the vertices of the hole feature or the end of visible portions of the edges for a corner feature should be checked. The measure for the field of view is given in Table 13.5. The field of view locus has been formulated in [20], e.g., an ellipsoid with its center at the feature center and its major axis coplanar with the feature plane, for a hole feature.

13.4.2.4 Depth of Field (Focus). For a point to be in focus, the diameter of its blur circle should be less than the minimum pixel dimension. It has been shown [21] that the feature point at distance D_f is in focus if it lies between maximum distance D_1 and minimum distance D_2, i.e.,

$$D_2 \leq D_f \leq D_1 \tag{13.17}$$

where D_1 and D_2 have been defined in terms of F, a (the diameter of the lens aperture), and c (the minimum of P_x and P_y), as

$$D_1 = \frac{D_f a F}{aF - c(D_f - F)} \quad \text{and} \quad D_2 = \frac{D_f a F}{aF + c(D_f - F)} \tag{13.18}$$

Since $(D_1 - D_2)$, *depth of field*, increases with D, focus constraint provides a lower bound for the distance of the camera from the target features. However, from the feature extraction point of view, unfocused features may be desirable. Moreover, in many assembly operations, fine motions require close operations on objects and it is difficult to have perfectly focused features in practice. Therefore, low weight has been given to this constraint in our implementation. The measure for focus constraint is given in Table 13.5 with its application to node 2 of our example trajectory.

13.4.2.5 Windowing. The windowing criterion measures the feasibility of the target feature(s) for windowing and for robust tracking of the selected windows during servoing. This is to make sure that the size of selected windows around the feature(s) is between the maximum and minimum allowable window sizes; the windows have enough clearance from the image plane boundaries; and they do not overlap other windows or portions of other features. The method given in [13] calculates the image coordinates of the window(s) center(s), the sizes of window(s) (W_x, W_y) in X^i and Y^i directions, respectively. The method has been developed for both single and double windowing of corner features. To ensure clearance from image boundaries, clearance factors, i.e., δx, δy, are taken into consideration. Also, to generate non-overlapped windows for different features, a clearance factor C_f is employed in our windowing algorithm. Finally the measure in Table 13.5 compares the selected window sizes with the minimum and maximum allowable window sizes in X^i and Y^i directions, i.e., $(S_{\min})_x$, $(S_{\min})_y$ and $(S_{\max})_x$, $S_{\max})_y$, respectively.

$$(S_{\min})_x \leq W_x \leq (S_{\max})_x \text{ and } (S_{\min})_y \leq W_y \leq (S_{\max})_y \tag{13.19}$$

13.4.3 Feature Extraction and Pose Estimation Criteria

The criteria introduced in this section improve the speed, ease, and robustness of feature extraction process, and also enhance the observability, accuracy, robustness, and sensitivity of EKF-based (Extended Kalman Filter) pose estimation process. Most of these constraints are defined for the set of features, hence called *feature set constraints* as opposed to *feature constraints*. To distinguish between these two, J_{f_i} and J_{s_i} will denote the measures associated with the feature and feature set constraints, respectively. Also, we define the notion of *q-COMB*ϑ.

Definition 2 Let ϑ be a set of m nonrepeated features ε and $q \leq m$ a positive integer number. A *q-COMB*ϑ is the set of all possible combinations, *q-COMB*ϑ_k, of m features taken q at a time, where the kth combination is *q-COMB*$\vartheta_k \triangleq \{\varepsilon_1^k, \varepsilon_2^k, \ldots, \varepsilon_q^k\}$.

The feature extraction and pose estimation constraints are generally *soft* because the features or feature sets that do not satisfy these constraints might not be rejected. A summary of the measures with examples is given in Table 13.6.

13.4.3.1 Type of Feature. The type of features affects the ease of identification and pose estimation. It is usually preferred to have features that are easily identifiable and different from other features in the image. Feddema et al. [15] define feature uniqueness measure for their feature selection algorithm under the image-based control structure. Corner and hole types of features are more common for visual servoing. A hole feature has some advantages—e.g., a hole feature is easier to identify; its center can be extracted with higher accuracy; measurement errors associated with a hole feature are less sensitive to the orientation of the hole on the object [11]; a mathematical expression can be used to update the measurement covariance matrix R in Kalman filtering for a hole feature; and finally the errors in the centroid of a circular hole can be approximated by zero mean Gaussian noise [9]. On the other hand, since the projection of two edges associated with a corner feature is sufficient to measure the location of its corner, corner features are more robust to the changes in the range of camera with respect to the object especially when the camera gets closer to the object. The measure for type constraint is given in Table 13.6 with its application to node 2 of our example trajectory. Here d_t is a threshold and D_f is the distance of focal point to the center of feature.

13.4.3.2 Size of Feature. It has been shown that the use of large holes minimizes the variance of errors in their centroids [9] reduces measurement errors [11], and hence increases feature extraction accuracy. Similarly, for the corners, it is preferred to have large edges associated with each corner. The feature set robustness measure defined in [15] gives priority to large features. On the other hand, from the computational point of view, smaller features are desirable. Therefore the computational-expense measure of Feddema et al. [15] favors smaller features. Due to the advances in image processing units, we have omitted the second criterion. However, it should be noted that at close ranges, large holes might move out of

TABLE 13.6 Feature extraction and pose estimation criteria applied to node 2 for hole and corner features, with $d_t = 15$ (cm), $o_t = -0.1$, $\rho_t = 0.06$ (cm), $\sigma_t = 0.5$ (cm), $Z_{ct} = 7$ (cm), $C_R = 0.9$, $\gamma_\vartheta = 1$, $\gamma_\varepsilon = 1/q$.

Criteria	Measure	Value
Type	$J_{f_6}(\varepsilon, \mathbf{T}) = \begin{cases} 1 & \text{if hole and } D_f > d_t \\ & \text{or corner and } D_f \leq d_t \\ 0.5 & \text{otherwise} \end{cases}$	1 for (5–8) and 0.5 for (31,32)
Size	$J_{f_7}(\varepsilon, \mathbf{T}, \Theta) = \begin{cases} 1 & \text{if } \varrho(\frac{\mathcal{A}(D_f)}{\mathcal{A}(r_f)}) \geq 0.75 \\ .5 & \text{otherwise} \end{cases}$	1 for (5–8) and 0.5 for (31,32)
Optical angle	$J_{f_8}(\varepsilon, \Theta) = \max_{RE_i \in RE} J_{f_8}(RE_i, \Theta)$	1 for (5-8,31,32)
	$J_{f_8}(RE_i, \Theta) = \begin{cases} 1 & \text{if} -1 \leq \mathbf{k} \cdot \mathbf{n}_\varepsilon \leq o_t \text{ for} RE_i \\ 0 & \text{otherwise} \end{cases}$	
Set	$J_{s_1}(\vartheta, \mathbf{T}, \Theta) = \prod_{k=1}^{n} J_{s_1}(\varepsilon_1^k, \varepsilon_2^k, \mathbf{T}, \Theta)$	1 for (5,6,7,8,31,32)
nonambiguity	$J_{s_1}(\varepsilon_1^k, \varepsilon_2^k, \mathbf{T}, \Theta) = \begin{cases} 1 & \text{if } \lvert x_1^i - x_2^i \rvert > \rho_t \\ & \text{or } \lvert y_1^i - y_2^i \rvert > \rho_t \\ 0 & \text{otherwise} \end{cases}$	
Set apartness	$J_{s_2}(\vartheta) = \begin{cases} \sigma_s = \frac{1}{q} \sum_{k=1}^{q} \sigma_k & \text{if } \sigma_s > \sigma_t \\ 0 & \text{otherwise} \end{cases}$	2.43 for (6,7,8,32) 2.32 for (5,7,31,32)
Set	$J_{s_3}(\vartheta) = \prod_{k=1}^{n} J_{s_3}(\varepsilon_1^k, \varepsilon_2^k, \varepsilon_3^k)$	1 for (5,6,7,8)
noncollinearity	$J_{s_3}(\varepsilon_1^k, \varepsilon_2^k, \varepsilon_3^k) = \begin{cases} 1 & \text{if (13.20) holds} \\ 0 & \text{otherwise} \end{cases}$	
Set	$J_{s_4}(\vartheta) = \frac{1}{m} \sum_{k=1}^{m} J_{s_4}(\varepsilon_1^k, \varepsilon_2^k, \varepsilon_3^k, \varepsilon_4^k)$	0 for (5,6,7,8)
coplanarity	$J_{s_4}(\varepsilon_1^k, \varepsilon_2^k, \varepsilon_3^k, \varepsilon_4^k) = \begin{cases} 0 & \text{if (13.21) holds} \\ 1 & \text{otherwise} \end{cases}$	0.87 for (5,6,7,8,31,32)
Set angle	$J_{s_5}(\vartheta) = 1 - \frac{\sum_{j=1}^{q} \lvert \kappa_{j,j+1} - \frac{360^\circ}{q} \rvert}{360}$	0.81 for (5,6,7,8,31,32)
Sensitivity	$J_{s_6}(\vartheta, \mathbf{T}, \Theta) == \frac{1}{q} \sum_{k=1}^{q} J_{s_6}(\varepsilon_k, \mathbf{T}, \Theta)$	0.52 for (5,6,7,8,31,32)
	$J_{s_6}(\varepsilon, \mathbf{T}, \Theta) = \begin{cases} \frac{\frac{1}{Z_\varepsilon^c} + R_\varepsilon^c}{\frac{1}{Z_{ct}} + \max_{Z_\varepsilon} R_{ct}} & \text{if } R_\varepsilon^c < R_{ct} \\ & \text{and } Z_\varepsilon^c > Z_{ct} \\ 0 & \text{otherwise} \end{cases}$	
Durability	$J_{s_7}(\vartheta) = \sum_{i=k}^{\ell} \gamma_\vartheta + \sum_{j=1}^{q} \sum_{i=k'}^{\ell'} \gamma_{\varepsilon_j} \ \forall \varepsilon_j \in \vartheta$	6.83 for (5,6,7,8,31,32)

field-of-view or their projection on the image plane might exceed window size limits. The measure in Table 13.6 reflects the above facts. In this measure, r_f denotes the radius of the feature circumscribing circle which can be calculated easily [22]. Also $\mathcal{A}(x)$ maps x to a value between 0 and 1 (e.g., a Sigmoid function with low slope as 0.1) and $\varrho(y)$ is a bell shaped function in the form of $\exp(-\frac{1}{2\sigma^2}(y - \mu)^2)$ with $\mu = 1.5$ and $\sigma = 0.25$.

13.4.3.3 Number of Features. In order to solve the set of photogrammetric equations for six unknown states, at least three noncollinear features should be included in the measurements [15]. However, inclusion of extra features will improve the performance of EKF, with the price of increasing the computational cost. It has been shown that the inclusion of more than six features will not significantly improve the performance of state estimation via EKF [8]. Therefore, we decide to have the number of set features q between four and six ($4 \leq q \leq 6$) with the preference given to higher number of features.

13.4.3.4 Optical Angle. It has been shown that features with surfaces close to coplanar with the camera optical axis lead to poor pose estimations by EKF algorithm [8]. This can be checked by measuring the optical angle, which is the angle between the unit vector along the optical axis $\vec{k}$ and the unit normal vector of the feature surface $\vec{n_\varepsilon}$. When this angle approaches 90°, the feature image will be ambiguous. The measure in Table 13.6 assigns a score to a feature for goodness of its optical angle. In the measure, o_t is a negative number close to 0.

13.4.3.5 Set Nonambiguity. An image will be ambiguous and tracking performance will degrade if the projections of two or more features coincide. Hence, the images of feature in the set should have enough clearance from each other. The set nonambiguity measure in Table 13.6 assigns the score to a feature set ϑ based on its features clearance, where ${\varepsilon_1}^k, {\varepsilon_2}^k \in$ 2-$COMB\vartheta_k$ of 2-$COMB\vartheta$ and $n = \frac{m!}{(m-2)!2!}$ for m features in the set. Also ρ_t is the nonambiguity threshold. For a relatively high speed scenario of about 20 cm/s and servoing with our camera [8] at frame rate of 61 Hz, using triangulation, it was found that $\rho_t = 0.06$ (cm) (10 pixels).

13.4.3.6 Set Apartness. To provide accuracy and robustness in both feature extraction and pose estimation, it is advantageous to have the features apart from each other [3]. The measure in Table 13.6 measures the apartness of the features in a set ϑ, where for ε_k, $\sigma_k \triangleq \min_j \|\varepsilon_k, \varepsilon_j\|_2$ with $\varepsilon_j \in \vartheta$, $j \neq k$, and $\|\varepsilon_k, \varepsilon_j\|_2$ denotes the Euclidean distance between the centers of these features. Also σ_t is the apartness threshold, e.g., in our experiments, it was observed that $\sigma_t = 0.5$ (cm) ensures enough minimum apartness requirements for tracking between frames.

13.4.3.7 Set Noncollinearity. In order to provide observability, accuracy, and robustness in the pose estimation process, the set of feature points should be noncollinear [8]. Also collinearity of the features will result in the singularity of control matrix **B** and tracking failure of a image based system as reported in [14]. To guarantee noncollinearity of the three points, the following condition must be satisfied:

$$\left| \begin{array}{cc} X_3^o - X_1^o & X_2^o - X_1^o \\ Y_3^o - Y_1^o & Y_2^o - Y_1^o \end{array} \right| \geq \epsilon \quad \text{and} \quad \left| \begin{array}{cc} Y_3^o - X_1^o & X_2^o - X_1^o \\ Z_2^o - Z_1^o & Z_3^o - Z_1^o \end{array} \right| \geq \epsilon \tag{13.20}$$

In the set noncollinearity measure of Table 13.6, $|.|$ is the absolute value of a matrix determinant, $\{\varepsilon_1^k, \varepsilon_2^k, \varepsilon_3^k\} =$ 3-$COMB\vartheta_k \in$ 3-$COMB\vartheta$ and ϵ is a small scalar.

13.4.3.8 Set Noncoplanarity. It has been shown that noncoplanar features outperform coplanar features in terms of accuracy and robustness in both direct solution of photogrammetric equations [3] and Kalman filter based pose estimation [8, 20]. The necessary and sufficient condition for four feature points ε_k ($k = 1, \ldots, 4$), with the homogeneous coordinates $[X_k Y_k Z_k w_k]^T$ defined in any coordinate frame, to be coplanar is that

$$\left| \begin{array}{cccc} X_1 & Y_1 & Z_1 & w_1 \\ X_2 & Y_2 & Z_2 & w_2 \\ X_3 & Y_3 & Z_3 & w_3 \\ X_4 & Y_4 & Z_4 & w_4 \end{array} \right| = 0 \tag{13.21}$$

where for the simplicity we can assume that $w_i = 1$ for $i = 1, 2, 3, 4$.

13.4.3.9 Set Angle. The set angle constraint requires q feature points in a feature set to be distributed at central angles of $\frac{360°}{q}$, where the *central angle* of two sequent features ε_i and ε_{i+1}, $\kappa_{i,i+1}$, is the smallest angle subtended at the center of features (or at the image plane center in image plane). This is to minimize the oscillations in state prediction error covariance matrix of EKF algorithm [9], and hence to improve the quality of pose estimation. If the features are ordered according to their central angle with respect to the center of the features in the preprocessing step, the measure for set angle can be written as in Table 13.6. Also, it is assumed that $\kappa_{q,q+1} \equiv \kappa_{q,1}$, and angles measured in degrees.

13.4.3.10 Sensitivity. Sensitivity is the measure of the change in the feature image positions for a change in the relative pose of the camera with respect to the object. In order to provide robust and accurate servoing, it is required to have substantially visible motions of the feature images due to the relative pose change. It is interesting to note that in an image-based control scheme, the inverse image Jacobian may be used [15] and updated each sample time. Therefore, from computational point of view, it is more desirable to minimize sensitivity and, hence, the variation of the elements of the Jacobian during the servoing, with the cost of less visible motions of the feature points on the image. Due to recent advancements of the computer technology, we have ignored this aspect in our AFS algorithm. Consequently, the features or feature sets that provide high sensitivity are desirable. In [13], it has been shown that sensitivity of a feature ε is a function of $R^c_\varepsilon \equiv \sqrt{X^{c2}_\varepsilon + Y^{c2}_\varepsilon}$, and $\theta_\varepsilon \equiv \arctan \frac{Y^c_\varepsilon}{X^c_\varepsilon}$. Also, the sensitivity will increase as the value of R^c_ε increases and/or the value of Z^c_ε decreases. At the neighborhood of a critical depth Z_{ct}, the slope of sensitivity S increase, due to R^c_ε increase, gets very sharp. For example, $Z_{ct} = 7$ (cm) for our experiments. Therefore, to provide high sensitivity, the depth of a feature Z^c_ε should be chosen as low as possible, yet Z^c_ε should be greater than Z_{ct}. Also the radius R^c_ε has to be chosen as high as possible. In [20], it was shown that this not only increases the sensitivity but also yields higher accuracy in (Z, *Pitch, Yaw*) parameters estimations. However, very high values of R^c_ε may lead to poor field of view and with small reduction in Z^c_ε parameter, features may move out of field of view. In order to avoid frequent switching of the features and hence improved durability, R^c_ε values should not be very close to an upper bound, i.e. R_{ct}. One can calculate R_{ct} from $R_{ct} = C_R Z^c_\varepsilon \tan(\frac{\alpha_f}{2})$ where $\alpha_f = 2 \arctan\left(\frac{\min(N_x P_x, N_y P_y)}{2F}\right)$ and C_R is a clearance factor which should be less than 1. Taking the above points into the consideration, the sensitivity measure for a set ϑ can be defined as in Table 13.6.

13.4.3.11 Durability. This criterion is to ensure the durability of a specific feature set during servoing in order to avoid frequent feature set switching during the servoing, and hence to provide faster and more accurate servoing. The measure for the durability constraint of a set ϑ with q features is given in Table 13.6, where γ_ϑ is the set durability factor (usually taken to be 1), k is the smallest node number such that the set ϑ is valid for all nodes $i = k, k+1, \ldots, \ell$, including the node under study. Similarly, γ_ε is the feature durability factor (usually $\gamma_\varepsilon \ll \gamma_\vartheta$ such as $1/q$), and the feature ε of ϑ is valid for all nodes $i = k', k'+1, \ldots, \ell'$. One way to apply the durability constraint is to backtrack from the last node of the relative trajectory and to increase incrementally the durability indices of the feature sets and features which are common with the sets in the previously examined node.

In addition to the above criteria, some criteria can be defined for image-based control scheme. Examples of those criteria are specified as follows. These criteria focus on image-based control issues.

13.4.3.12 Observability. In order to find a unique solution to gradient search method of Feddema et al. [15], they suggest a measure that uses error function as a means of evaluating the observability of the object's pose with respect to the camera. For instance, a fourth point in the middle of an equilateral triangle that is centered about the focal axis will not provide a unique solution.

13.4.3.13 Controllability. The resolved motion rate control of Feddema et al. [15] uses the inverse image Jacobian matrix to estimate the change in the object's pose from the change in three feature points. Therefore, the pose of the object will be controllable if the inverse Jacobian exists and is nonsingular. Hence, they define a measure (based on the condition of a Jacobian matrix, $c(J) = (\|J\|\|J^{-1}\|)$ to favor those feature sets that minimize the condition of J. Similarly, for a robust control, Papanikolopoulos [14] suggests the features that do not cause singularity of a matrix $\mathbf{B}$ for his image-based control law. For instance when the three feature points are collinear, the matrix $\mathbf{B}$ will be singular. Since the position-based techniques do not use inverse Jacobian control law, they do not need to deal with this issue.

13.4.4 Off-Line Feature/Trajectory Planning

The complexity of feature selection and planning is highly dependent on the dimensionality of features and feature sets used for the feature selection analysis. This is because for a set of n features, it is possible to form $\sum_q \frac{n!}{(n-q)!q!}$ feature sets each with q nonrepeated features. In particular, the study of many industrial parts shows that these objects usually have considerable number of features. A good planning strategy must be based on reducing the dimensionality of feature and feature set analyses. Dimensionality reduction can be achieved by partitioning the whole selection problem into the following subproblems. Here we assume that a relative trajectory has already been generated by our automatic trajectory planner in the off-line phase of operation. A detailed version of the following discussions can be found in [20].

1. *Features Representation*: This is done by selecting and representing the features that are allowed by the problem formulation, feature extraction, and pose estimation processes. For instance, in our current implementation, hole and corner features are used. In our AFS (Automatic Feature Selection) package, a solid model of the object is used to detect the candidate features $\Upsilon = \{\varepsilon_1, \varepsilon_2, \ldots, \varepsilon_n\}$.
2. *Feasible Features Selection*: The objective of this step is to test the set of candidate features Υ using the feature constraints J_{f_i} $(i = 1, 2, \ldots, 8)$, and to remove nonfeasible features in the early stages of calculation, before they increase the dimensionality of computation. This will determine the set of feasible features Γ for a node $\mathcal{T} : (\mathbf{T}, \Theta)$ of a trajectory. Examples of feasible features for node 2 of our trajectory were given in Tables 13.5 and 13.6. Similarly, the feasible features for the whole trajectory, i.e., nodes 1–6, can be calculated as (3, 5, 6, 7, 8, 27, 28, 31, 32); (5, 6, 7, 8, 31, 32); (5, 6, 7, 8, 27, 28, 31, 32); (5, 6, 7, 8, 9, 10, 27, 28, 31); (27, 28); and (5, 6, 31), respectively. The windowing information associated with the feasible features are also determined at this step.
3. *Feasible Feature Sets Formation*: This is achieved by forming all q-$COMB\Gamma_k$ sets from Γ according to the number of features constraint, $q = 4, 5, 6$. Each of these sets will be denoted by ϑ_k. In spite of the reduced number of features in step (2), the number of combinations would still be very large (e.g., 216 combinations for node 1).
4. *Admissible Feature Sets Selection*: This is done by applying the feature set constraints J_{s_i} $(i = 1, 2, \ldots, 6)$ (as the secondary filter) to feasible feature sets, with the attempt of removing inadmissible feature sets in the early stages of computation. The result will be a set of admissible feature sets Λ_a associated with a node $\mathcal{T} : (\mathbf{T}, \Theta)$ of

a trajectory. It should be noted that J_{s_4}, J_{s_5}, J_{s_6} are not associated with the hard constraints and 0 values for these measures of a feature set will not exclude that set from further considerations. Examples of feature set constraints applications were again given in Table 13.6.

5. *Optimal Feature Set Selection for Nodes of Relative Trajectory*: This is accomplished by first imposing the durability constraint over the admissible feature sets in $\Lambda_a(\mathbf{T}, \Theta)$. Next the overall optimality measure will be calculated and optimal feature set χ for a node will be obtained by

$$J_{op}(\chi, \boldsymbol{T}, \boldsymbol{\Theta}) = \max_{\vartheta_k \in \Lambda_a} (q_k \sum_{j=1}^{7} J_{s_j}(\vartheta_k, \boldsymbol{T}, \boldsymbol{\Theta}) + \sum_{i=1}^{q_k} \sum_{j=6}^{8} J_{f_j}(\varepsilon_j^k, \mathbf{T}, \Theta)) \tag{13.22}$$

where q_k is the number of features in the set ϑ_k. For example, in node 2 of our trajectory relative to the object, the feature set $(5, 6, 7, 8, 31, 32)$ scores the highest ($J_{op} = 78.65$) among the other features, indicating its optimality. On the other hand, sets such as $(5, 6, 31, 32)$ score low mainly because of their coplanarity, low sensitivity, and poor durability. Similarly, the optimal feature sets for nodes 1–6 of our trajectory will be: $(3, 5, 7, 8, 27, 28)$; $(5, 6, 7, 8, 31, 32)$; $(5, 6, 7, 8, 27, 32)$; $(6, 7, 8, 9, 27, 28)$; (NIL); and (NIL).

In [13] it has been shown that the time complexity of the above algorithm is $O(N(s + s' + n))$, for N nodes, s feasible feature sets, and s' admissible feature sets. One, however, could achieve faster processing time by parallel processing of AFS algorithm. For instance, partitioning steps (2)–(4) among N processors (one for each node of the trajectory) is possible and would decrease the running time by $O(N)$ for the above stages. Also it is possible to further partition the task of step (2) among eight parallel processors (one for each J_{f_i}) and that of step (4) among six parallel processors (one for each J_{s_i}), reducing the execution time to $1/8 \times 1/6$ of that with sequential processing. Further parallel processing is possible by allocating a processor to each feasible feature set during the admissible feature sets selection. This will further reduce the time by $O(s)$. Hence, the full parallel processing task will require $\max(8N, 6N, 6Ns) = 6Ns$ parallel processors.

One problem with the AFS strategy is that the approximate relative trajectory must be known in advance. Also, running time of the above algorithm would be long [13]. In the next section we will discuss possible techniques for on-line feature selection.

13.4.5 On-Line Feature Selection

In some on-line operations, there is no access to a relative trajectory and a point-to-point motion is carried out. Here, the system must plan for the next point of motion according to predictions made by Kalman filter. Many parts of the algorithm can be processed in parallel, and hence shorter processing time. In this case, since $N = 1$, at most only $6s$ processors will be necessary. However, some other strategies might be applied to reduce the processing time. In this section we propose two techniques for real-time feature selection.

1. *Reduced-Constraints Method*: Here, some of the the soft constraints can be dropped from the calculations with the cost of reduced robustness and accuracy. Examples of these constraints are: optical angle, depth of field, sensitivity, set noncoplanarity, set apartness, set angle, and durability. Also the loci of some constraints such as visibility can be determined in advance. Therefore, during the on-line phase of operation, the satisfaction of the constraints can be checked easily by checking the loci of the corresponding constraints.

2. *AFspace Method*: Here $AFspace_{\mathcal{C}}$ $(\mathcal{P})$ of the object is determined. Such space would be used as a look-up table during the on-line phase of the operation. In order to form $AFspace_{\mathcal{C}}$ $(\mathcal{P})$, one has to calculate *Fspace* of an object which is defined as follows.

   Definition 3 The space of feasible features of an object $\mathcal{P}$ viewed from all possible poses **x** of sensor $\mathcal{C}$ is called the feasible feature space of the object and is denoted by $Fspace_{\mathcal{C}}$ $(\mathcal{P})$. That is,

$$Fspace_{\mathcal{C}}(\mathcal{P}) = \{\varepsilon \in \Gamma_{\mathrm{x}}^{\mathcal{P}} \mid \mathrm{x} \in Cspace_{\mathcal{C}}^{\mathcal{P}}\} \tag{13.23}$$

   where $\Gamma_{\mathrm{x}}^{\mathcal{P}}$ is the set of feasible features of object $\mathcal{P}$ viewed from **x**, and $Cspace_{\mathcal{C}}^{\mathcal{P}}$ denotes the configuration space of $\mathcal{C}$ with respect to the frame of object $\mathcal{P}$. Similarly, the space of admissible feature sets $AFspace_{\mathcal{C}}$ $(\mathcal{P})$ can be calculated. If the operational zones of $AFspace_{\mathcal{C}}$ $(\mathcal{P})$ can be calculated, the real-time feature selection can be achieved in a very short time. However, $AFspace_{\mathcal{C}}$ $(\mathcal{P})$ is at least 7-dimensional for the general relative motion of the end-effector with respect to the object. To reduce the complexity of $AFspace_{\mathcal{C}}$ $(\mathcal{P})$ calculation, one may make assumptions about the relative orientation of the sensor with respect to the object. A good practical case is when the relative orientation of the camera is fixed by assuming that the optical axis passes through the center of the features. In order to further reduce the computational cost, the 3D relative position of the end-effector (or the camera) could be divided into patches. One efficient way to achieve this is to define the concentric viewing spheres around the object, as defined in [13]. After the discretization process, the set of admissible (or optimal) feature sets associated with each patch can be calculated using the AFS algorithm. This completes the $AFspace_{\mathcal{C}}$ $(\mathcal{P})$ calculation algorithm. During the execution, the relative pose of the sensor can be determined from the pose estimations obtained from SESEFS (State Estimation and Sensor Fusion System) [23] and the admissible (or optimal) feature set associated with that patch can be read from $AFspace_{\mathcal{C}}$ $(\mathcal{P})$ map accordingly.

13.5 INFORMATION REDUNDANCY AND SENSOR INTEGRATION

13.5.1 Role of Additional Sensor Information

There are three main sources of information that can be utilized by a sensor-based tracking system: observations obtained from a sensor, a priori information such as a description of the object model, and derived information such as the predicted object pose at a given time step. This section describes how information from multiple sources can be combined to develop a robust visual servoing system. The use of multiple sources of information has both a direct and an indirect impact on the robustness of the system's image processing. The resulting ability to detect and reject outlier measurements has a direct impact on the robustness of the image processing. The resulting improvement in estimation quality and in sample rate indirectly improves the image processing by improving the image windowing and feature selection processes.

Information that describes the same features within the environment is considered to be competitive information. For example, measurements of a hole centroid and a prediction of the centroid location based upon derived information are considered to be competitive. Another example involves two cameras that obtain measurements of the same object corner location with respect to the robot end-effector. The use of competitive information can increase the accuracy and robustness of the estimation process, which in turn increases the robustness of the image processing as the resulting window locations and sizes used for directed image processing improve.

Competitive information provides a "second opinion" about object feature measurements. Integrating competitive data from multiple information sources can reduce the effect of measurement noise, the effect of sensor limitations, and the dependence of the system on possibly invalid model assumptions or ad-hoc constraints [24]. The availability of competitive information permits the detection and rejection of measurement outliers. This improves the system's robustness in the event of object measurement degradation due to environmental conditions such as shadows, object feature occlusion, and sensor degradation and failure. While competitive information can be obtained from a single camera and other types of information sources, the system will still fail in the event of camera failure as a priori and derived information is not sufficient to estimate the pose of an object that is following an unknown trajectory. The addition of a second camera or other sensors increases the system robustness in the event of the failure of one sensor.

Information about different features in the environment is classified as complementary. Complementary information can include measurements of different features obtained from a single camera, or information from multiple sensors. For example, a second camera will perceive different aspects of the environment due to differences in its location, and possibly due to differences in the type of camera. The integration of the complementary information allows features in the environment to be perceived that are impossible to perceive using just the information from each individual sensor [25]. For example, stereo (or trinocular) vision can be used to determine scene depth information [26–33]. Barnard and Fischler [34] and Dhond and Aggarwal [35] survey the different approaches to computational stereo. When multiple cameras are used, a larger set of object features is available for feature selection at any given time step, allowing the system to use "better" features (as described in Subsection 13.4.1) and/or to use more features.

A sensor utilizes cooperative information when it relies upon another information source for information in order to obtain observations. This information can be used to improve the quality of the observations. For example, depth information can be used to set the focal length of a camera lens. Cooperative information can also be used to increase the overall system sample rate. The system described in this chapter uses a prediction of the feature locations in order to perform directed image processing, thus reducing the amount of image processing required to obtain information about the features.

The use of multiple cameras increases a system's tolerance toward poor image measurements through the fusion of competitive information to increase the overall state estimation accuracy, and by enabling the system to better perform outlier rejection. In addition, each individual camera does not need to provide complete information about the robot's environment, which makes the system more tolerant of real-world problems such as sensor failure and feature occlusion. Finally, cooperative information can be used to improve the quality and robustness of each camera's image processing.

Similarly, the use of other types of sensor's can improve the robustness of image processing for visual servoing. Other types of sensors can be used to provide a more accurate and robust measurement of a particular parameter. For example, a range sensor is less likely to provide erroneous depth information than a stereo camera system which must match image features. In addition, the range sensor may be able to provide the depth information faster, which improves the overall sample rate of the system. The advantages of using different types of sensors include the availability of different sensing modalities such as tactile information, and improvements in the accuracy and timeliness of measurements. The disadvantage is that the sensory system is more complex as it must be able to integrate information from diverse sensors. The following sections explore the problem of integrating data from multiple sources of information in more detail.

13.5.2 Interaction of Sensors and Sensor Integration

The terms multisensor fusion and multisensor integration are still ambiguous. Luo and Kay [25] state the following definitions. Multisensor integration refers to "the synergistic use of the information provided by multiple sensory devices to assist in the accomplishment of a task by the system." The notion of multisensor fusion is more restrictive. It refers to "any stage in the integration process where there is an actual combination (or fusion) of different sources of sensory information into one representational format." In this chapter, the notion of multisensor fusion is widened, to include parts of the data combination process, such as performing data alignment and association.

The sensor fusion process involves the combination of sensor data from multiple sources into a coherent framework. This process consists of several steps. First, the data from individual sensors are processed into the desired format. Second, the formatted data are aligned into a common spatial-temporal frame. Then, corresponding measurements are grouped, or associated, with each other. Finally, the data is combined, and the desired types of inferences are drawn using the fused data. Figure 13.19 shows a block diagram of the multisensor fusion process.

The sensor preprocessing that is performed by the system is dependent on the types of sensors employed by the system, and on the desired level of data abstraction. For example, feature extraction can be performed upon a camera image as part of the sensor preprocessing stage. The model is also used to determine the uncertainty associated with a particular sensor observation.

13.5.3 Fusion of Measurement Data from Different Sensors

The sensor fusion approach implemented as part of the relative position sensing visual servoing system described in this chapter is based on an object model reference approach using ray tracing [36]. The approach provides a practical, computationally tractable, means of designing and implementing a sensor fusion system that is able to fuse data that are temporally and spatially distributed, and is obtained from diverse sensors. For the class of problems of interest in this research work, extended Kalman filtering was selected as the most suitable

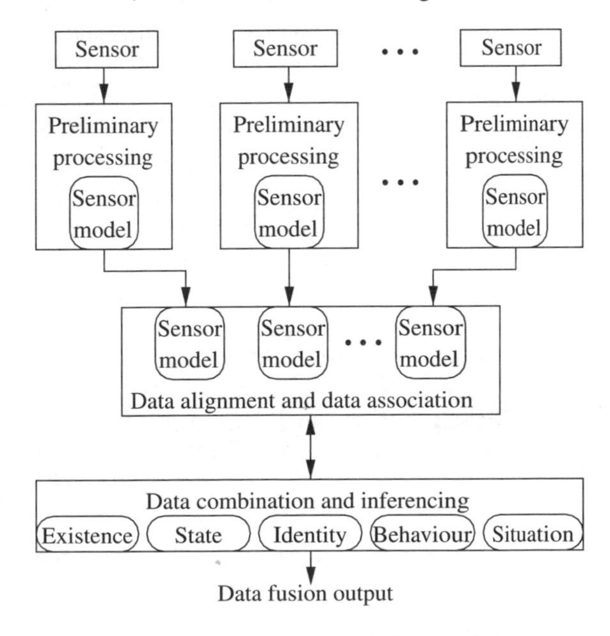

Figure 13.19 Functional diagram of a multisensor fusion system.

method for performing the actual fusion of the sensor measurements [36]. It provides near-optimal estimates, is suitable for a real-time implementation, and can perform dynamic sensor fusion. The main drawbacks of Kalman filtering are that Gaussian noise is required, and that a model describing how the system states change over time is required. However, these problems do not pose a great problem. The assumption of Gaussian noise is often well approximated by real-world conditions [36]. The lack of a system model can be dealt with by estimating the model, or by assuming a particular model and allowing any variations to be modeled as noise. The use of a constant velocity model has been shown to be suitable for the type of visual servoing described in this chapter [4]. Thus, (13.8), the discrete time system dynamic model is written as

$$\mathbf{x}_k = A\mathbf{x}_{k-1} + w \tag{13.24}$$

where w is a zero mean Gaussian disturbance noise vector with covariance Q, and A is block diagonal with 2×2 blocks of the form $\begin{bmatrix} 1 & T \\ 0 & T \end{bmatrix}$ for a sample period T. The system state vector is defined as

$$\mathbf{x} = \begin{bmatrix} x & \dot{x} & y & \dot{y} & z & \dot{z} & \phi & \dot{\phi} & \theta & \dot{\theta} & \psi & \dot{\psi} \end{bmatrix}^{\mathrm{T}} \tag{13.25}$$

where x, y, z is the relative position, ϕ, θ, ψ is the relative orientation described by roll, pitch, and yaw angles, and $\dot{x}, \dot{y}, \ldots, \dot{\psi}$ are the relative velocities. The structure output equation (13.9), is derived using models of the sensor observation behavior.

13.5.4 Data Association and Outlier Rejection

As part of the sensor fusion process, competitive information is aligned into a common spatial-temporal frame. Observations corresponding to the same aspect of the environment are then grouped together. Measurement outliers can then be detected and rejected. Given a number of observations from different sensors, the data association problem is to determine which observations represent the same entity, and thus should be grouped together. In the image processing community this problem is known as the *correspondence problem*. Most of the methods used to solve the correspondence problem for multiple image sensors are based on constraints that are image-specific, and thus are not suitable for data association in the general case. Ayache [7] presents a good summary of these constraints.

To quantify the similarity between a given observation and a group, an association metric must be defined. Association measures include: correlation coefficients such as Pearson's product-moment and angular measures; distance measures such as the Euclidean, Manhattan, Minkowski, and Mahalanobis distances; association coefficients based on simple matching, Jaccard's coefficient and Gower's coefficient; and probabilistic similarities such as maximum likelihood [37]. Of these association measures, the Mahalanobis distance is the one most commonly used in multisensor fusion systems for data association and to reject outlier measurements [7, 38, 39, 40]. It gives the weighted Euclidean distance with weight equal to the inverse of the observations' covariance matrix. The Mahalanobis distance has been shown to be effective for grouping data and rejecting outlier measurements [7]. Also, it is easier to calculate than the other association measures with the exception of the nonweighted Euclidean distance [37].

The generalized Mahalanobis distance for a new observation vector z_k and predicted state vector $\hat{\mathbf{x}}_{k,k-1}$ is given by [7]

$$\delta(\mathbf{x}_k, \hat{\mathbf{x}}_{k,k-1}) = R_k \quad - \quad \frac{\partial g}{\partial \mathbf{x}} P_{k,k-1} \frac{\partial g}{\partial \mathbf{x}}^T \tag{13.26}$$

where $g(\cdot)$ is the nonlinear output system used by the Extended Kalman filter which replaces (13.9),

$$\mathbf{z}_k = g(\mathbf{x}_k) + \eta_k \tag{13.27}$$

An acceptable distance threshold is determined using the χ^2 distribution. For the system described in this chapter, the generalized Mahalanobis distance is used to accept or reject the camera measurement pairs (x, y) describing the location of the object hole centroid or corner location. It is also used to perform outlier rejection of range sensor measurements.

There are two approaches to associating data with a group [41]. The nearest-neighbor approach determines a unique pairing, so that at most one observation from a given sensor at a particular time can be paired with a previously established group. The observation can only be used once, either with an existing group or to start a new group. This method is implemented by minimizing an overall distance function based on all of the observations. The all-neighbors approach incorporates all observations within a certain distance of a group into that group. A given observation can be used as part of many groups. This approach allows an averaging of the data over multiple hypotheses. While this method is effective for single entities and multiple hypotheses, it is less effective for multiple entities [41].

13.5.5 Camera Sensor Redundancy and the Effect on Image Processing

Sensor data fusion can involve the fusion of multiple measurements from a single sensor such as a camera, as well as measurements from multiple sensors. This section looks at the improvements in accuracy, and hence robustness of the directed image processing, when additional measurements beyond the three noncollinear feature measurements that are required to estimate the object pose are integrated. First, the case of redundant information obtained from a single camera is considered. Then, the case involving multiple sensors is explored. A second camera and a range sensor were added to a simulation of the system, while only a range sensor was added to the experimental system.

When additional feature point measurements from a single camera are used as part of the pose estimation process, the accuracy of the state estimates improves. Table 13.7 shows the state error variances obtained for different numbers of camera features using the experimental system [36]. The experimental system utilizes a five degree-of-freedom robot, and so the error variances for the yaw are not considered.

As the number of features included in the measurement set increases, the overall accuracy of the pose estimation increases. However, the results show that increase in accuracy diminishes as additional good feature measurements are included. There is a significant difference between utilizing three features, which provide six measurements, and utilizing four. There is less difference between using four and five features. Beyond five, there is little improvement. As the use of additional measurements increases the processing time it becomes counterproductive to utilize additional image features. In the case of poor quality measurements, some of the measurements are likely to be rejected as outliers thus reducing the resultant number of features used. In this situation the use of more than five features can be productive.

When a second camera is added to a visual servoing system, redundant measurements of a feature point's location can be obtained. The first camera was assumed to have measurement noise variances of 0.06 pixel^2, which corresponds to those obtained experimentally [9]. Table 13.8 shows the error variances for the single camera case, and the cases where a noisier

TABLE 13.7 State error variances for different numbers of camera features.

# Features	x	y	z	ϕ	θ
5	4.91−06	4.60−07	3.76−07	1.05−05	9.99−05
4	5.80−06	2.13−07	4.20−07	5.20−06	1.05−04
3	7.50−05	5.31−05	7.36−06	3.34−04	1.64−03

TABLE 13.8 Complementary information error variances.

	x	y	z	ϕ	θ	ψ
Single camera	1.07−7	8.59−7	3.67−6	5.39−5	3.59−4	6.45−5
Add better camera	1.95−8	6.05−7	7.44−7	1.15−5	7.12−5	8.65−6
Add same camera	5.80−8	7.03−7	2.11−6	3.41−5	2.06−4	2.37−5
Add noisy camera	7.02−8	7.35−7	2.56−6	4.06−5	2.47−4	3.03−5

(variances of 0.1 pixel^2), less noisier (variances of 0.01 pixel^2), and comparably noisy camera is added to the system.

As expected, the addition of better or comparable information results in a noticeable improvement in the quality of the state estimates. Even the addition of poorer information results in some improvement in the accuracy of the state estimates. Richardson and Marsh [42] have shown, in theory, that the performance (defined in terms of a statistical ensemble of test data) of a system can never be degraded when the information from additional sensors is properly integrated into a system, and that in most cases the performance improves. However, under real-world conditions it can be difficult or impossible to ensure the proper integration of additional information.

The use of a second camera can also provide complimentary information. This case is the same as the use of additional feature points from a single image that is discussed above. Similarly, when a range sensor is used to provide complimentary information the accuracy of the state estimates improved. For experiments using three camera features and one laser range sensor measurement, there was a 23% improvement in the estimation accuracy of the z parameter, and moderate improvement in the estimation accuracy of the other parameters [36].

Each camera feature provides two independent measurements, while the range sensor provides one additional measurement. When two camera features are used by the system, there is not enough information to fully determine the object pose. This, combined with the fact that the system model is unknown, result in a reduction of the quality of the estimates with each passing time step, and eventual failure of the system.

Figure 13.20 shows the z parameter estimates when two camera features are used. As expected, for the camera only case, even with an extremely good initial state estimate, the system fails almost immediately. When the camera-range sensor fusion system is used, there is a significant increase in the length of time until the system fails.

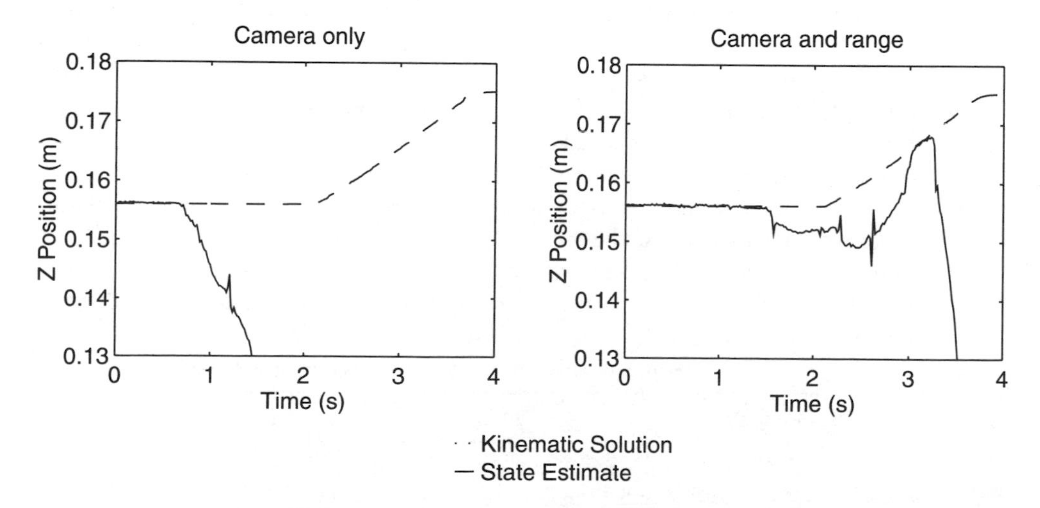

Figure 13.20 Estimate errors for two camera features.

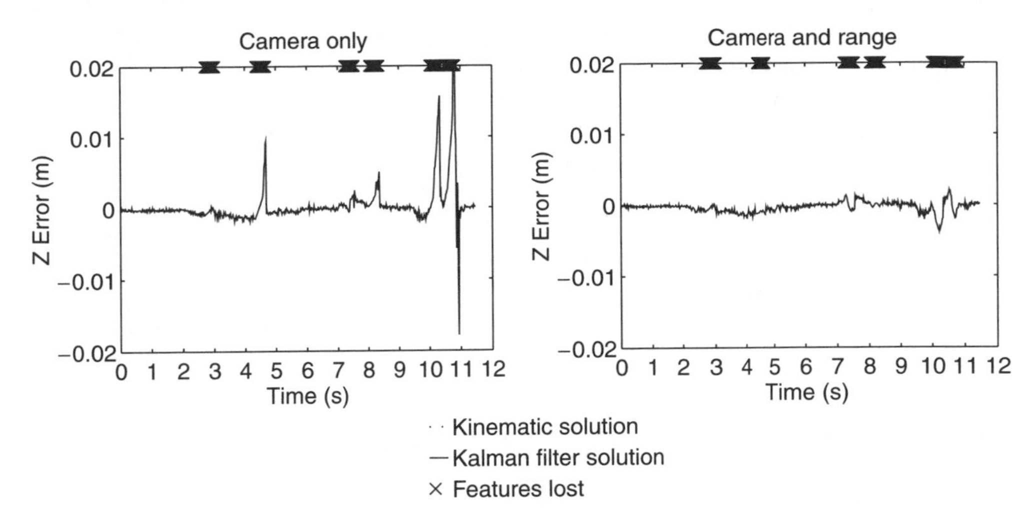

Figure 13.21 Estimate error during intermittent feature loss.

Figure 13.21 shows the results of estimating the z parameter when four of five features are lost intermittently. As expected, there is a significant improvement in the estimation process. The use of the additional range information is particularly important when the trajectory does not correspond to the assumed behavior as described by the constant velocity system model.

13.6 CONCLUSIONS

This chapter has focused on image processing techniques as they relate to the requirements of position-based visual servoing. To achieve reasonable dynamic control of the robot trajectories using vision feedback, sample rates of 50–100 Hz are required. This means that all of the image processing required at each sample must be completed in less than 10–20 ms. At the same time, it is imperative to reliably measure multiple feature locations on the image plane as accurately as possible over a wide range of relative motion of the camera with respect to the target object. This presents a challenging image processing problem.

To perform these image processing functions successfully, *directed* image processing techniques must be applied. In this directed image processing, it is important to use all available information regarding the expected image to be seen at any individual sample time. When implementing position-based vision servoing, this implies strong interaction between the image processing functions and the pose estimation functions. This interaction occurs in several ways. First, the pose estimation provides the best estimate of the relative pose of the target object with respect to the camera, and when used in conjunction with a description of the target object, this leads to a high quality estimate of the feature locations on the image plane. This information is used to predict the location and size of image plane windows required to isolate the features of interest.

A second effect of the interaction between the image processing and the pose estimation is that it leads to a well-defined environment for feature planning and real-time feature switching. As a result of a wide range of relative motion of the camera with respect to the target object, features may become occluded, become too large or too small, move out of the camera field of view, or mearly become poor quality features for visual servoing.

The third affect of this interaction results from the ability of the Kalman filter, which is used for the pose estimation to integrate measurements from a variety of sensors, such as multiple cameras and range sensors. This leads to improve robustness and accuracy of the pose estimation, so that image processing problems from one source do not destroy the pose estimation, and therefore, the visual servoing. It also leads to the identification of outlier measurements which allows corrective action to be taken, such as the removal of the offending feature measurements.

The chapter has demonstrated that through the use of binary images and *directed* image processing techniques, sample rates of at least 50–100 Hz can be achieved with accuracies and robustness that lead to good quality visual servoing. It has also shown that focusing is not a major issue when applying these methods. However, there is still potential for improvement with the possible use of grey scale images and/or adaptive threshold levels, which may lead to better isolation of the features and improved robustness and accuracy of the feature measurements.

REFERENCES

[1] S. Hutchinson, G. D. Hager, and P. I. Corke. "A tutorial on visual servo control." *IEEE Transactions on Robotics and Automation*, vol. 12, no. 5, pp. 651–670, October 1996.

[2] W. J. Wilson. "Visual servo control of robots using Kalman filter estimates of robot pose relative to work-pieces." *Visual Servoing: Real-Time Control of Robot Manipulators Based on Visual Sensory Feedback*, K. Hashimoto, ed., pp. 71–104. Singapore, World Scientific, 1993.

[3] J. S.-C. Yuan. "A general photogrammetric method for determining object position and orientation." *IEEE Transactions on Robotics and Automation*, vol. 5, no. 2, pp. 129–142, April 1989.

[4] W. J. Wilson, C. C. Williams Hulls, and G. S. Bell. "Relative end-effector control using cartesian position based visual servoing." *IEEE Transactions on Robotics and Automation*, vol. 12, no. 5, pp. 684–696, October 1996.

[5] A. Rosenfeld and A. C. Kak. "Digital Picture Processing." New York, Academic Press, 2nd edition, 1982.

[6] H. Lee and W. Lei. "Region matching and depth finding for 3D objects in stereo aerial photographs." Pattern Recognition, vol. 23, no. 1/2, pp. 81–94, 1990.

[7] N. Ayache. "Artificial Vision for Mobile Robots: Stereo Vision and Multisensory Perception." Cambridge, MA: The MIT Press, 1991.

[8] J. Wang. "Optimal Estimation of 3D Relative Position and Orientation for Robot Control." M.S. thesis, University of Waterloo, 1991.

[9] C. Madhusudan. "Error Analysis of the Kalman Filtering Approach to Relative Position Estimations Using Noisy Vision Measurements." M.S. thesis, University of Waterloo, 1990.

[10] J. S. Wong. "Close Range Object Operations using a Robot Visual Servoing System." M.S. thesis, University of Waterloo, 1996.

[11] R. Smith. "Relative Position Sensing Using Kalman Filtering of Vision Data." M.S. thesis, University of Waterloo, 1989.

[12] B. Bishop, S. Hutchinson, and M. Spong. "On the performance of state estimation for visual servo systems." *Proceedings of IEEE International Conference on Robotics and Automation*, vol. 1, pp. 168–173, 1994.

[13] F. Janabi-Sharifi and W. J. Wilson. "Automatic selection of image features for visual servoing." *IEEE Transactions on Robotics and Automation*, vol. 13, no. 6, pp. 890–903, December 1997.

[14] N. P. Papanikolopoulos. "Controlled Active Vision." PhD thesis, Carnegie Mellon University, Pittsburgh, PA, 1992.

[15] J. T. Feddema, C. S. G. Lee, and O. R. Mitchell. "Weighted selection of image features for resolved rate visual feedback control." *IEEE Transactions on Robotics and Automation*, vol. 7, no. 1, pp. 31–47, February 1991.

[16] L. E. Weiss, A. C. Sanderson, and C. P. Neuman. "Dynamic sensor-based control of robots with visual feedback control." *IEEE Journal of Robotics and Automation*, vol. RA-3, pp. 404–417, October 1987.

[17] B. Espiau, F. Chaumette, and P. Rives. "A new approach to visual servoing in robotics." *IEEE Transactions on Robotics and Automation*, vol. 8, pp. 313–326, 1992.

[18] S. Yi, R. M. Haralick, and L. G. Shapiro. "Automatic sensor and light source positioning for machine vision." *Proceedings of the 1990 IEEE International Conference on Pattern Recognition*, Los Alamitos, CA, 1990.

[19] R. Tsai and K. Tarabanis. "Occlusion-free sensor placement planning." *Machine Vision for Three-Dimensional Scenes*, H. Freeman (Ed.), San Diego,CA, Academic Press, 1990.

[20] F. Janabi-Sharifi. "A Supervisory Intelligent Robot Control System for a Relative Pose-Based Strategy." PhD thesis, University of Waterloo, 1995.

[21] E. P. Krotkov. "Focusing." Tech. Rep. MS-CIS-86-22, University of Pennsylvania, 1986.

[22] D. J. Elzinga and D. W. Hearn. "The minimum covering sphere problem." *Management Science*, vol. 19, no. 1, pp. 96–104, September 1972.

[23] F. Janabi-Sharifi and W. J. Wilson. "An intelligent assembly robotic system based on relative pose measurements." *Intelligent and Robotic Systems*, vol. 11, no. 1, 1995.

[24] J. J. Clark and A. L. Yuille. "Data Fusion for Sensory Information Processing Systems." Boston, Kluwer Academic Publishers, 1990.

[25] R. C. Luo and M. G. Kay. "Multisensor integration and fusion in intelligent systems." *IEEE Transactions on Systems, Man and Cybernetics*, vol. 19, no. 5, pp. 901–931, September/October 1989.

[26] H. H. Baker and T. O. Binford. "Depth from edge and intensity-based stereo." *Proceedings of the International Joint Conference on Artificial Intelligence*, pp. 631–636, 1981.

[27] A. Gerhard, H. Platzer, J. Steurer, and R. Lenz. "Depth extraction by stereo triples and a fast correspondence estimation algorithm." *Proceedings of the International Conference on Pattern Recognition*, 1986.

[28] C. Hansen and N. Ayache. "Efficient depth estimation using trinocular stereo." Sensor Fusion: Spatial Reasoning and Scene Interpretation, P. S. Schenker (ed.), pp. 124–131. *Proceedings SPIE 1003*, 1988.

[29] M. Ito and A. Ishii. "Range and shape measurement using three-view stereo analysis." *Proceedings of the 1986 IEEE Conference on Computer Vision and Pattern Recognition*, pp. 9–14, 1986.

[30] Y. C. Kim and J. K. Aggarwal. "Finding range from stereo images." *Proceedings of the 1985 IEEE Conference on Computer Vision and Pattern Recognition*, pp. 289–294, 1985.

[31] M. Pietikainen and D. Harwood. "Depth from three camera stereo." *Proceedings of the 1986 IEEE Conference on Computer Vision and Pattern Recognition*, pp. 2–8, 1986.

[32] M. Shao, T. Simchony, and R. Chellappa. "New algorithms for reconstruction of a 3-D depth map from one or more images." *Proceedings of the 1988 IEEE Conference on Computer Vision and Pattern Recognition*, pp. 530–535, 1988.

[33] G. B. Smith. "Stereo reconstruction of scene depth." *Proceedings of the 1985 IEEE Conference on Computer Vision and Pattern Recognition*, pp. 271–276, 1985.

[34] S. T. Barnard and M. A. Fischler. "Computational stereo." *ACM Computing Surveys*, vol. 14, no. 4, pp. 553–572, December 1982.

[35] U. R. Dhond and J. K. Aggarwal. "Structure from stereo - A review." *IEEE Transactions on Systems, Man, and Cybernetics*, vol. 19, no. 6, pp. 1489–1510, November/December 1989.

[36] C. C. Williams Hulls. "Dynamic Real-Time Multisensor Fusion Using an Object Model Reference Approach." PhD thesis, University of Waterloo, 1996.

[37] D. L. Hall. "Mathematical Techniques in Multisensor Data Fusion." Boston, Artech House, 1992.

[38] J. L. Crowley and F. Ramparany. "Mathematical tools for representing uncertainty in perception." *Spatial Reasoning and Multi-Sensor Fusion: Proceedings of the 1987 Workshop*, A. Kak and S. Chen (eds.), pp. 293–302, Los Altos, CA, Morgan Kaufman Publishers, 1987.

[39] H. F. Durrant-Whyte. "Integration, Coordination and Control of Multi-Sensor Robot Systems." Boston: Kluwer Academic Publishers, 1988.

[40] J. K. Hackett and M. Shah. "Multi-sensor fusion: A perspective." *Proceedings of the 1990 IEEE International Conference on Robotics and Automation*, vol. 2, pp. 1324–1330, 1990.

[41] S. S. Blackman. "Theoretical approaches to data association and fusion." Sensor Fusion, C. B. Weaver (Ed.), pp. 50–55. *Proceedings SPIE 931*, 1988.

[42] J. M. Richardson and K. A. Marsh. "Fusion of multisensor data." *The International Journal of Robotics Research*, vol. 7, no. 6, pp. 78–96, December 1988.

Chapter 14

VISION-BASED OBJECTIVE SELECTION FOR ROBUST BALLISTIC MANIPULATION

Bradley E. Bishop
United States Naval Academy
Mark W. Spong
University of Illinois

Abstract

The use of visual measurement for high-speed and high-precision robotic tasks requires careful analysis of the fundamental properties of the sensing paradigms chosen. In this work we analyze the information generated by a state estimator for a target moving object that is to be intercepted by a robotic manipulator. Selection of potential interaction objectives is carried out through analysis of the data stream from the visual system. The underlying sensing problem is designed for a principally unstructured environment under fixed-camera visual measurement. The test domain is chosen to be ballistic manipulation of a sliding object on a low-friction planar surface: the air hockey problem.

14.1 INTRODUCTION

The use of visual feedback for robotic interaction with changing environments is strongly motivated by the primacy of vision among human senses for planning and executing interaction tasks. It is therefore unfortunate that digital cameras and computers lack the functionality, resolution, and encoded image processing techniques available to humans. The fundamentally discrete nature of computer vision (in time, space and resolution) as well as the inherent lack of high-level real-time visual processing techniques lead to significant difficulties for dynamic visual servo applications.

One of the most challenging cases of visually driven robotics is manipulation of rapidly moving objects in an unstructured environment. High environmental object speeds require the use of visual techniques that can be carried out in real time, generally limiting the amount of information attainable from the visual process. Significant delay times must be avoided to guarantee the success of the manipulation task and stability of any controller that includes the visual process in a feedback loop.

Any visual measurement techniques that are utilized for high-speed and high-precision robotic tasks must be carefully analyzed from a performance standpoint. Both timing issues and robustness properties of a selected sensing paradigm must be carefully considered when integrating vision into a robotic system. In fact, speed and accuracy of visual techniques are often competing criteria, requiring a delicate balance of processing delay and visual acuity. System objectives must be selected such that the inherent delay and error in visual measurement does not significantly affect the task.

In order to study the interaction of robotic systems with high-speed objects through application of computer vision, we have chosen to study the air hockey problem. Fundamental to the air hockey problem is ballistic manipulation of a sliding, circular puck on a low-friction surface through impulsive impacts. In the framework of a competitive game, in which the robot seeks to score goals on an opponent while simultaneously protecting its own goal, the task requires not only rapid response but also highly accurate state estimation given a sparse data set. Using the visually derived motion of the puck, the robot must plan and execute trajectories of the circular mallet (attached to the tip of the last link in the manipulator) such that the outgoing puck velocity after impact matches some desired profile.

A number of interesting visual servo applications for target tracking and interception have been generated. In [1], an eye-in-hand robot uses motion estimation from optical flow to generate estimates of target object trajectories for an interception and grasping task that is carried out after the estimates have converged sufficiently. The task domain of air hockey precludes, at the most basic level, the use of trajectory filtering techniques that must converge before any action can be taken. In air hockey, a decision regarding what action the robot should take must be made before the puck reaches the robot's end of the table, regardless of the number of vision samples received or the accuracy of the estimate. Additional work that relates to visual interception and manipulation tasks in dynamic environments can be found in [2, 3], although the task domains are significantly different for these works, and target velocities tend to be lower than those encountered in air hockey. Fundamental work in vision-driven ballistic robot games can be found in [4].

The remainder of this paper is organized as follows: Section 14.2 includes details of the visual measurement scheme selected for this manipulation task. Details of data analysis used for prediction and estimation of motion can be found in Section 14.3. The full ballistic manipulation scheme, with objective selection criteria, is outlined in Section 14.4, together with experimental results. Conclusions and future directions can be found in Section 14.5.

14.2 VISUAL MEASUREMENT SCHEME

In order to utilize visual data, it is necessary to extract those portions of the scene that are of interest from the background data. There are many scene partitioning methods available, but only a few of these are typically used in dynamic visual servo systems carrying out visual processing in real time.

Simple scene partitioning methodologies are often task specific. We will consider the problem of locating a single (moving) object on a static background. Many common methods of scene partitioning for extraction of moving objects rely on the use of *thresholding*. Thresholding refers to the mapping of a grey scale image to a binary image: pixels with intensity above a certain value, or threshold, are mapped to one, while those below that threshold are mapped to zero. In visual servo literature, the scene is often set so that the object of the vision process is easily extracted by means of simple intensity variation; for example, a white ping-pong ball on a black background [5].

The thresholding approach to scene partitioning has the advantage of simplicity, but may require either special intensity conditions or additional processing. If the histogram of image intensities is not clearly bifurcated between object and background intensities, the thresholding algorithm will offer multiple regions of value one, or possibly none at all, when only one object of interest is present in the image plane.

A straightforward method of dealing with background intensity profiles is to generate a difference image by subtracting a current image from a pregenerated image of the static background. The difference image is then thresholded. This approach works moderately well for the air hockey case, as the object of interest (the puck) is uniform in intensity and darker than most of the background profile. The outcome of the subtract-and-threshold algorithm is a binary image in which, ideally, the single region of value one corresponds to the puck's trajectory during the sample interval. The location of the puck is assumed then to be the centroid of the region.

We have chosen to leave the commercial air hockey table used in our experiments unmodified (Fig. 14.1). This choice is made primarily for purposes of studying the use of visual servo in unstructured environments. Due to the subtract-and-threshold scheme utilized, locations at which the table is dark (low original image intensity) will never be mapped to a value of one in the thresholded image. This causes a systematic disturbance of the centroid when the puck passes over these regions.

A number of additional factors influence the accuracy of the extracted centroid, including lens distortions, specularities and shadows, and blur. Additionally, random noise can result in inappropriate segmentation if the threshold is not set appropriately. These issues are discussed in a general framework, as well as with specific application to the thresholding algorithm, in [6] and [7].

A primary disadvantage of the thresholding algorithm, from the perspective of sensing for ballistic manipulation, is the effect of blur on the thresholded image. Let the puck have intensity equal to a grey scale level of zero (out of 256) while the background (for sake of discussion) has a uniform intensity value of I_b. Further, allow that the camera's exposure time is approximately equal to the sample period T. A pixel onto which the image of the puck projects for some portion of the exposure interval will have a grey scale value *higher* than zero. The smaller the portion of the exposure interval over which the puck images onto the pixel, the higher the pixel value. Specifically, if the given pixel is covered by the image of the puck for nT, with $0 < n < 1$, the grey scale level of the pixel will be approximately $(1-n)I_b$. Thus, when the current image is subtracted from the background image, the pixel's value will be nI_b.

Given a threshold value I_t, there is always a velocity v_b such that, for any velocity $v \geq v_b$, no pixel is covered by the image of the puck for more than $\frac{I_t}{I_b}T$. In fact, as the velocity increases toward v_b, the region of the final binary image corresponding to the puck shrinks in size.

As the threshold level increases with the background intensity constant, v_b *decreases*. Even when we have optimized the I_b and I_t values, we are still velocity limited for purposes of sensing: there is always a finite v_b that is generally quite achievable.

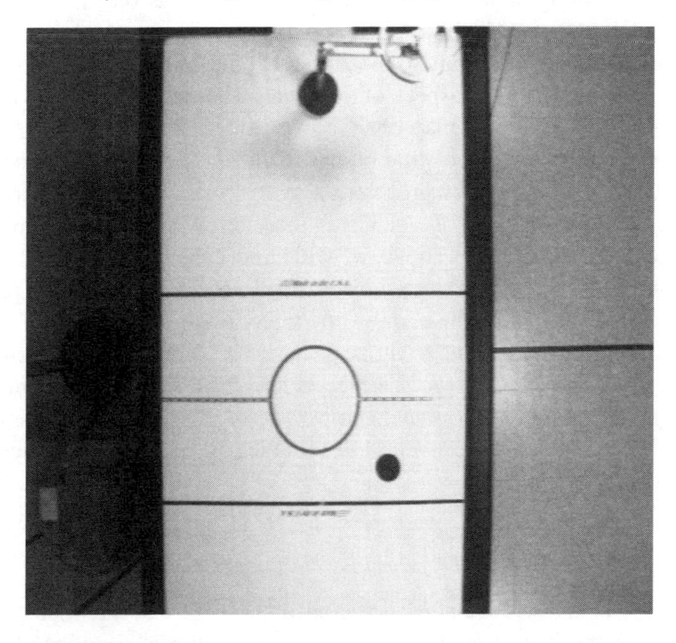

Figure 14.1 Image of commercial air hockey table, puck and robot end-effector from overhead CCD camera.

The sensing paradigm that we utilize is end-effector open loop for purposes of visual servo control (see Section 14.4). As such, considering that the robot has not been painted to match the background, the workspace of the robot is exclusive of the sensing area. Poor transient behavior of the robot system will result in a disturbed centroid as the robot enters the image plane region of attention. Other difficulties include the opponent impinging on the play area, table misalignment, etc.

In order to deal with spurious data caused by the difficulties mentioned above or by the communication between various pieces of hardware in our system (see Section 14.4), we have implemented a data validation scheme. This scheme consists of considering the smallest bounding box on the thresholded region and the $0th$ order moment of the region (the area). We discuss the use of these measures in Section 14.3.

All of the visual processing techniques used for sensing of ballistic trajectories are written using ImageFlow code for the DataCube MAXVIDEO 20 system, and are achievable at frame rate.

14.3 STATE ESTIMATION AND PREDICTION

Air hockey is an example of a ballistic control problem, in which the robot must intercept a moving object in such a way as to modify the object's trajectory in a desirable fashion. In order for a ballistic control system to approach peak performance when dealing with rapidly moving target objects, fast and accurate prediction of target object motion must be achieved.

For the air hockey system, a number of physical models can be used for the purpose of state estimation and prediction. Selection of an appropriate model should be based on the desired accuracy of prediction and the amount of computational power and time available. The most simple model of the air hockey process considers the puck to be an extended particle, all collisions to be fully elastic, and admits no forces tangent to impact surfaces (thus the puck's rotation is unimportant). The most complicated practical model includes dynamic effects of spin, friction, inelastic collisions, and fluid dynamics (for the air flow). Work has been done on the modeling of ice hockey pucks [8] as well as on the study of the dynamics of impulsive manipulation [9]. Both of these works are computationally intensive and include some dynamics that do not apply to the air hockey problem.

The most problematic aspect of physical modeling for the air hockey system is the possibility that the dynamics may change over time. The air supply for the table varies in pressure over time and with position on the surface. The table itself may be roughed by the puck or mallets, resulting in a change in its already position-dependent friction characteristics. Additionally, the introduction of a new puck may result in significantly different dynamics.

Due to the considerations above, we will choose the most simple physical process model, relying on analysis of the data stream for estimation of the accuracy of prediction. We model the impact of the puck with the edge of the table as follows. Let $b(t)$ be the puck position, with $\dot{b}_t(t)$ the velocity tangent to an impact surface and $\dot{b}_n(t)$ the velocity normal to that surface. Let $\dot{b}(T^+)$, $\dot{b}(T^-) \in \mathcal{R}^2$ be the velocity of the object just after and just before an impact, respectively. Assuming that the impact surface is immobile and immovable, the simple coefficient of restitution law (proven to be reliable for dynamic tasks such as juggling [10]) is given by

$$\dot{b}_n(T^+) = -\alpha_r \dot{b}_n(T^-) \tag{14.1}$$

$$\dot{b}(T^+) = \dot{b}(T^-) - (1 + \alpha_r) n n^T \dot{b}(T^-) \tag{14.2}$$

where $\alpha_r \in [0, 1]$ and n is the normal vector to the surface at the impact point.

Modeling the behavior of the puck including spin is significantly more difficult than modeling the simple impacts discussed above [9]. Further, we do not have the ability to sense the spin on the puck except by a comparison of velocities before and after impacts. To further complicate matters, the coefficient of restitution and the friction coefficients of the table edges with respect to the puck are seen to be position dependent. As such, we have chosen to not model the spin explicitly in the puck dynamics, but rather to account for our lack of knowledge concerning the system behavior at edge impacts in the control algorithm (see Section 14.4). This negates the need for consideration of the friction coefficients on the mallet, puck, and table edges.

For ballistic manipulation, it is necessary to acquire not only the target object's current position and velocity, but also to predict the most likely trajectory of the object in the future. By estimating the accuracy of this prediction, we will be able to choose control objectives in a robust and intelligent manner.

Prediction theory has a broad base in literature (e.g., [11]). Many techniques exist for performing filtering of signals to compensate for noise and disturbances in a control theoretic framework [12]. However, in the case of simple linear motion, most of these techniques become little more than weighted averaging methods. Further, the systems are designed to achieve their objectives with a large sample base. In the case of air hockey, we must perform prediction as rapidly as possible in order to achieve the goals of the control system. As such, we will consider straightforward prediction techniques.

Before we consider the algorithms for prediction in depth, we must outline some of the major difficulties with analysis of the raw data. In addition to the problems discussed in Section 14.2, difficulties also arise when the puck undergoes an impact with the table edge. Specifically, consider the case of two sample periods with centroids as shown in Fig. 14.2. While the actual trajectory is angled at 45°, the centroids indicate a trajectory straight down the table. Additionally, as we consider each maximal bounce-free trajectory segment to be an independent data entity (due to problems with spin, as outlined above), we must attempt to estimate between which two samples a bounce has occurred.

Wall impacts are seen to have an effect not only on the inter-sample prediction, but also on the accuracy of the extracted centroid for samples including a bounce. Experimental investigations have indicated that the vision sample at an impact point is significantly disturbed, and can result in poor estimation. The cause of this can be seen in Fig. 14.3. The image taken

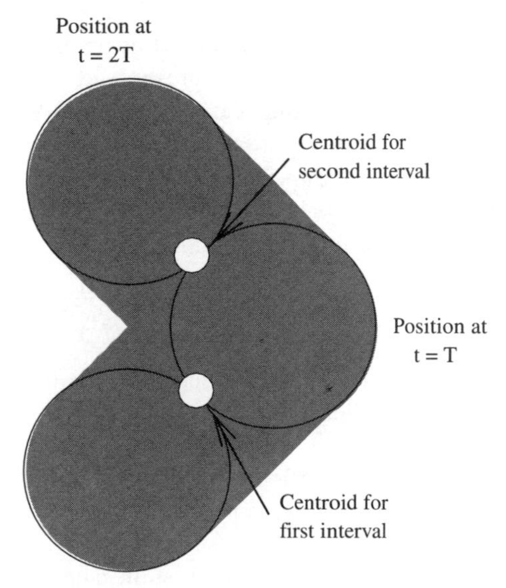

Figure 14.2 Ambiguity for centroid-based prediction under the thresholding algorithm.

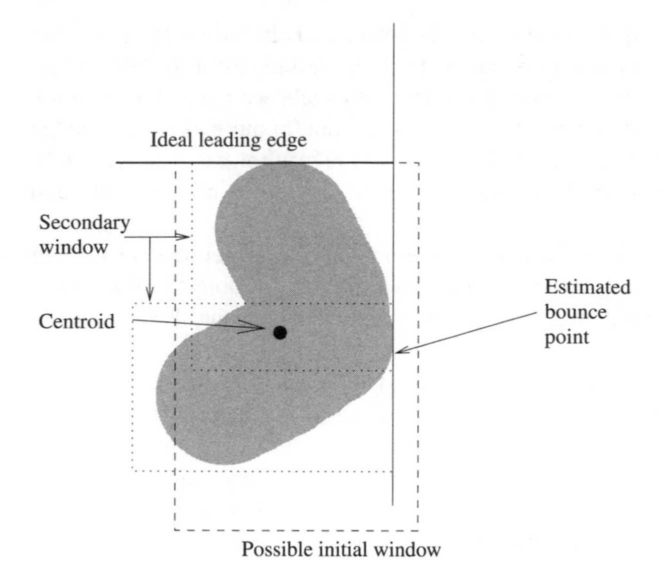

Figure 14.3 Example of an intra-sample wall impact (bounce), including an example of windowing.

at a bounce point does not accurately lie in either the pre- or post-impact trajectory segments. One solution to this problem is to window the image and extract information for each of the trajectory segments (as suggested in Fig. 14.3), but this is a difficult task to accomplish in real time. Therefore, we have chosen to ignore the vision sample that includes a bounce point. Determination of the inclusion of a bounce in an image is accomplished by prediction (for the pre-bounce estimates) and simple deduction (for the post-bounce estimates). While electronic camera shuttering could alleviate some of these difficulties, we have chosen to study the case where the camera's imaging dynamics are nontrivial in order to emphasize the robustness of the developed schemes.

After the thresholded region has been generated, a data validation step is utilized to attempt rejection of exceptionally poor data points. In addition to concerns discussed above, we must be careful of data generated when the processing window includes additional objects other than the puck. The size of the smallest bounding box (as well as its edge locations) can be used in most circumstances to detect errors introduced by spurious objects in the field of view. Additionally, the area of the thresholded region is useful for similar cases. If all of the validation steps are successfully carried out, the centroid of the thresholded image is generated.

The algorithm that we utilize to generate estimates of the puck motion is a simple least-squares approach in two dimensions. We assume that, for sample i, the centroid is $(u(i), v(i))$ and the time is $t(i)$. For each bounce-free trajectory segment $(u(i), v(i), t(i))$, $i = 1, \ldots, N$, we determine the best fit lines for $(t, u(t))$ and $(t, v(t))$ of the form

$$\hat{u}(i) = A_u(N) + B_u(N)t(i) \tag{14.3}$$

$$\hat{v}(i) = A_v(N) + B_v(N)t(i) \tag{14.4}$$

where $\hat{u}(i)$, $\hat{v}(i)$ are the best linear estimates at time $t(i)$. The index on the A and B values indicates the best fit using data up to step N, and all A and B values are updated at each sample beyond the second.

The calculation of a least-squares linear constants for data of the form $(t(i), x(i))$, $i = 1, \ldots, N$ is straightforward [13]. The benefit of using this form of data fitting is that the resulting linear equations can be easily used to perform prediction. Further, we can analyze the statistical nature of the data set if we assume a particular distribution on the noise process. For simplicity, and from experience, we have selected a Gaussian distribution for the noise. We also note that the Central Limit theorem [14] supports the use of a Gaussian random variable for the noise on a complicated signal. We have assumed that the noise in our process is zero mean with some unknown standard deviation, and uncorrelated in the two coordinates.

Given A, B for the data set (t_i, x_i), $i = 1, \ldots, N$, we determine the estimated variance for each of the coefficients using standard methods [13]. We then perform the following calculations for each bounce-free trajectory segment at each time instant:

- Determine σ_u and σ_v, the standard deviations.
- Determine σ_{A_u}, σ_{B_u}, σ_{A_v}, σ_{B_v}.
- Choose a probability p.
- Determine $A_{u_{\min}}$ and $A_{u_{\max}}$ s.t. $P(A_{u_{\min}} \leq A_u \leq A_{u_{\max}}) = p$, assuming a Gaussian distribution of the A_u about the best A_u. Do the same for A_v, B_u and B_v.
- Determine the following trajectories:
  1. Most likely 2D trajectory (defined by A_u, B_u, A_v, B_v)
  2. 1D trajectory determined by $A_{u_{\min}}$, $B_{u_{\max}}$
  3. 1D trajectory determined by $A_{u_{\max}}$, $B_{u_{\min}}$
  4. 1D trajectory determined by $A_{v_{\min}}$, $B_{v_{\max}}$
  5. 1D trajectory determined by $A_{v_{\max}}$, $B_{v_{\min}}$
  6. 2D trajectories determined by combinations of the above 1D trajectories

This results in a most likely trajectory and a probability-based possible trajectory set. We define $\Gamma(i)$ as the set of trajectories generated by the worst-case 1D trajectories at sample instant i.

An example using experimental data is given in Figs. 14.4–14.7. We have chosen $p = 0.6826$, equal to the probability that a Gaussian random variable X lies within σ_X of its mean. The coefficient of restitution for the wall-puck impacts was experimentally estimated at an average value of 0.725. The trajectory is shown in Fig. 14.4, with the centroid of the

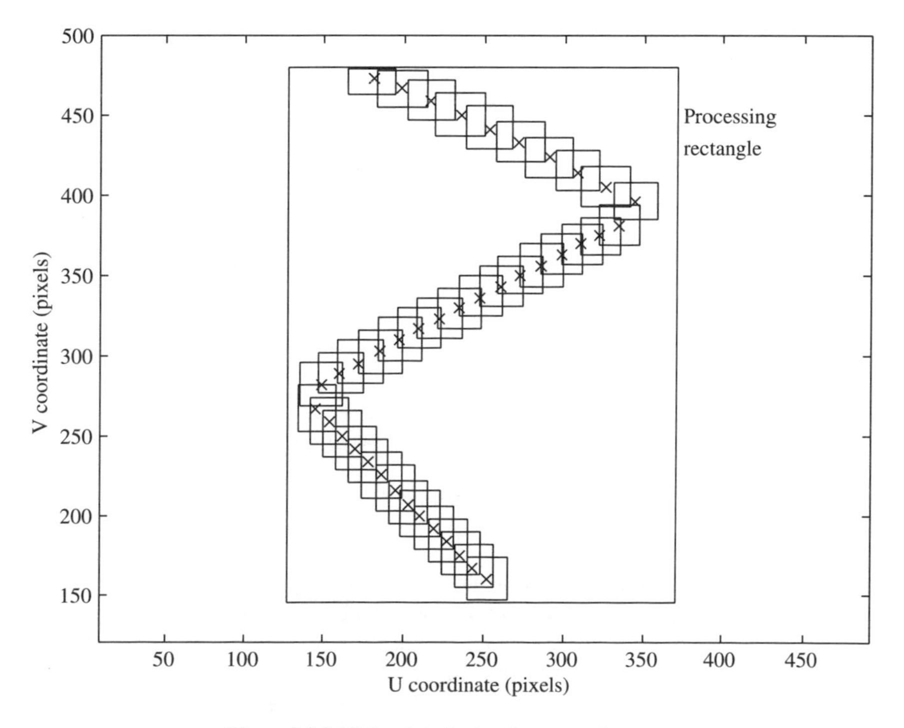

Figure 14.4 Vision data for two-bounce trajectory.

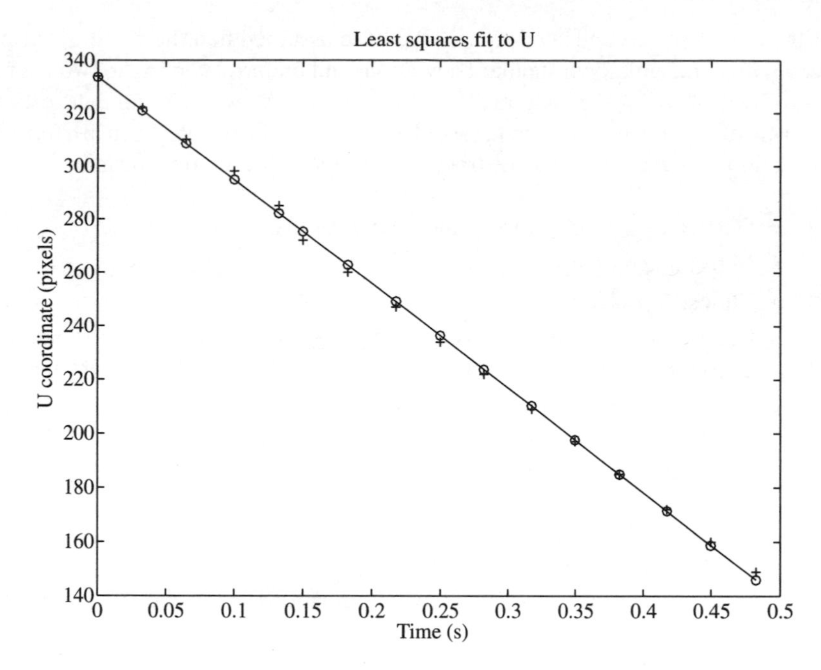

Figure 14.5 Least-squares fit for the u coordinate of segment two of the two-bounce trajectory [data points (+) and best fit points (o)].

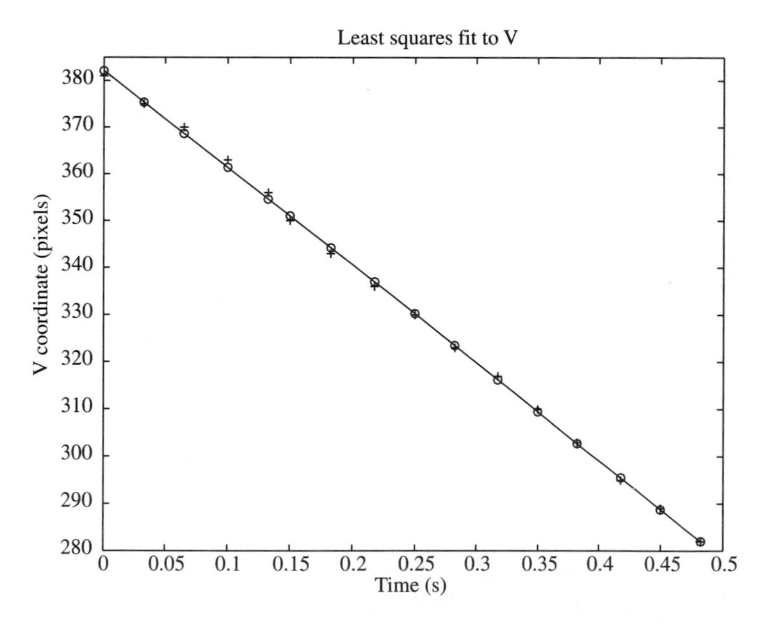

Figure 14.6 Least squares fit for the v coordinate of segment two of the two-bounce trajectory.

thresholded image indicated by an 'x' and the bounding box on the thresholded region drawn for each sample. For this trajectory, there are three bounce-free trajectory segments over 1.2 s. We will consider each trajectory segment to be a unique data set and normalize time such that the first sample of each segment occurs at $t = 0$ s.

The least squares fit to the data for the second of the trajectory segments is shown in Figs. 14.5–14.6. This data fit was generated using all of the data on the trajectory segment.

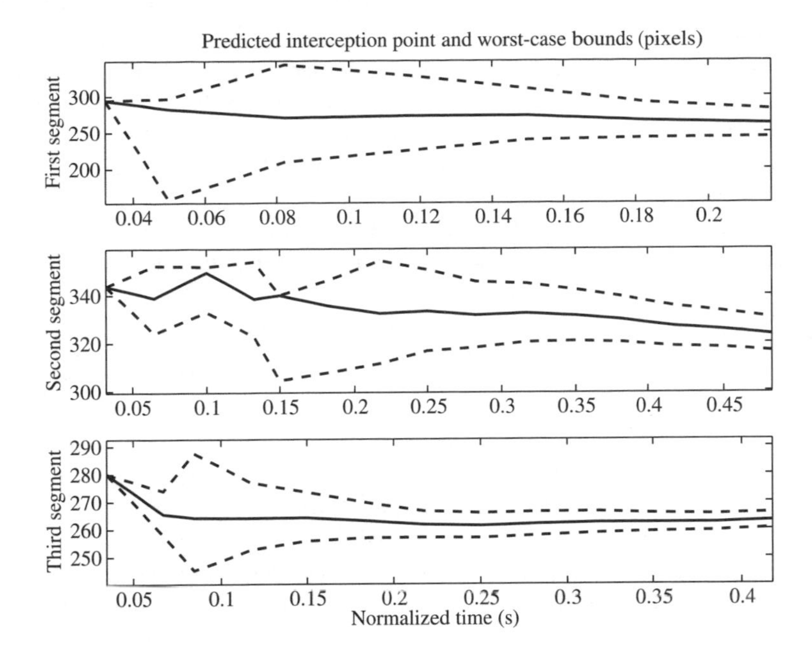

Figure 14.7 Predicted (solid) and worst-case (dashed) intersection points for the two-bounce trajectory.

The least-squares fit at each sample instant was used to predict the intersection of the centroid with the image plane line $v = 147$ pixels. The predicted intersection point at each sample instant is shown in Fig. 14.7, together with the worst-case intersection points for the trajectories in Γ. Timing information, which can be equally important for ballistic control applications, can be seen in Fig. 14.8. Note that the predicted interception times are referenced to the normalized time for the trajectory segment over which they are defined.

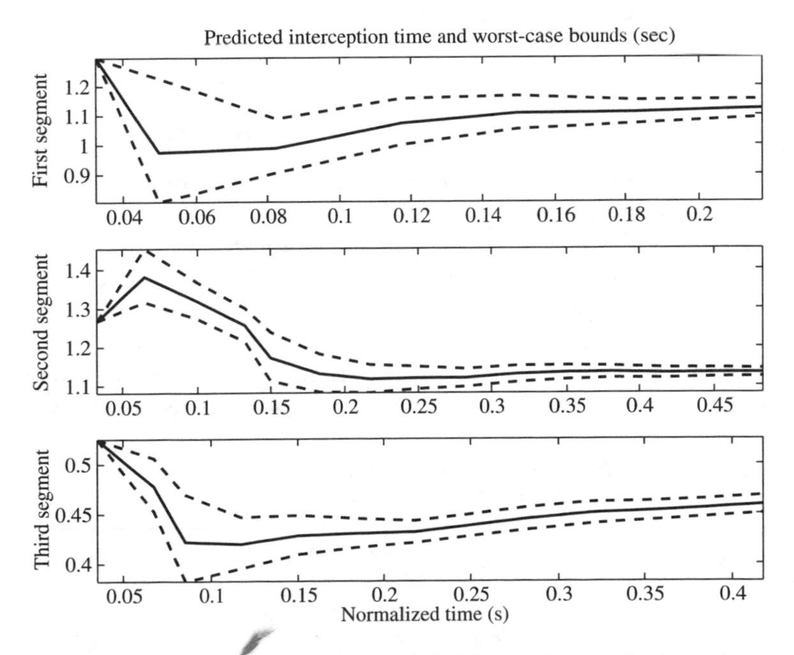

Figure 14.8 Predicted (solid) and worst-case (dashed) intersection times for the two-bounce trajectory.

The estimate of the intersection point on the third segment is seen to be very accurate, indicating a likely error of no more than three pixels in the intersection point as generated from 13 samples. An interesting note concerning these data is that the prediction for the first segment converges to a value closer to the third segment estimate than does the second segment's prediction. The most likely cause of this difficulty is spin on the puck, resulting in poor prediction at the wall impacts.

In Section 14.4, we will consider the use of our statistical analysis of visual data in selection of appropriate robust objectives for ballistic manipulation.

14.4 AIR HOCKEY

The object of air hockey is to produce desired puck motions through impacts between a circular mallet and a circular puck. As a competitive game, it is necessary to choose appropriate puck trajectories such that the puck enters the opponent's goal and not a defended goal. We will consider the strategy of selecting appropriate desired puck trajectories as determination of a system *objective*, defined over the set of possible actions.

The physical setup for our experiments is as shown in Fig. 14.1, and consists of a 3 DOF (degree of freedom) planar direct drive robot arm and a CCD camera. Visual processing is carried out on the DataCube MAXVIDEO 20 system, while robot control is achieved using a TMS320 floating point DSP and dedicated motor driver cards (see [15] for more details). Communication is achieved across the backplane of a VME cage.

The underlying controller for our robot will be PID with partial inverse dynamics, in which a full inverse dynamics scheme will be utilized for the first two links assuming that the third link is negligible, and the third link controller will be a simple PID [15]. The PD gains of the joint-level controllers are given in Table 14.1. The integral gain k_{i_3} used for link 3 is 0.5. No other integral gain was utilized.

The camera-robot system for these experiments was calibrated off-line. For simplicity, the camera was placed with its optical axis perpendicular to the playing surface. The calibration parameters for the camera-robot system are θ, $\frac{\alpha_1 f}{Z}$, $\frac{\alpha_2 f}{Z}$ and O_r, where a point x_r in the robot plane projects onto the image plane of the camera at point p_r given by:

$$p_r = \begin{bmatrix} \frac{\alpha_1 f}{Z} & 0 \\ 0 & \frac{\alpha_2 f}{Z} \end{bmatrix} \begin{bmatrix} \cos\theta & -\sin\theta \\ \sin\theta & \cos\theta \end{bmatrix} (x_r - O_r) \tag{14.5}$$

The values of the calibration data and the puck measurements of interest for planning actions are given in Table 14.2.

Selection of a system objective is achieved through a heuristic decision process carried using a number of primitives defined over the set of all possible incoming puck trajectories. The principle set of objectives for air hockey can be broken down into classes that we will denote `Strike`, `Block`, and `Ignore`. The `Strike` objective is designed to attempt to generate a specific outgoing trajectory after impact, while the `Block` objective is aimed at guaranteeing that a specific range of outgoing trajectories are *not* generated. A lower-level objective is

TABLE 14.1 Controller parameters for the experimental air hockey system.

Link	τ_i^{max}	Parm.	Value	Parm.	Value
1	50 N–m	k_{p_1}	121.0	k_{d_1}	22.0
2	30 N–m	k_{p_2}	72.6	k_{d_2}	17.04
3	15 N–m	k_{p_3}	4.0	k_{d_3}	0.4

TABLE 14.2 Measurements of interest for experimental investigation of the air hockey problem.

Parameter	Value
Diameter of the puck	0.065 m
Diameter of the mallet	0.1 m
Width of the goal	0.23 m
Opponent goal location	$x = [2.0, 0.133]^T$ m
θ	1.571 rad
$\frac{\alpha_1 f}{Z}$	282.3 pixels/m
$\frac{\alpha_2 f}{Z}$	348.0 pixels/m
O_r	$[0.971, 0.103]^T$ m

`Ignore`, in which the current incoming puck trajectory is such that the system simply allows it to evolve on its own.

We define a number of quantities based on the analysis of centroid data streams. These definitions will be useful for determination of a best objective given the incoming trajectory. For this discussion, we will ignore the effects of manipulator dynamics on objective selection, although this topic is covered in detail in [15]. For the measures below, we will assume that the table is oriented so that the nongoal table edges lie along the x_1 direction, corresponding to the v direction in the image plane.

The measures that we will be concerned with are as follows:

P_s: Measure of the probability of a score on the home goal (to be defended) based on the distribution of the data ($P_s \in \mathcal{R}^+$).

V_m: The (workspace) velocity of the trajectory in Γ with maximum $|B_v|$ ($V_m \in \mathcal{R}^2$).

V_s: The (workspace) velocity of the trajectory in the subset of Γ that projects into the home goal with maximal $|B_v|$ ($V_s \in \mathcal{R}^2$).

N_b: Projected number of wall impacts of the most likely trajectory before the desired interception point ($N_b \in \mathbf{N}$).

ρ: Width of Γ.

We select the desired objective of the system based on these heuristic measures. To assist in this goal, we will define the following value:

δ: Heuristic limit on ρ determined by the value of P_s. May take on two values: δ_{low} and δ_{high}.

There are any number of methods of selecting an appropriate objective given various values of the measures defined above. The V_m and V_s values are primarily useful for control-based objective decisions (carried out in consort with the decision process associated with visual measurement) and will not be utilized herein (see [15] for details). A simple example of a vision-based objective selection scheme that utilizes many of the measures is as follows:

If ($P_s > P_{s_{\min}}$) OR ($N_b > 0$)
    $\delta = \delta_{\text{low}}$
Else
    $\delta = \delta_{\text{high}}$
If ($\rho > \delta$)
    objective = `Block` (or `Ignore` if $P_s = 0$)
Else
    objective = `Strike`

This flow of definitions is designed to incorporate our knowledge that impact dynamics, especially when the puck is spinning, do not match our model. Essentially, if the confidence of the current prediction is high and there are no wall impacts between the puck and the desired interception point, we allow a `Strike` behavior to dominate. Otherwise, we must allow that adjustment of the end-effector position for the `Block` maneuver is more important.

Once a specific objective has been selected, we must generate the desired end-effector motion and controller parameters. Briefly, we must solve the following problems (where x^i is the desired interception point):

> `Block`: Interception problem (e.g., see [16]). We choose x^i such that the end-effector intercepts the maximal range of trajectories in Γ that project onto the home goal, thereby minimizing P_s after placement of the end-effector. The velocities V_m and V_s determine minimum time for placement.
>
> `Strike`: We must carefully instantiate the end-effector desired trajectory. This requires us to choose a x^i and a desired outgoing puck trajectory.

The choice of x^i for `Block` is very straightforward, while the `Strike` action requires a great deal of care for end-effector motion planning. We must be concerned with the dynamics of the puck–mallet interaction (see [9, 10]) as well as the robustness and timing issues associated with an impact defined by a desired outgoing puck trajectory.

The first two links of the manipulator are used for gross positioning, tracking a joint-space trajectory interpolated from the initial to final configurations using a cubic polynomial. The third link is used for specific interaction tasks based on the selected objective. We will discuss commands for the third link as we cover each objective.

We will begin our discussion of the objectives with `Strike`. Instantiation of a `Strike` requires knowledge of the dynamics of the end-effector and the puck–mallet interaction. We can again use a simple coefficient of restitution law, similar to that discussed in Section 14.3, together with our knowledge of the end-effector dynamic response to execute an appropriate `Strike`. We must choose x^i and the desired outgoing puck trajectory in order to fully instantiate the `Strike`.

A straightforward methodology for selecting x^i is to choose to intercept the puck when its centroid reaches a horizontal line on the table. In the image plane, this corresponds to a line of constant v. We may modify the x^i when the interception point on the line of constant depth is too near the table bounding edge by simply allowing a secondary interception line.

We study the step response of the end-effector under PID control in order to generate an appropriate `Strike` instantiation. Strikes consist of appropriately timed and oriented "slap shots" using the third link only. We study the efficacy of various `Strike` instantiations using the same end-effector step size (thereby effectively removing our ability to regulate velocity, but isolating our capability of generating a desired outgoing angle). The reasoning for this simplification is to study accuracy of the basic `Strike` behavior.

Instantiation of the `Block` objective is straightforward. We can utilize the same technique as selected for `Strike`, but allow the third link to lie in a neutral position to perform the block (i.e., no "slap" step response).

Let us now consider a very simple objective selection scheme based entirely on estimated accuracy of prediction, the fundamental basis of our scheme. Given the width ρ of the projection of Γ on the horizontal line we have chosen in the image plane, we choose an objective as follows: we select a limit ρ_{limit} and choose `Strike` when $\rho \leq \rho_{\text{limit}}$, `Block` otherwise.

An instantiation of a `Strike` objective includes the desired outgoing trajectory. As we have limited our available control, we will attempt to dictate only the outgoing puck angle with respect to the table, denoted θ_{out}. In fact, we will attempt to strike a target (the opponent's goal) with the puck. As such, a specific instantiation of `Strike` will be denoted as `Strike[X]` where X is the number of wall impacts desired before the target goal is to be entered. We will denote shots going to the opponent's left-hand table edge first with a minus sign, so that a `Strike[-2]` would indicate a desired outgoing trajectory that struck the left-hand wall, followed by the right-hand wall followed by the goal.

Effectively, only `Strike[0]`, `Strike[-1]`, and `Strike[1]` are reliable objectives, as the buildup of uncertainty and modeling errors over two wall impacts results in poor performance. We choose to instantiate the `Strike[±1]` that minimizes the difference between $\theta_{in} + \pi$ and θ_{out}, indicating that the `Strike` will not "crosscut" the incoming trajectory unless necessary. This decision was made to reduce the error in θ_{out} resulting from inaccurate impact modeling of the puck–mallet interaction. We exclude the option of selecting `Strike[0]` as success using `Strike[±1]` indicates an ability to accurately control puck angle.

The algorithm used to determine an appropriate `Strike` instantiation utilizes a nominal coefficient of restitution for puck–mallet interactions of 0.4 and a carefully characterized "slap shot" using the third link (with a nominal velocity at impact time of 0.6354 m/s, normal to the third link). The desired outgoing trajectory is to enter the center of the target goal, as defined in Table 14.2. The equation for the impact is

$$\dot{b}(T^+) = \dot{b}(T^-) + (1 + \alpha_r)nn^T(v_p - \dot{b}(T^-)) \tag{14.6}$$

where the velocity of the puck is $\dot{b}(T^-)$ before the impact and $\dot{b}(T^+)$ after the impact, v_p is the velocity of the mallet at T, $\alpha \in [0, 1]$ is the coefficient of restitution and n is the normal to the mallet at the point of impact. During the `Strike` (not during repositioning, but only during the actual "slap"), the gains for the third link were increased to $k_{p_3} = 7.0$ and $k_{d_3} = 0.7$. Total angle covered by q_3 across the "slap" was 0.4 rad, equally distributed across pre- and post-strike. Timing of the impact was calculated under the assumption that contact occurred for 0.005 s and that the end-effector would reach the interception point in 0.0675 s. The nominal coefficient of restitution for puck–wall impacts was again taken to be 0.725.

The Newton–Raphson method [17] was utilized to solve the nonlinear problem of determining the desired end-effector angle at the point of impact, given the desired outgoing puck angle. An initial guess determined using the relation that the incoming angle relative to the striking surface was equal to the outgoing angle relative to that same surface. This resulted in a very rapid convergence of the N–R method (2–4 iterations, typically).

Below are three example trajectories and the resulting control switching signals for experimental evaluation of the performance of the system discussed above. In each case, the objective was attained and θ_{out} (if defined) was achieved to an acceptable degree, resulting in the puck entering the (unguarded) goal on the opposite end of the table.

The four example cases detailed below (Case 1–Case 4) include instantiations of `Strike[1]`, `Strike[-1]`, and a pure `Block`. All utilize $\rho_{limit} = 20$ pixels. An incoming puck trajectory has negative velocity in v. The interception line (in the image plane) was taken to be $v = 110$ pixels. The limits on acceptable puck centroids (i.e., the region of interest) are the limits of the centroid graphs (excluding those in Fig. 14.11). Note that for the coordinate systems associated with our CCD camera, a low value of u is the opponent's left-hand wall. In all cases, raw centroid data is shown (i.e., the optical axis is located at $(u, v) = (256, 240)$). All measured puck centroids are shown by an 'X.' Notice that this information is raw data, including several spurious data points introduced by difficulties in the VME implementation. These obvious pieces of spurious data are easily extracted using timing information, but have been included to demonstrate the system's robustness.

The raw centroid data and objective selection history for Case 1 are shown in Fig. 14.9. This case displays a switch from `Block` to `Strike[1]` as Γ becomes tighter with more samples. The width of the error cone is defined only after three centroid samples, at which point the statistical analysis of the data has meaning. Before three samples become available,

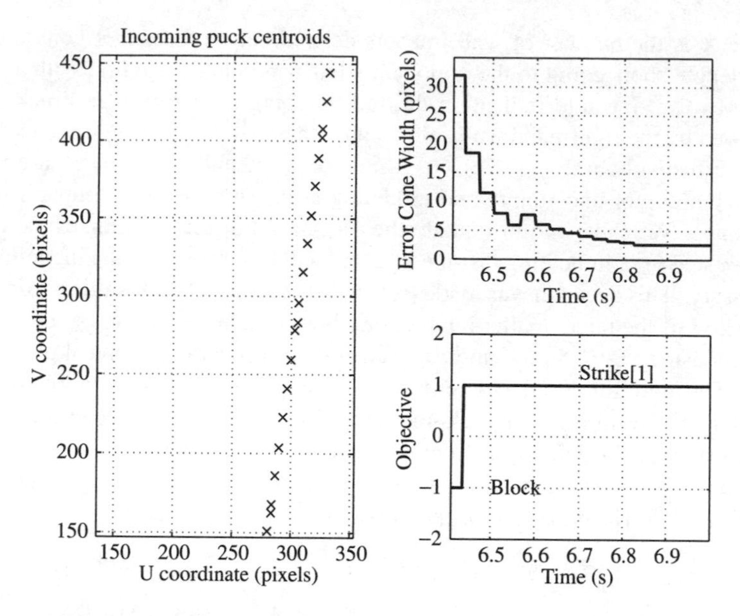

Figure 14.9 Raw centroid data, ρ, and objective selection for Case 1.

we always utilize the `Block` objective. Note that, were we to have set $w_{\text{limit}} = 7$ pixels, we would have seen an objective switching history of the form

$$\texttt{Block} \rightarrow \texttt{Strike[1]} \rightarrow \texttt{Block} \rightarrow \texttt{Strike[1]}$$

Case 2, utilizing `Strike[-1]`, is seen in Fig. 14.10. Note that the initial portion of the trajectory is only slightly disturbed from the ideal, linear case, but a bad sample just after $t = 13.4$ s causes a jump in the uncertainty cone. Nevertheless, the objective `Strike[-1]` is always instantiated.

Figure 14.11 shows the results of the instantiated `Strike` objectives of Cases 1 and 2. In order to achieve the objective, the puck must enter the opponent's goal at any point, not hit

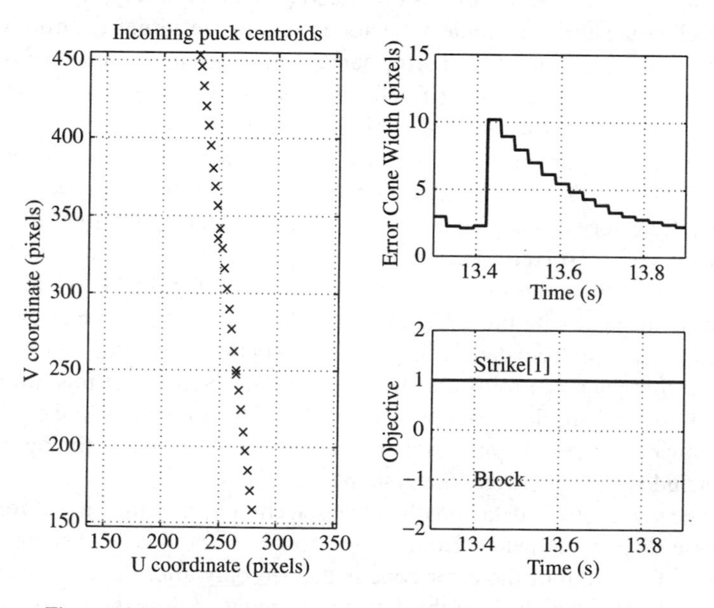

Figure 14.10 Raw centroid data, ρ, and objective selection for Case 2.

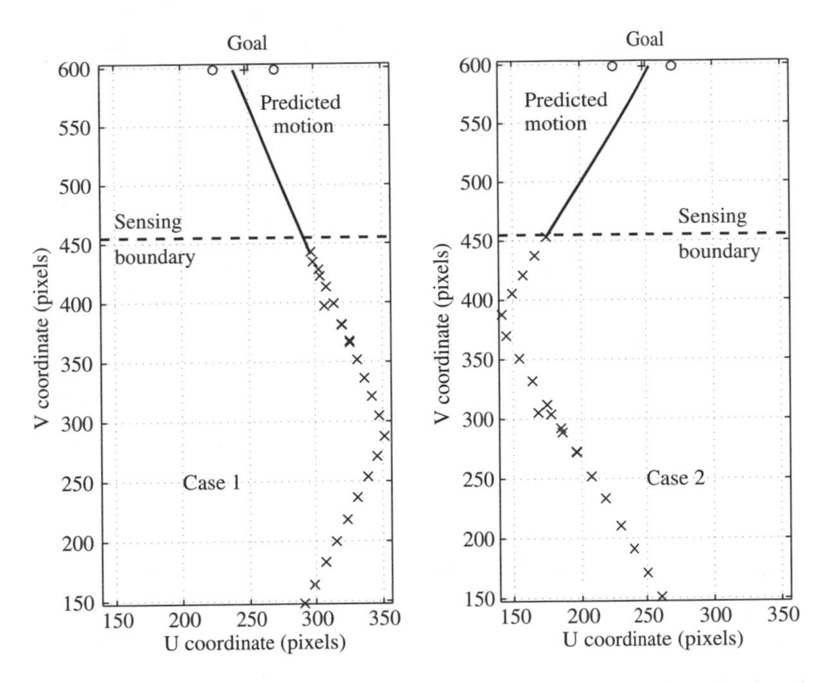

Figure 14.11 Outgoing puck centroid history and projected puck motion for evaluation of the `Strike` objective.

the center exactly. In Fig. 14.11 we have made a phantom extension of the image plane and transformed the actual goal coordinates into image plane coordinates. The edges of the goal (in the configuration space of the puck defined by the centroid) are marked with 'o,' while the center and actual target of the outgoing puck is marked with a '+.' Predicted motion of the puck after it leaves the sensing region is shown with a solid line. We see that both `Strike` objectives are successfully instantiated. Note that, in Case 1, we see the friction characteristics of the table come into play dramatically near the sensing boundary, where the centroids become very closely spaced as the puck slows down.

Figure 14.12 (Case 3) shows a `Block` action. The puck was successfully intercepted in this case, but we have no way of determining whether or not a `Strike` action might have been successful. Note that it is the errors in centroid location and not the speed of the incoming shot that resulted in the instantiation of the `Block` objective. Figure 14.13 (Case 4, objective `Strike`[1]) demonstrates the distinction.

We see that the objectives in each case were achieved. We note here that the successful implementation of a `Strike` requires that the system be able to reach the desired pre-strike location with sufficient time to begin the action. This puts an effective limit on switches from `Block` to `Strike` based on the time required to reposition. Additional factors influencing the behavior of the system include friction, spin, and timing. In this particular instantiation of the control algorithm, we have chosen to ignore timing difficulties, as our choice of `Strike` instantiations are implemented only when the Γ cone is small, which typically coincides with small timing errors.

We see from these examples that our implementation of a vision-based objective selection scheme achieves the desired goals. In many circumstances, the `Strike` objective will be achievable even for very high ρ. In circumstances where failure to achieve the `Strike` is not of consequence to the primary objective (i.e., P_s is small), we would be able to relax the selection criteria along the lines of the early discussions in this section.

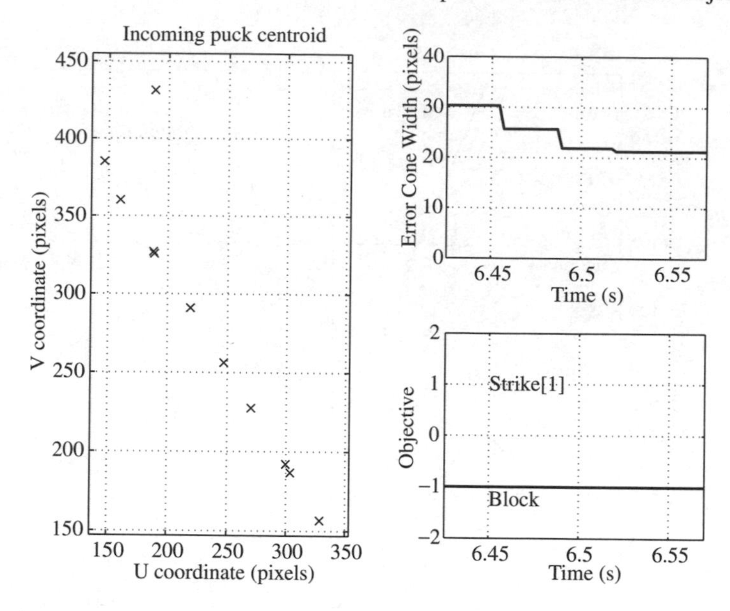

Figure 14.12 Raw centroid data, ρ, and objective selection for Case 3.

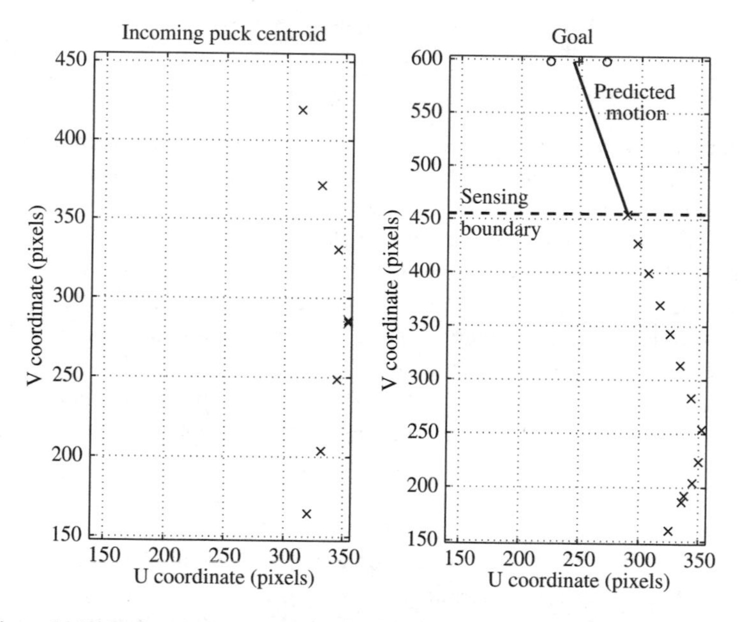

Figure 14.13 Puck centroid history and projected puck motion for Case 4 (`Strike`[1]).

14.5 CONCLUSIONS AND FUTURE WORK

This work has focused on the use of real-time visual feedback for selection of robustly achievable control objectives in the case of ballistic manipulation in two dimensions. The primary contribution of this effort is to suggest that careful analysis of visual measurement and prediction techniques is necessary for appropriate behavior generation when multiple possible actions can be taken. The work in [15] approaches the problem by including dynamic control in the decision-making process, thereby enhancing the achievability properties of selected objectives.

Future directions relating to this work include the synthesis of sensing and control strategies in a unified framework when multiple controllers, objectives, and sensing algorithms are available. Further, work on real-time single-step velocity estimation and associated prediction schemes is currently being undertaken. The goal of these endeavors is to generate tools and design methodologies that will enhance the flexibility of robotic manipulators using visual feedback for interaction with uncertain and dynamic environments. Many exciting avenues of research remain open.

14.6 ACKNOWLEDGMENTS

This research was supported by the National Science Foundation and the Electric Power Research Institute under the joint NSF/EPRI Intelligent Control Initiative, NSF Grant number ECS-9216428 and EPRI Grant number RP8030-14. Additional support was provided by the Naval Academy Research Council and the Office of Naval Research under grant N0001498WR20010.

REFERENCES

[1] P. Allen, A. Timcenko, B. Yoshimi, and P. Michelman. "Automated tracking and grasping of a moving object with a robotic hand–eye system." *IEEE Transactions on Robotics and Automation*, vol. 9, no. 2, pp. 152–165, IEEE, 1993.

[2] G. C. Buttazzo, B. Allotta, and F. P. Fanizza. "Mousebuster: A robot for real-time catching." *Control Systems Magazine*, vol. 14, no. 1, pp. 49–56, IEEE, 1994.

[3] E. Grosso, G. Metta, A. Oddera, and G. Sandini. "Robust visual servoing in 3–d reaching tasks." *IEEE Transactions on Robotics and Automation*, vol. 12, no. 5, pp. 732–742, IEEE, 1996.

[4] R. L. Andersson. "Dynamic sensing in a ping–pong playing robot." *IEEE Transactions on Robotics and Automation*, vol. 5, pp. 728–739, IEEE, 1989.

[5] A. A. Rizzi and D. E. Koditschek. "An active visual estimator for dextrous manipulation." *IEEE Transactions on Robotics and Automation*, vol. 12, pp. 697–713, IEEE, 1996.

[6] B. E. Bishop. "Real Time Visual Control With Application to the Acrobot." Master's thesis, University of Illinois at Urbana–Champaign, Coordinated Science Laboratory, 1994.

[7] B. E. Bishop, S. A. Hutchinson, and M. W. Spong. "Camera modelling for visual servo applications." *Mathematical and Computer Modelling*, vol. 24, no. 5/6, pp. 79–102, 1996.

[8] K. Voyenli and E. Eriksen. "On the motion of an ice hockey puck." *American Journal of Physics*, vol. 53, pp. 1149–1153, 1985.

[9] W. H. Huang, E. P. Krotkov, and M. T. Mason. "Impulsive manipulation." *Proceedings of the 1995 IEEE International Conference on Robotics and Automation*, pp. 120–125, Nagoya, Japan, May 1995.

[10] A. A. Rizzi and D. E. Koditschek. "Progress in spatial robot juggling." *Proceedings of the 1992 IEEE International Conference on Robotics and Automation*, pp. 775–780, Nice, France, May 1992.

[11] B. D. O. Anderson and J. B. Moore. "Optimal Filtering." Englewood Cliffs, NJ, Prentice–Hall, 1979.

[12] A. P. Sage and C. C. White. "Optimum Systems Control." Englewood Cliffs, NJ, Prentice–Hall, 1977.

[13] J. R. Taylor. "An Introduction to Error Analysis: The Study of Uncertainties in Physical Measurements." Mill Valley, CA, University Science Books, 1982.

[14] P. Z. Peebles. "Probability, Random Variables, and Random Signal Principles." New York, McGraw Hill, 1987.

[15] B. E. Bishop. "Intelligent Visual Servo Control of an Air Hockey Playing Robot." PhD dissertation, University of Illinois at Urbana–Champaign, Coordinated Science Laboratory, 1997.

[16] A. E. Bryson Jr. and Y. C. Ho. "Applied Optimal Control: Optimization, Estimation and Control." New York, Halsted Press, 1975.

[17] W. H. Press, B. P. Flannery, S. A. Teukolsky, and W. T. Vetterling. "Numerical Recipies in C: The Art of Scientific Computing." Cambridge, Cambridge University Press, 1988.

Chapter 15

VISION-BASED AUTONOMOUS HELICOPTER RESEARCH AT CARNEGIE MELLON ROBOTICS INSTITUTE (1991–1998)

Omead Amidi, Takeo Kanade, and Ryan Miller
Carnegie Mellon Robotics Institute

Abstract

This chapter presents an overview of the Autonomous Helicopter Project at Carnegie Mellon Robotics Institute. The advantages of an autonomous vision-guided helicopter for a number of goal applications are enumerated through possible mission scenarios. The requirements of these applications are addressed by a central goal mission for the project. Current capabilities, including vision-based stability and control, autonomous take off, trajectory following, and landing, aerial mapping, and object recognition and manipulation are presented. In conclusion, the project future directions are discussed.

15.1 INTRODUCTION

Precise maneuverability of helicopters makes them useful for many critical tasks ranging from rescue and security to inspection and monitoring operations. Helicopters are indispensable air vehicles for finding and rescuing stranded individuals or transporting accident victims. Police departments use them to find and pursue criminals. Fire fighters use helicopters for precise delivery of fire extinguishing chemicals to forest fires. More and more electric power companies are using helicopters to inspect towers and transmission lines for corrosion and other defects and to subsequently make repairs. All of these applications demand dangerous close proximity flight patterns, risking human pilot safety. An unmanned autonomous helicopter will eliminate such risks and will increase the helicopter's effectiveness.

Typical missions of autonomous helicopters require flying at low speeds to follow a path or hovering near an object of interest. Such tasks demand accurate helicopter position estimation relative to particular objects or landmarks in the environment. In general, standard positioning equipment such as inertial navigation systems or global positioning receivers can not sense relative position with respect to particular objects. Effective execution of typical missions is only possible by on-board relative sensing and perception. Vision, in particular, is the richest source of feedback for this type of sensing.

At Carnegie Mellon Robotics Institute, we have been investigating the applicability of on-board vision for helicopter control and stability since 1991. Through an incremental and step-by-step approach, we have developed a number of vision-guided autonomous helicopters. We started our development by building a number of indoor testbeds for calibrated and safe experimentation with model helicopters. Using these testbeds, we developed and verified different system components individually before deploying them on-board free flying autonomous helicopters. Over the years, critical components have matured enough to form a framework for several autonomous systems currently in service. These systems can autonomously fly midsized (14 ft long) helicopters using on-board vision, inertial sensing, global positioning, and range sensing. The latest system can autonomously take off, follow a prescribed trajectory, and land. In flight, the system can build aerial intensity and elevation maps of the environment, scan and locate objects of interest by using previously known appearance or color, and track the objects if necessary.

This chapter presents an overview of the Autonomous Helicopter Project at Carnegie Mellon Robotics Institute. The chapter discusses the project's goals, current status, and future plans.

15.2 GOALS

Autonomous helicopters guided by on-board vision can carry out a wide range of useful tasks. In particular, the goal applications we focus on include: search and rescue, law enforcement, inspection, aerial mapping, and cinematography. Figures 15.1 and 15.2 show scenarios involving autonomous helicopters in these applications.

15.2.1 Goal Applications

Search and Rescue: A group of autonomous helicopters can collaborate to quickly and systematically search a very large area to locate victims of an accident or a natural disaster. They can then visually lock on to objects or stranded victims at the site to guide rescue forces to the scene. The helicopters can help focus the efforts of search and rescue crews on the rescue operation instead of the time-consuming search operation. They can be more readily deployed

Figure 15.1 Rescue and crime-fighting goal applications.

Figure 15.2 Inspection and cinematography goal applications.

in weather conditions which would normally prevent human piloted search and rescue. They can be sacrificed in very dangerous conditions to save human lives. Typical tasks may include flying close to a forest fire to look for stranded individuals, searching in contaminated areas, and identifying potential radioactive leaks after a nuclear reactor accident.

Law enforcement: Vision-guided robot helicopters can fly overhead to aid the police in dangerous high-speed chases or criminal search operations. Stationed on top of buildings in urban areas, they can be dispatched in seconds to take off and relay images from trouble spots. This real time imagery is crucial to the tactical assessment of the situation by human experts who dispatch police units to the area.

Inspection: Vision-guided robot helicopters can inspect high voltage electrical lines in remote locations. They can inspect large structures such as bridges and dams cost effectively. They can be quickly called upon to inspect buildings and roads for potential damage after an earthquake. They can locate hazardous materials in waste sites by providing aerial imagery to human experts or by automatically identifying waste containers or materials by on-board vision.

Aerial mapping: Vision-guided robot helicopters can build more accurate topological maps than conventional aircraft at a substantial cost savings. Unlike airplanes, they can fly close to the ground while carrying cameras or range sensors to build high resolution 3D maps. They can fly in smaller and more constrained areas to build highly detailed elevation maps.

Cinematography: Vision-guided robot helicopters can be a director's eye-in-the-sky camera. Because they can fly precisely under computer control, the need for skilled human pilots for aerial photography is eliminated. They can automatically track subjects with their on-board vision-based object trackers. They can fly a prescribed path over and over again to help plan shots or to aid in producing special effects.

15.2.2 Goal Mission

We are pursuing a prototype goal mission which addresses the crucial technologies required by all of the above goal applications. This goal mission requires a robot helicopter to:

1. Automatically start operation and take off.
2. Fly to a designated area on a prescribed path while avoiding obstacles.
3. Search and locate objects of interest in the designated area.
4. Visually lock onto and track or, if necessary, pursue the objects.
5. Send back images to a ground station while tracking the objects.
6. Safely return home and land.

It is important to realize that accomplishing this goal mission is only the first step toward an ultimate goal of building reliable and useful intelligent aerial vehicles. Many important issues ranging from safety and fault tolerance to vision algorithm robustness require strict testing and evaluation in controlled environments before deployment of such vehicles in real life applications.

15.3 CAPABILITIES

Over the past few years, we have been developing a number of basic capabilities to accomplish the project's goal mission. These capabilities include: vision-based stability and control;

autonomous take off trajectory following and landing; aerial mapping; and object recognition and manipulation. This section presents an overview of these capabilities.

15.3.1 Vision-Based Stability and Position Control

The most basic capability our goal mission requires is robust autonomous flight. A truly autonomous craft can not completely rely on external positioning devices such as GPS satellites or ground beacons for stability and guidance. Rather, the craft must sense and interact with its environment. We chose to experiment with on-board vision as the primary source of feedback for this interaction.

Our initial efforts toward vision-based flight has produced a "visual odometer" [1] which tracks helicopter position based on visual feedback. The odometer first determines the position of objects appearing in the camera field of view in relation to the helicopter, and thereafter tracks the objects visually to maintain an estimate of the helicopter's position. As the helicopter moves, older objects may leave the field of view, but new objects entering the scene are localized to continue tracking the helicopter position.

The odometer tracks objects by image template matching. Image templates are rectangular windows containing the objects of interest. Template matching between consecutive images provides lateral and longitudinal image displacement which may result from both helicopter translation and rotation. Template matching in two images, taken simultaneously by a pair of stereo cameras, measures helicopter range. The odometer estimates 3D helicopter motion by combining the lateral and longitudinal image displacements and range estimates with helicopter attitude, measured by on-board synchronized angular sensors.

The visual odometer relies on a "target" template initially taken from the center of an on-board camera image. This process is shown in Fig. 15.3. The 3D location of the target template, P_T, is first determined by sensing the camera range vector, V_o, at the image center given current helicopter position, P_o, and attitude. Relying on the target template's sensed position, the odometer can estimate the helicopter position vector, P_1, by visually locking on to the target template to estimate the new template vector, V_1. In anticipation of losing the current target template, the odometer selects and localizes a new candidate target in every cycle to guarantee uninterrupted operation.

Because the target template will change in appearance as the helicopter rotates and changes its altitude, the odometer tracks pairs of target templates with a small baseline. Before matching templates, the odometer scales and rotates the target template pair using the

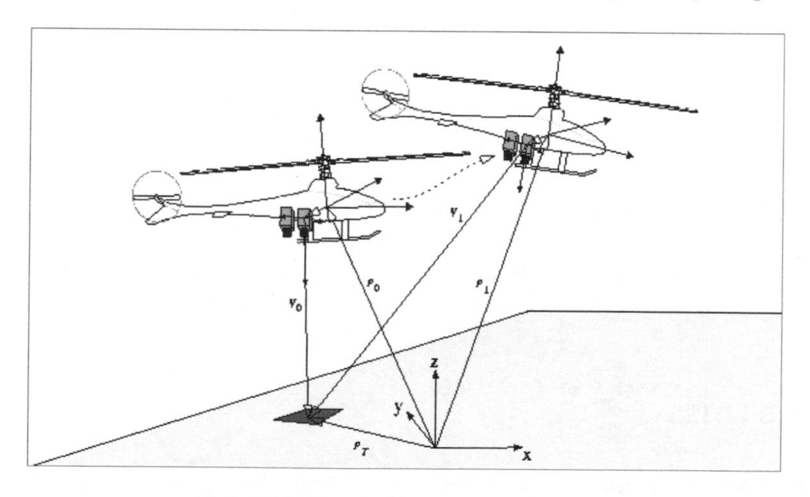

Figure 15.3 Visual odometer positioning cycle.

magnitude and angle of the baseline from the previous match. The current implementation of the odometer tracks pairs of (32×32) pixel templates using a custom-made TI C44-based vision system. The system can support up to six C44s, each tracking one pair of templates at 60 Hz with 24 ms latency. Images are preprocessed by an 8×8 convolution ASIC (GEC Plessey) before matching. The current visual odometer, integrating a pair of b/w video cameras, a set of inexpensive angular sensors (Gyration gyro-engines and KVH digital compass), has been shown to successfully stabilize and maneuver (< 15 mph) small to midsized model RC helicopters. See Appendix for more details. Figure 15.4 shows the odometer's aerial view during autonomous flight over a grassy field. Figure 15.5 compares the odometer's positioning accuracy (solid) to ground truth (dashed) during indoor and outdoor helicopter flight.

15.3.2 Precise Autonomous Flight: Takeoff, Trajectory Following, and Landing

Our goal mission cannot be accomplished without highly precise helicopter maneuvers. Precise helicopter maneuvering is a game of accurate force control. Inherently, vision alone cannot provide the necessary feedback for this control mode. This is because vision can sense only the motion created by applied forces, not the forces themselves. It is impossible for a controller to completely eliminate undesired movements due to disturbances after they are sensed. Precise helicopter maneuvers such as takeoff, trajectory following, and landing require inertial sensing.

Figure 15.4 Odometer's view

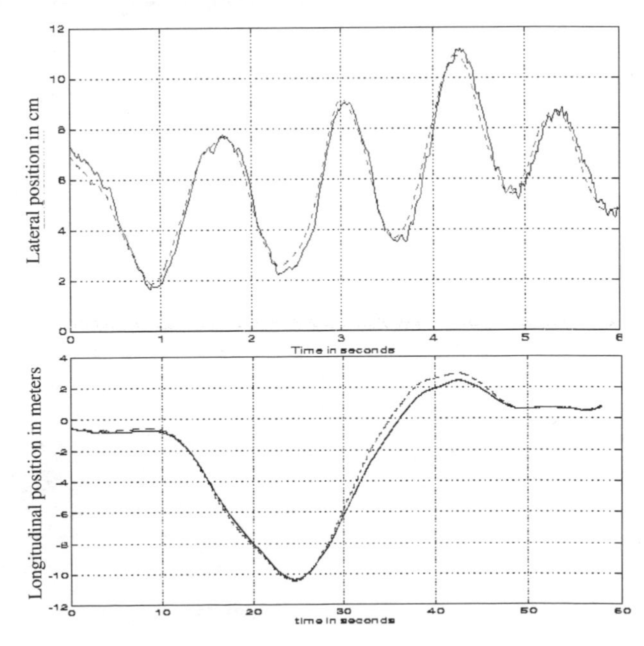

Figure 15.5 Visual odometer positioning accuracy indoor (top) and outdoor (bottom).

Combining vision with other low-level sensors such as accelerometers and range sensors can produce very robust autonomous navigation systems. A notable example is the work of Dickmanns [2] who applies an approach that exploits spatiotemporal models of objects in the world to control autonomous land and air vehicles. He has demonstrated autonomous position estimation for an aircraft in landing approach using a video camera, inertial gyros, and an air velocity meter.

We have begun following this approach by integrating an array of on-board sensors with vision to form a robust state estimator. The state estimator will ultimately fuse data from the visual odometer, an inertial measurement unit (IMU), a GPS receiver, a flux-gate compass, and a laser range finder (see Fig. 15.6). These sensors form an integrated vision-based navigation system for our autonomous helicopters. The current state estimator integrates an IMU (Litton LN-200, 3-axis Silicon accelerometers, and angular rate sensors), a GPS receiver (NovAtel MillenRt2 dual-frequency carrier-phase unit, 2 cm accuracy positioning using nearby (< 20 miles) ground differential correction stations), and a digital compass (KVH Industries).

The systems employs lat-lon mechanization (see [3] for details) for inertial navigation. The data from the sensors are fused by a 12th order Kalman Filter which keeps track of latitude, longitude, height, 3-axis velocities, roll, pitch, yaw, and accelerometer biases (Gauss-Markov model).This data fusion is demonstrated in Fig. 15.7. The jagged graph (marked by x) is GPS data, the solid line is ground truth, and the third (marked by +) is the Kalman filter output.

An on-board classical (feed-forward PD) controller controls the helicopter on smooth (cubic spline) trajectories planned by a flight path editor. As displayed in Fig. 15.8, the flight path editor's console shows the planned trajectory between goal points supplied by a human operator. The helicopter can be programmed to maintain its heading tangent to the path or to always orient itself toward a point in space. The console also displays 3D range data collected by the mapping system (see Section 3.4) to help in selecting goal points.

15.3.3 Aerial Mapping

Our goal mission requires 3D range sensing for autonomous obstacle detection and aerial mapping. In general, aerial vehicles are particularly well-suited for 3D mapping for the following main reasons:

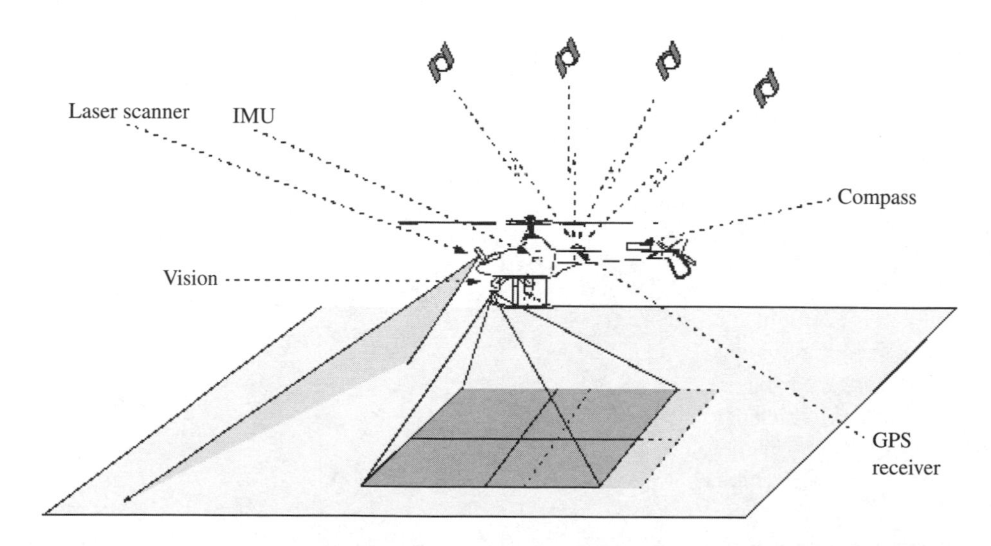

Figure 15.6 Integrated vision-based navigation.

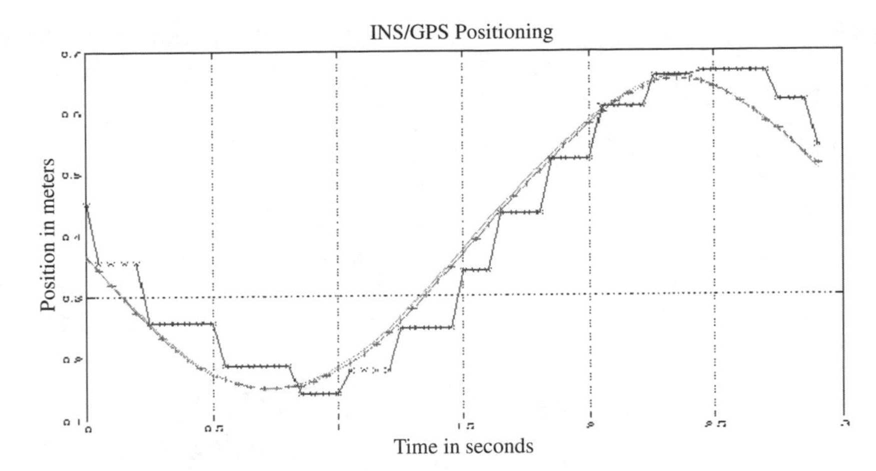

Figure 15.7 IMU/GPS fusion result.

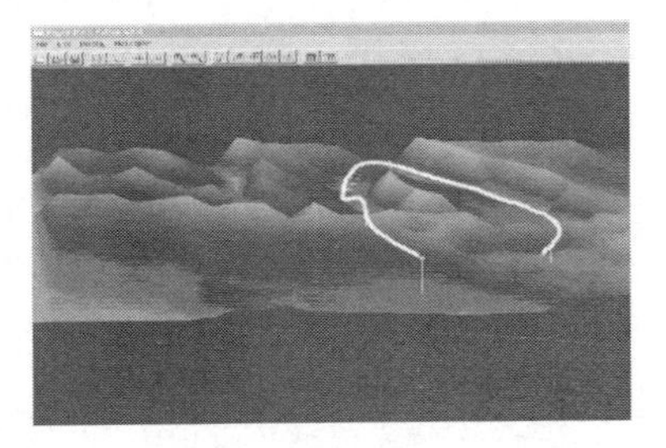

Figure 15.8 Flight path editor.

1. While carrying on-board active sensors, aerial vehicles can quickly and efficiently scan very large areas.
2. The performance of active sensors such as laser rangefinders on-board aerial vehicles is superior to that of ground-based systems due to the fact that aerial sensors receive better signal returns than sensors on-board ground vehicles.

Our first prototype mapping system integrates a laser rangefinder (Reigl LD90), custom synchronization hardware, and our inertial state estimator on-board an autonomous helicopter. The system scans the environment line by line using a spinning mirror during helicopter flight. Each line, consisting of an array of range measurements, is tagged with helicopter state and is registered in memory to incrementally build a 3D map. Figure 15.9 compares an aerial photograph with its corresponding 3D elevation map built by the mapping system.

15.3.4 Object Recognition and Manipulation

Our goal mission requires detecting, tracking, and possibly manipulating objects during autonomous flight. Our current prototype systems can detect and track objects based on color and appearance.

The color-based detector is, in essence, a color discriminator. Built by high-speed digitally controllable analog hardware, the discriminator can be configured at NTSC field rate (60 Hz) to look for as many RGB color combination as necessary in sequential image fields. The discriminator normalizes RGB intensity levels to eliminate the effects of lighting, determines the difference between each pixel color from the target color, and penalizes each pixel based on this distance. Most recently, for the 1997 Unmanned Aerial Robotics Competition [4], the dis-

Figure 15.9 Aerial 3D elevation map.

criminator was used to pick up an orange disk from a barrel on the ground. The image sequence below, taken prior to the contest, displays a successful pickup attempt (Fig. 15.10). The pickup system tracked a blue magnet and aligned it with the orange disk as the helicopter descended to the estimated range to the disk. (Helicopter range to the disk was measured by triangulation.)

The template-based detector locates objects based on appearance. To find a potential match, the detector locates image regions which resemble a picture of the object. Because searching every image for a the object's picture or template in any orientation requires enormous computational power, the detector exploits principal component methods to reduce its work load. The detector employs Karhunen-Loeve [5] expansion to reduce computational complexity and storage of necessary templates. Rotated template images differ slightly from each other and are highly correlated. Therefore, the image vector subspace required for their effective representation can be defined by a small number of eigenvectors or eigenimages. The eigenimages which best account for the distribution of image template vectors can be derived by Karhunen-Loeve expansion [6].

The detector has been used to carry out a number of tasks. Most notable is the successful identification of all warning barrel labels for the 1997 Unmanned Aerial Competition. The detector successfully identified the location, orientation, and type of three different labels. A picture of each label was supplied to the detector prior to flight. Figure 15.11 shows the these labels. They are typically affixed on radioactive, biological hazard, and explosive containers.

Figure 15.10 Autonomous vision-based object pickup.

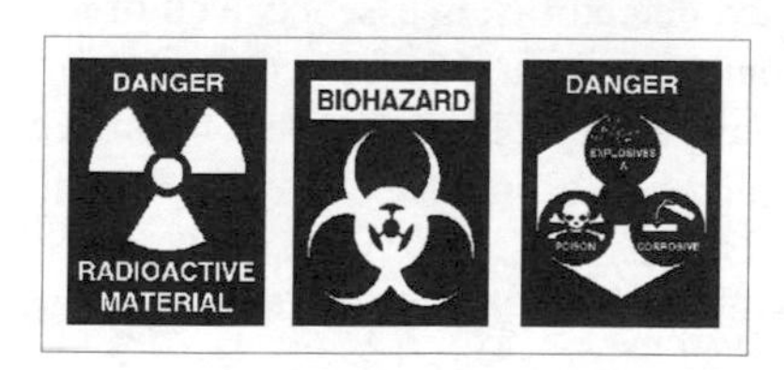

Figure 15.11 Hazard labels.

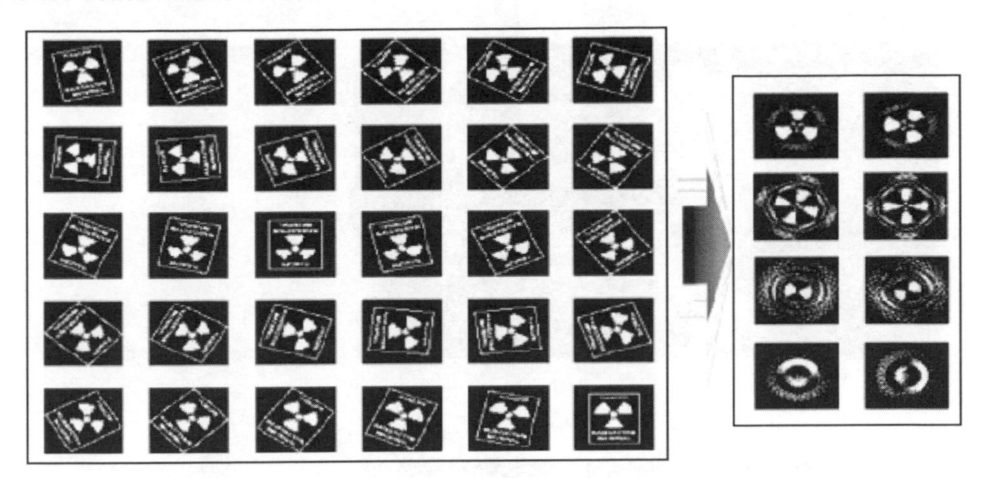

Figure 15.12 Principle component analysis.

Figure 15.12 compares 25 radioactive hazard symbol templates, rotated at 15 degree intervals, to the eight principal templates used by the detector for matching. In the competition trials, the eight principal templates never failed to detect the radioactive hazard label, provided the helicopter height was kept close to (within 1 m) the target camera range when the original template was recorded.

15.4 FUTURE WORK

The current prototype flight systems have demonstrated the potential of vision-based autonomous flight. Current prototype systems have shown useful capabilities such as autonomous take off, trajectory following and landing, aerial mapping, and object recognition and tracking. These capabilities form the building blocks of future research on aerial robots.

We plan to continue this research by improving the current template-based positioning strategy as well as by developing new optical flow-based systems for flight. The list of current activities include: texture analysis to automatically locate and select high contrast image areas for the target template to improve the current visual odometer performance; fusion of optical flow estimates in the Kalman filter with inertial and global positioning data for more robust position estimation; development of high resolution laser scanners to detect overhead electrical lines; and system identification to model helicopter dynamics.

Appendix: Carnegie Mellon Autonomous Helicopter Product History

September 1991: Initial attitude control experiments (Fig. 15.13).

- Developed to test and tune attitude control system. Electrical model helicopter mounted on a swiveling arm platform.
- Optical encoder mounted with a frictionless bearing measures ground-truth angles in real time.
- Configurable for roll, pitch, and yaw.

February 1992: Free flight and vision-based state estimator (Fig. 15.14).

- Six-degree-of-freedom testbed developed for evaluating various position estimation and control systems.

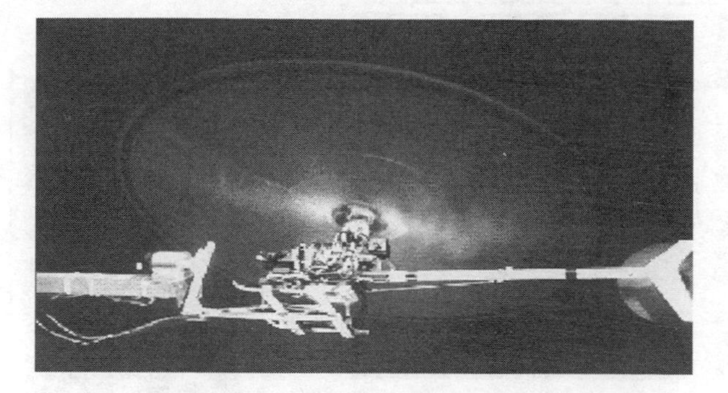

Figure 15.13 Initial helicopter experiments.

Figure 15.14 Helicopter with free flight and vision-based state estimator.

- Electrical model helicopter attached to poles by graphite rods for safety and helicopter ground-truth position estimation.
- Lightweight composite material and custom-designed frictionless air bearings allow unobtrusive helicopter free flight in a cone shaped area.
- Mechanical stops prevent the helicopter from crashing or flying away.

September 1994: First autonomous platform (Fig. 15.15).

- Indoor testbed developed as an step toward autonomous operation.
- Used for testing the vision system, control/sensor platform, power system, RF interference, and overall system integrity.
- Allows relatively large (1.5 m) longitudinal travel. Severely limits helicopter travel laterally and vertically.
- Helicopter is tethered with ropes which are fastened to the ground and poles positioned on either side of the platform.
- Steel rod with hooks on either end connects the ropes to the helicopter. Steel rod is secured to the helicopter's center of gravity to eliminate any torques from restraining forces which could cause dangerous rotations.

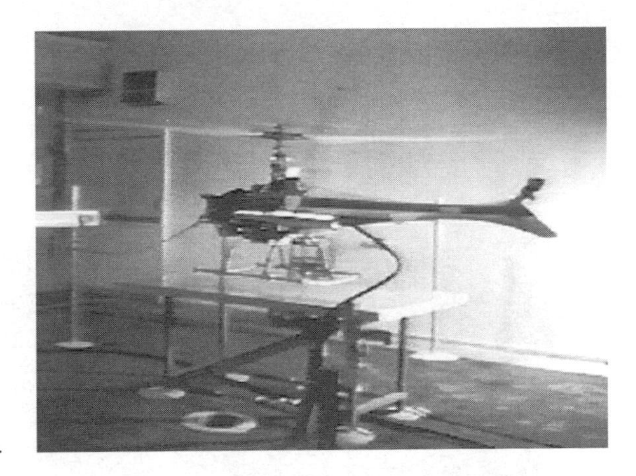

Figure 15.15 First autonomous platform.

October 1995: Autonomous Helicopter #1 (Fig. 15.16).

- Visual odometer (4 cm accuracy, 60 Hz), tracks image patches and templates with helicopter motion.
- Initial computer control trials performed at relatively high (15 m) altitudes to allow safety pilot time to override computer.
- Latitudinal and longitudinal controls were first tested by mixing human control for height and heading with the computer commands.
- GPS used for ground-truth measurements.

August 1996: Autonomous Helicopter #2 (Fig. 15.17).

- Control system for autonomous takeoff, landing, and smooth trajectory following.
- System tested in harsh conditions (40–45 mph wind gusts).
- State estimator fusing data from a dual-frequency carrier-phase GPS receiver, 3-axis angular rate and inertial sensors, and field-rate vision-based odometry.
- Custom-designed vision system capable of field-rate position sensing, multiple object tracking, color discrimination, and aerial intensity map building. Custom-designed camera stabilization system.
- 3D laser line scanner.
- Power system for up to 33 minutes autonomous operation.
- Winning entry in the 1997 International Aerial Robotics Competition.

Figure 15.16 Autonomous helicopter #1.

Figure 15.17 Autonomous helicopter #2.

REFERENCES

[1] O. Amidi, "An Autonomous Vision-Guided Helicopter." Ph.D. thesis, Electrical & Computer Engineering Department, Carnegie Mellon, August 1996.

[2] E. Dickmanns, "A General Dynamic Vision Architecture for UGV and UAV." *Journal of Applied Intelligence*, New York, vol. 2, pp. 251–270, 1992.

[3] *Aerospacs Avionics Systems*. New York, Academic Press, 1993.

[4] R. Michelson, "Aerial Robotics Competition Rules." Technical report, Smyma, Georgia, Georgia Tech Research Institute, 1995.

[5] S. Yoshimura and T. Kanade, "Fast template matching based on the normalized correlation by using multiresolution eigenimages." *IEEE/RSJ International Conference on Robotics and Systems*, Munich, August 1994.

[6] M. Uenohara nd T. Kanade, "Vision-based object registration for real- time image overlay." *Computer Vision, Virtual Reality and Robotics in Medicine. First International Conference*, CVR Med-95, Nice, April 1995.

INDEX

ABOUT THE EDITORS

Marcus Vincze was born in Salzburg, Austria in 1965. He has worked in the field of vision for robotics since 1990.

In 1965 he received his "Diplomingenieur" in mechanical engineering from Vienna University of Technology and two years later a M.Sc. from Rensselaer Polytechnic Institute. He finished his Ph.D. at the Vienna University of Technology in 1993. From 1995 to 1998, he was awarded an APART research grant from the Austrian Academy of Sciences and spent a year as guest researcher at HelpMate Robotics, Inc. and Robot and Vision Lab of Professor Gregory D. Hager at Yale University. Since 1998 he has led national and European research projects in the area of service robotics and robust vision for industrial applications at the Vienna University of Technology.

He received in Barcelona 1992 the JIRA-Award for the best paper in Research and Development at the 23rd International Symposium of Industrial Robots (ISIR) Conference from the Japan Industrial Service Robot Association (now JRA). He is a member of IEEE, ISRA (International Service Robot Association), and OAGM (Austrian Pattern Recognition Association). His special interests include cue integration, visual servoing, calibration, and service robotics.

Gregory D. Hager received the B.A. degree in computer science and mathematics from Luther College in 1983, and the M.S. and Ph.D. in computer science from the University of Pennsylvania in 1985 and 1988, respectively. From 1988 to 1990, he was a Fulbright Junior Research Fellow at the University of Karlsruhe and the Fraunhofer Institute IITB in Karlsruhe, Germany. From 1991 through 1999, he was on the faculty of the Computer Science Department at Yale University. In July 1999, he accepted a position as full professor in the Computer Science Department at The Johns Hopkins University.

Professor Hager is a member of IEEE and AAAI. He is on the editorial boards of the *IEEE Transactions on Robotics and Automation* and *Pattern Analysis and Applications*, and he is currently co-chairman of the Robotics and Automation Society Technical Committee on Computer and Robot Vision. His research interests include visual tracking, hand-eye coordination, medical robotics, sensor data fusion and sensor planning. He is the author of *Task-Directed Sensor Fusion and Planning* published by Kluwer Academic Publishers, Inc., and he is also coeditor of *The Conference of Vision and Control* published by Springer-Verlag.

ABOUT THE EDITORS